Lecture Notes in Physics

New Series m: Monographs

Managing Editor

W. Beiglböck
Assisted by Mrs. Sabine Landgraf
c/o Springer-Verlag, Physics Editorial Department II
Tiergartenstrasse 17, D-69121 Heidelberg, Germany

The Editorial Policy for Monographs

The series Lecture Notes in Physics reports new developments in physical research and teaching - quickly, informally, and at a high level. The type of material considered for publication in the New Series m includes monographs presenting original research or new angles in a classical field. The timeliness of a manuscript is more important than its form, which may be preliminary or tentative. Manuscripts should be reasonably self-contained. They will often present not only results of the author(s) but also related work by other people and will provide sufficient motivation, examples, and applications.

The manuscripts or a detailed description thereof should be submitted either to one of the series editors or to the managing editor. The proposal is then carefully refereed. A final decision concerning publication can often only be made on the basis of the complete manuscript, but otherwise the editors will try to make a preliminary decision as definite as they can on the basis of the available information.

Manuscripts should be no less than 100 and preferably no more than 400 pages in length. Final manuscripts should preferably be in English, or possibly in French or German. They should include a table of contents and an informative introduction accessible also to readers not particularly familiar with the topic treated. Authors are free to use the material in other publications. However, if extensive use is made elsewhere, the publisher should be informed. Authors receive jointly 50 complimentary copies of their book. They are entitled to purchase further copies of their book at a reduced rate. As a rule no reprints of individual contributions can be supplied. No royalty is paid on Lecture Notes in Physics volumes. Commitment to publish is made by letter of interest rather than by signing a formal contract. Springer-Verlag secures the copyright for each volume.

The Production Process

The books are hardbound, and quality paper appropriate to the needs of the author(s) is used. Publication time is about ten weeks. More than twenty years of experience guarantee authors the best possible service. To reach the goal of rapid publication at a low price the technique of photographic reproduction from a camera-ready manuscript was chosen. This process shifts the main responsibility for the technical quality considerably from the publisher to the author. We therefore urge all authors to observe very carefully our guidelines for the preparation of camera-ready manuscripts, which we will supply on request. This applies especially to the quality of figures and halftones submitted for publication. Figures should be submitted as originals or glossy prints, as very often Xerox copies are not suitable for reproduction. For the same reason, any writing within figures should not be smaller than 2.5 mm. It might be useful to look at some of the volumes already published or, especially if some atypical text is planned, to write to the Physics Editorial Department of Springer-Verlag direct. This avoids mistakes and time-consuming correspondence during the production period.

As a special service, we offer free of charge $\text{L}^{\text{A}}\text{T}_{\text{E}}\text{X}$ and $\text{T}_{\text{E}}\text{X}$ macro packages to format the text according to Springer-Verlag's quality requirements. We strongly recommend authors to make use of this offer, as the result will be a book of considerably improved technical quality.

Manuscripts not meeting the technical standard of the series will have to be returned for improvement.

For further information please contact Springer-Verlag, Physics Editorial Department II, Tiergartenstrasse 17, D-69121 Heidelberg, Germany.

M. Dineykhan   G.V. Efimov   G. Ganbold
S.N. Nedelko

Oscillator Representation in Quantum Physics

Springer

Authors

M. Dineykhan
G. V. Efimov
G. Ganbold
S. N. Nedelko
Bogoliubov Laboratory of Theoretical Physics
Joint Institute for Nuclear Research (JINR)
141980 Dubna, Russia

ISBN 978-3-662-14063-5   ISBN 978-3-540-49186-6 (eBook)
DOI 10.1007/978-3-540-49186-6

CIP data applied for.

Originally published by Springer-Verlag Berlin Heidelberg New York in 1995
Softcover reprint of the hardcover 1st edition 1995

Typesetting: Camera-ready by the authors
SPIN: 10127309   55/3142-543210 - Printed on acid-free paper

Contents

Part II. The Gaussian Equivalent Representation of Functional Integrals in Quantum Physics

1. Introduction

The investigation of most problems of quantum physics leads to the solution of the Schrödinger equation with an appropriate interaction Hamiltonian or potential. However, the exact solutions are known for rather a restricted set of potentials, so that the standard eternal problem that faces us is to find the best effective approximation to the exact solution of the Schrödinger equation under consideration. In the most general form, this problem can be formulated as follows. Let a total Hamiltonian H describing a relativistic (quantum field theory) or a nonrelativistic (quantum mechanics) system be given. Our problem is to solve the Schrödinger equation

$$H\Psi_n = E_n\Psi_n,$$

i.e., to find the energy spectrum $\{E_n\}$ and the proper wave functions $\{\Psi_n\}$ including the ground state or vacuum $\Psi_0 = |0\rangle$. The main idea of any approximation technique is to find a decomposition

$$H = H_0 + H_I$$

in such a way that H_0 describes our physical system in the "closest to H" manner, and the Schrödinger equation

$$H_0\Psi_n^{(0)} = E_n^{(0)}\Psi_n^{(0)}$$

can be solved exactly. The interaction Hamiltonian H_I is supposed to give small corrections to the zero approximation which can be calculated.

In this book, we shall consider the problem of a strong coupling regime in quantum field theory, calculations of path or functional integrals over the Gaussian measure and spectral problems in quantum mechanics. Let us consider these problems briefly.

The real problems of quantum field theory which will be studied in this book can be exemplified by a one-component scalar field. The total Hamiltonian is

$$H = \int dx \left[\pi^2(x) + (\nabla\phi(x))^2 + m^2\phi^2(x) + g\phi^4(x) \right].$$

If the coupling constant g is small enough one can choose in the decomposition $H = H_0 + H_I$

$$H_0 = \int dx \left[\pi^2(x) + (\nabla\phi(x))^2 + m^2\phi^2(x) \right], \quad H_I = g \int dx \phi^4(x).$$

Aside from all mathematical problems H_0 describes scalar particles with mass m, and the interaction Hamiltonian H_I gives small corrections. The point is what will this system be if the coupling constant g begins to grow, $g \to \infty$? Another question is what happens with this system if the temperature T is introduced?

The same kind of problems arise when we wish to calculate a functional or a path integral of the Gaussian type

$$I(g) = N \int \delta\phi \exp\left\{ -\frac{1}{2} \int \int dx dx' \phi(x) D^{-1}(x, x')\phi(x') + gW[\phi] \right\},$$

where $W[\phi]$ is a functional of $\phi(x)$. Evidently, for small g we can obtain the expansion of the function $I(g)$ in a series over g

$$I(g) = \sum_{n=0}^{\infty} g^n I_n.$$

In principle, each term I_n can be calculated, although practically only several of the lowest terms can be computed. As above the same point is how to calculate this integral if the coupling constant grows $g \to \infty$.

In quantum mechanics the standard Hamiltonian has the form

$$H = \frac{p^2}{2m} + V(q),$$

and the problem is to find the energy spectrum $\{E_n\}$ and the proper wave functions $\{\Psi_n\}$. Again, we face looking for a decomposition $H = H_0 + H_I$ in which the Hamiltonian H_0 keeps the main properties of the total Hamiltonian H.

Thus, we have two problems:

- What type of operators should be chosen as H_0? We are looking for a representation of H_0 for which the solutions of the Schrödinger equation can be found in the explicit form, and subsequent perturbation calculations should be as simple as possible.
- What principles should be formulated to take into account the main quantum contributions arising from the interaction described by the total Hamiltonian?

Our goal is to obtain the representation $H = H_0 + H_I$ in which H_0 describes the physical system closest to H and the interaction Hamiltonian H_I gives small perturbation corrections only.

The first step is to choose the Hamiltonian of the harmonic oscillator for H_0. The harmonic oscillator is the simplest quantum system both from a physical and a mathematical point of view. One can say that the oscillator accumulates the essence of quantum physics in the most pure form. Moreover, it is no exaggeration to say that our notions about quantum systems, especially in quantum field theory, have been formed under the crucial influence of ideas associated with quantum oscillators. This influence was very fruitful and we can see that the idea of harmonic oscillators spans very different areas of quantum physics.

Roughly speaking, the expansion of the oscillator picture on most quantum systems is realized *via* a very simple scheme: *a stable quantum system in an appropriate representation can be described by some set of harmonic oscillators with suitable frequencies.* If we are able to find this *appropriate representation* then we can expect that the difference between the quantum system under consideration and its oscillator representation is not too large and can be taken into account by perturbation calculations. Thus, the mathematical point is to find and construct these representations.

Of course, this prescription taken as a rigid rule works successfully only in the simplest cases and we should exhibit our inventiveness to get *the appropriate representation* for nontrivial Hamiltonians.

Our optimism is based on the following observations. In quantum field theory (QFT) a quantum field is treated as an infinite set of oscillators with a frequency defined by a mass parameter of a field. One of the most important features of QFT is that no interaction of quantum fields changes the oscillator nature of the quantum fields but only redefine their masses and other physical characteristics.

The same situation takes place for path integrals over the Gaussian measure in most interesting physical problems. In these cases path integrals of the type mentioned above contain the interaction functionals $gW[\phi]$, which do not change the Gaussian character of the measure. So that we can expect that in the strong coupling regime $g \to \infty$ the main contributions from interaction lead to modification of the operator $D^{-1}(x, x')$ defining the Gaussian measure.

In quantum mechanics the situation is more difficult because in most cases the asymptotic behavior of the wave functions for large and short distances is quite far from the behavior of functions of the oscillator basis. However, it is possible to change the variables in the Schrödinger equation to coordinate the asymptotic behavior of exact and oscillator wave functions. Mathematically these transformations lead to equivalence of quantum problems in spaces of different dimensions. The well-known Kustaanheimo-Stiefel transformation is a bright example of this kind. We shall follow this idea.

Thus, we can see that for quite a wide range of quantum problems there exist reasons to look for the initial approximation in the form of an oscillator basis.

The second step is to find a universal method to extract the leading quantum contributions of an interaction and to include them in H_0, i.e. in the formation of basic states of a quantum system. For this aim, let us pay attention to the following result of the QFT formalism which indicates how to do this. Namely, the total Hamiltonian $H = H_0 + H_I$ of a quantum field system is usually represented in the form in which all field operators in the Hamiltonian are written in the normal product and the interaction Hamiltonian contains field operators in powers more than 2. For example, in the simplest case (2) and (3) the total Hamiltonian in the formalism of creation and annihilation operators $a_{\mathbf{k}}^+$ and $a_{\mathbf{k}}$ looks like

$$H_0 = \int d\mathbf{k}\omega(\mathbf{k})a_{\mathbf{k}}^+ a_{\mathbf{k}},$$

$$H_I \sim g \prod_{j=1}^{4} \int \frac{d\mathbf{k}_j}{\sqrt{2\omega(\mathbf{k}_j)}} \sum_{m=0}^{4} \delta(\mathbf{k}_1 + \ldots + \mathbf{k}_m - \mathbf{k}_{m+1}\dot{s} - \mathbf{k}_4)$$

$$\times a_{\mathbf{k}_1}^+ \cdots a_{\mathbf{k}_m}^+ a_{\mathbf{k}_{m+1}} \cdots a_{\mathbf{k}_4}.$$

We shall call this representation *the correct form* of the total Hamiltonian. The total Hamiltonian is written in *the correct form* if

- all field operators in the total Hamiltonian H are written in *the normal product*,
- the Hamiltonian H_0 is quadratic over the field operators,
- the interaction Hamiltonian H_I contains field operators in powers more than 2.

Mathematically, the requirement of the normal product means the following. For example, in quantum field theory of scalar fields of the type (2) and (3), noninteracting particles are described by a Hamiltonian representing a set of oscillators and the interaction is described by the product of field operators of the type $g\phi^4$, which contains diverging contributions related to the divergence of the scalar field propagator at zero

$$\langle\phi(x)\phi(x)\rangle = D(0) \sim \hbar \int \frac{d^4k}{m^2 + k^2} \sim \infty.$$

This term is the highest divergence in renormalizable theories of the ϕ^4 type. This singularity can be removed by renormalizing the mass m and vacuum energy. This renormalization procedure is equivalent to the postulate that field operators in the interaction Hamiltonian have to be written in terms of normal products over particle creation and annihilation operators. Thus one can say that

The normal ordering of the Hamiltonian means essentially that the main quantum contributions to the ground state or vacuum of the system are taken into account.

We wish to use this result and our idea is to represent the Hamiltonian for a quantum system:

- the Hamiltonian is written in the formalism of creation and annihilation operators of an oscillator basis with an appropriate frequency,

- the Hamiltonian is written in the correct form.

We shall call this representation *the oscillator representation* (OR). The Hamiltonian in the oscillator representation is very convenient for any calculations and this form implies the main quantum contributions to be taken into account.

In this book, we use the oscillator representation method in three areas of quantum physics: quantum field theory, computation of the path integrals over a Gaussian measure, and the calculation of spectra for different potentials in quantum mechanics. The results are presented in three parts of this book, respectively. Application of the general idea of the oscillator representation method to the particular quantum systems requires some specific tools. The main instruments are: in the case of QFT, the canonical transformations and the renormalization group; in the case of calculations of path integrals, the conception of the normal product with respect to a given Gaussian measure; in quantum mechanics, the transformations of the Kustaanheimo-Stiefel type and representations of the Schrödinger equations in spaces of different dimensions.

Primarily, the method was applied to the problem of the strong coupling regime and phase structure of the φ_2^4 quantum field theory. It has been found that an approximate solution of the problem can be obtained by combining the canonical transformations of field variables and normal ordering of the creation and annihilation operators.[1] In the lowest approximation, the solution obtained there is identical to the result of the variational method of the Gaussian effective potential[2]. Meanwhile, in contrast with the variational method the new technique ensures a canonical structure of the theory and, hence, provides calculation of perturbation corrections over effective coupling constants which can also be used to control an accuracy of approximation. Later on, the method was generalized and applied to renormalizable quantum field theories, path integrals, and quantum mechanics of bound states. Thus, the oscillator representation method arises in QFT and is based on the ideas of QFT. That is why we begin the book with the part devoted to the QFT problems.

The OR can be considered as a kind of the extended variational technique, but in contrast with the variational method, it is applicable to the QFT models with ultraviolet (UV) divergencies and to theories with nonhermitian and complex action (stochastic and dissipative processes). The OR is characterized by a high accuracy of the lowest approximation, which can be obtained by simple and rapid calculations. It gives a regular prescription for the calculation of the highest order corrections to the lowest approxima-

---

[1] G.V. Efimov, Int.J.Mod.Phys. **A4** (1989) 4977. A similar technique was also developed by Chang and Magruder in earlier papers: S.-J. Chang, Phys.Rev.**D13** (1976) 2778; S.F. Magruder, Phys.Rev. **D14** (1976) 1602.

[2] P.M. Stevenson, Phys. Rev. **D32** (1985) 1389.

tion. Detailed investigations in the following problems of quantum physics are represented in this book:

- phase sructure and phase transitions (strong coupling regime) in the QFT systems at arbitrary coupling constants and temperature;
- ground-state energy of the d-dimensional polaron for any values of the coupling constant;
- wave propagation in stochastic or random mediums;
- spectra of two-body quantum systems described by the Schrödinger equation for any potentials;
- bound states and stability of three-body Coloumb systems;

and some others.

We would like to stress that the three parts of this book are rather independent one from other. Each part contains all the necessary information needed for understanding the generalities and examples. The references between the parts are reduced to a minimum. Therefore, readers can start without any trouble with the parts most interesting to them.

Part I

The Phase Structure of
Quantum Field Systems

2. Formulation of the Method

We start this book with the realization of the oscillator representation method in quantum field theory and the application of this method to the problems of a strong coupling regime and phase transitions in various quantum field models. Many readers might think that this is an unusual manner in which to describe some new calculational techniques taking as an example the most difficult part of quantum physics. However, we hope that astonishment will be dispelled as soon as the main ideas of the OR method are explained. The central point here is that according to the conventional scattering picture in QFT the initial and final states correspond to the physical free particles. The problem is how to describe these particles for different values of the renormalized coupling constants or temperature. This means that interactions of quantum fields do not change the oscillator nature of the states of the fields. As a result the realization of the OR method in QFT is more direct and its algorithm is more universal for different QFT models than is the quantum mechanical systems.

The most subtle point of investigation of the phase transitions and strong coupling regime in QFT is a careful formulation of the problem. What are the free parameters of a field system? What is the mathematical definition of the different phases of a field system? How should we manipulate the ultraviolet divergences, which are important ingredient of most QFT models? The current chapter is devoted just to answers to these questions.

2.1 Breaking Symmetry in Quantum Field Theory

About twenty years ago, when the systematic investigation of phase structure of quantum field systems was begun, S. Coleman and E. Weinberg [1] found that quantum corrections can initiate spontaneous symmetry breaking (SSB) in the field theories symmetric on the classical and quasi-classical level.

Approximately at the same time D.A. Kirghnits and A.D. Linde showed [2, 3] that the symmetry is restored at high temperature in some field theories with SSB postulated at zero temperature (see also [4]–[7]).

These conclusions are based on the idea that a constant classical scalar field or condensate can arise in all space-time. The appearance of such a condensate means reconstruction of the ground state of a system. This reconstruction results in the changing of masses of fields and the character of field interactions. In other words, it has turned out that quantum field systems have a complicated vacuum or phase structure and phase transitions take place for specific values of dynamical parameters (coupling constants) and external parameters (temperature, external fields).

The methods, used in [1, 3], are based on the loop expansion of the effective potential, so that their applicability is restricted by the weak coupling regime. Further progress in studying the phase structure of quantum field models was conditioned by the development of nonperturbative methods.

Most investigators concentrated on φ^4-type scalar field theories with the following classical Lagrangians:

$$L(x) = \frac{1}{2}\varphi(x)\left(\Box - m^2\right)\varphi(x) - \frac{g}{4}\varphi^4(x), \tag{2.1}$$

$$L(x) = \frac{1}{2}\varphi(x)\left(\Box + \frac{1}{2}m^2\right)\varphi(x) - \frac{g}{4}\varphi^4(x), \tag{2.2}$$

$$L(x) = \frac{1}{2}\sum_i^N \varphi_i(x)\left(\Box - m^2\right)\varphi_i(x) - \frac{g}{4}\left(\sum_i^N \varphi_i^2(x)\right)^2 \tag{2.3}$$

in space-time R^{d+1} ($d = 1,\ 2,\ 3$) at zero and finite temperature T. Here $x = (\mathbf{x}, t)$.

Lagrangians (2.1) and (2.2) describe the one-component scalar field. They are invariant under the transformation $\varphi \to -\varphi$. Lagrangian (2.3) describes the $O(N)$-multiplet of scalar fields and is invariant under $O(N)$-transformations and under $\varphi_i \to -\varphi_i$. The parameters m and g are positive.

If the dimensionless parameters

$$G = \frac{g}{2\pi m^{3-d}} \quad \text{and} \quad \theta = \frac{T}{m}$$

are small enough, then Lagrangians (2.1) and (2.3) describe in QFT symmetric interaction, while Lagrangian (2.2) corresponds to spontaneously broken symmetry.

Scalar fields play a fundamental role in the unified theories of strong, electromagnetic and weak interactions. The mathematical structure of scalar field theory is simpler than that of the vector and spinor field theories. At the same time, the mechanisms of symmetry rearrangement specific for scalar fields occur in realistic four-dimensional field theories as well (e.g., [3]). Thus models (2.1–2.3) look attractive (simple, but nontrivial) objects for research into dynamical mechanisms of symmetry rearrangement.

Within the framework of so-called constructive QFT B. Simon and R. Griffitths [8, 9], J. Glimm and A. Jaffe [10, 11], O. Mc Bryan and J. Rosen [12] have proven a set of general theorems which establish the existence of nontrivial two-dimensional theories of self-coupling scalar field. These theorems prove an existence of the phase transition in two- [9, 10, 11] and three-dimensional [12] φ^4 theories at zero temperature. It has been argued that these transitions are of the second order, although the proof is not complete [11, 12]. Unfortunately constructive quantum field theory gave no effective method (like Feynman diagrams) for the calculation of matrix elements and other physical characteristics of the QFT models. In particular, any information about the critical value of the coupling constant or the dependence of the mass and order parameter on the coupling constant was not obtained within constructive QFT.

Investigation of the so-called triviality problem of the $(\varphi^4)_{d+1}$ theory has shown that for $d > 3$ this theory turns out either to be noninteracting (the bare coupling constant is equal to zero when ultraviolet (UV) regularization is removed) or unstable (the coupling constant becomes negative) [13]. For the realistic case $d = 3$ complete solution of the triviality problem has not been achieved up to now [14].

Very attractive approach to the problem under discussion is the variational method of the Gaussian effective potential (GEP). This approach in QFT was initiated by the investigations of T. Barnes and G.I. Chandour [15], W.A. Bardeen and M. Moshe [16], P.M. Stevenson [17], M. Consoli [18] and others. Original investigations in the same direction were made by S.-J. Chang [19], S.F. Magruder [20], G. Baym and G. Grinstein [21], F. Grassi, R. Hakim and H.D. Sivak [22]. These papers differ from one another in renormalization prescriptions, methods of taking into account the temperature effects, using or not using the $1/N$-expansion for studying $O(N)$-invariant systems and so on. L. Polli and U. Ritchel [23] researched into the nature of phase transitions in the $(\varphi^4)_2$-theory and undertook an attempt to go beyond the GEP (the so-called post-Gaussian approximation).

Within the GEP approach critical values of the coupling constant and temperature were calculated approximately for various theories of self-interacting scalar fields. A dependence of the masses and order parameter in the broken symmetry (BS) phase on the coupling constant and temperature were obtained as well. L. Polli and U. Ritchel [23] found the second order phase transition to be in the $(\varphi^4)_2$ theory in accordance with the above-mentioned general theorems (see also [66]).

At the same time, specific features of the variational approach in QFT make their results unreliable if a theory has UV divergencies in the highest perturbation orders (see Feynman's paper in [24]). Such a situation is realized in the models (2.1–2.3) for $d > 1$. Many authors have stressed that this difficulty is a direct result of the application of the variational method to the Hamiltonians which are not operators in a mathematical sense [25]-[27]. The variational principle is applied to a fixed Hamiltonian *with given set of bare parameters* (masses and coupling constants). On the other hand, an effective potential *with given set of renormalized parameters*, that is important from physical viewpoint. After renormalization the basic inequality of the variational approach

$$U_{\text{eff}}^{(+)}(\varphi) \equiv \min_{\psi}\langle\psi|H|\psi\rangle \ \geq \ U_{\text{eff}}(\varphi) \tag{2.4}$$

turns out to be useless, because UV divergencies of variational estimation $U_{\text{eff}}^{(+)}$ and exact effective potential U_{eff} are different and, as a consequence, the difference between $U_{\text{eff}}^{(+)}$ and U_{eff} is infinitely large [25]. Inequality (2.4) is reliable only for the Hamiltonians which are well-defined operators on the Hilbert space of states. Such a situation takes place in QFT models containing normally ordered divergences only (e.g., (2.1–2.3) for $d = 1$).

Another problem relates to the impossibility of controlling the accuracy of approximation in the variational method even in the cases where inequality (2.4) is applicable [28].

A way to overcome these difficulties of the GEP approximation was proposed in the original approach of S.-J. Chang [29] and S.F. Magruder [20]. Their method is based directly on mass renormalization within perturbation theory (PT). An analogous technique was developed in the paper [30] for the finite temperature case. Unlike the GEP these methods take into account mass renormalization in the highest perturbation orders. S.-J. Chang and S.F. Magruder have researched the phase structure of model (2.2) in R^3 and have shown that the symmetry is restored in strong coupling regime. This conclusion contradicts the result of the GEP approximation [21].

The approach of S.-J. Chang and S.F. Magruder is quite close in spirit to the OR method, which is the main subject of this book. In this part we shall discuss and illustrate by examples an application of the oscillator representation method in QFT.

The OR is a specific modification of the canonical transformation method providing the regular prescriptions for dealing with the highest order renormalization and giving a possibility to control the accuracy of approximation.

This modification was proposed in the paper [32], where the phase structure of two-dimensional models (2.1,2.3) was investigated. Extension to more complicated three- and four-dimensional theories and field systems at finite temperature was performed in papers [33]-[37]. The most consistent formulation of the method is given in [36].

Our approach to the problem consists in combination of the canonical transformation method and renormalization group. The idea of such a combination originates in the fundamental properties of the local QFT: existence of nonequivalent representations of the canonical (anti)commutation relations (CR) and UV divergences (e.g., see [48, 60] and references therein). Each of these properties relates closely to the problem of phase structure. Let us clarify what we mean.

– Description of any quantum system requires choosing the Hilbert space of states on which the canonical variables are defined as operators. Equivalently, one can say that definite representation of the canonical (anti-)commutation relations has to be chosen. For the case of quantum systems with a finite number of degrees of freedom all representations are unitary equivalent to each other.[1] This is not true for the QFT and quantum statistical systems with infinite number of degrees of freedom. There are a lot of unitary nonequivalent CR representations. Each representation corresponds to a specific physical picture that is characterized, in particular, by symmetry properties. From a physical viewpoint, the existence of the nonequivalent representations means that the field system may have a set of vacua. The internal dynamics of a system chooses among them the vacuum, which has minimal free energy and, hence, is an actual ground state of the system for given values of dynamical and external parameters of the model. At the new values of these parameters the old vacuum may become unstable and the ground state would be realized with another vacuum from the above-mentioned set, so that nonequivalent representations give a method for the classification and description of the different phases. We will identify a *phase structure* of given QFT model with a *set of nonequivalent CR representations* realized in this model for different values of dynamical and external parameters. The transformations of the canonical variables preserving the canonical relations (canonical transformations) are a convenient tool for construction of nonequivalent representations.

– The ultraviolet divergencies give the main quantum contribution to physical quantities like masses, effective coupling constants and so on. Renormalization removes the divergencies and actually means taking into account the leading quantum corrections, which have a crucial influence on the formation of different phases of a system. A way to recover this influence is provided by the formalism of the renormalization group.

These two more or less evident statements leads to the following scenario. Let some QFT model in the form of a Lagrangian and (or) Hamiltonian is given in the space-time R^{d+1}. First of all, the ultraviolet divergences should be classified and removed by an appropriate renormalization procedure. The renormalization counter-terms should be introduced into the Hamiltonian. This procedure as well as the canonical formalism of QFT indicate the **correct form** of the total Hamiltonian:

$$H = H_0 + H_I + H_{ct} + VE. \tag{2.5}$$

The standard free part H_0 describes a ground state of the field system and therefore determines the Fock space, i.e., a definite representation of the CR. The interaction Hamiltonian H_I does not contain terms linear or quadratic

---

[1] Although all representations are equivalent in quantum mechanics, an appropriate representation should be chosen for each concrete system in order to provide the simplest solution of the problem under consideration. Specific aspects of application of the OR to QM systems are discussed in the last part of the book.

in fields and describes corrections to H_0. The counter-term operator H_{ct} removes all UV divergencies in each perturbation order over the interaction Hamiltonian H_I. The form of H_{ct} is determined by H_0, H_I and renormalization scheme. The constants V and E mean the "volume" of the space R^d and the density of the vacuum energy respectively. The vacuum energy E is connected with the free energy density according to the equation

$$F = E - TS,$$

where S is the entropy density. Usually the energy E is chosen to be equal to zero in the initial representation. One can say that the QFT model is defined: the Hamiltonian and appropriate representation of the CR are chosen.

It is important to stress that the division (2.5) of the total Hamiltonian has a physical sense if the coupling constants in the operator H_I are small.

If the parameters of a model are changed, the effective coupling constants can become large, quantum corrections induced by the operator H_I become large either and the initial representation of the total Hamiltonian loses its physical sense. This means that we should look for an another CR representation. The problem is formulated as follows: *what representation of the canonical (anti)commutation relations is suitable for different values of the dynamical and external parameters of the model and what physical picture corresponds to this representation?*

The OR method for investigation of this problem is based on two steps. First, the total Hamiltonian of a field system should be written in the **correct form** in the representation which seems to be suitable for specific values of the dynamical and external parameters (e.g., in the weak coupling regime at zero temperature). The renormalization scheme should be fixed.[2] Second, the **canonical transformations** of the field variables should be performed. The requirement that the Hamiltonian expressed in new variables has the **correct form** leads to equations defining unitary nonequivalent representations of the CR for given values of dynamical and external parameters. Each representation is characterized by the effective coupling constants and free energy density. A system is considered to be in a definite phase if the effective coupling constants and free energy density in the representation describing this phase are the smallest ones. The effective coupling constants are used to control the accuracy of the approximation.

This part of the book has the following structure. Chapter 2 contains a brief review of the canonical quantization method and the renormalization group (RG) technique, which are important ingredients of the OR method in QFT. The phase structure problem in QFT and nonequivalent CR representations are discussed and the most important examples of nonequivalent representations are considered in section 2.4. Different aspects of quantization,

---

[2] Renormalization schemes in different representations should be equivalent. This requirement follows from the statement of the problem and is explained in details in sections 2.3 and 2.4.

canonical transformations and renormalization group discussed in sections 2.2–2.4 indicate the detailed formulation of the OR method in QFT. Such formulation is given in section 2.5. Other chapters are devoted to research into the phase structure of concrete quantum field systems.

2.2 Canonical Quantization and the S-matrix

Research into the phase structure of the QFT system can be understood as an investigation of the behaviour of the field system when input parameters characterizing it are changed. It is quite obvious that such an investigation needs to know at least what the system under consideration is at some fixed values of the parameters. One can say that the problem of phase structure in QFT requires quite definite "initial conditions". In this section we clarify this point and show that such "initial conditions" can be extracted from consideration of the QFT scattering problem.

As has already been mentioned, an investigation of relativistic quantum field systems comes across two main difficulties:

– nonequivalent representations of the canonical relations and, as a result, the mathematical unfoundedness of the interaction representation;
– ultraviolet (UV) divergences.

Mostly because of these difficulties the general theory of interacting quantum fields has not been formulated up to now, so that we are forced to reduce the complete problem to a more specific scattering problem. Namely, we are able to describe processes for which particles at the beginning (time $t \rightarrow -\infty$) and the end (time $t \rightarrow +\infty$) are widely separated from each other and can be considered free. The sole objective is to calculate the probabilities of various transitions between these asymptotically free states. The object characterizing an interacting quantum field system and containing all probability amplitudes for elastic and inelastic scattering and for mutual transformations of particles is the scattering matrix or the S-matrix. One can say that the scattering problem in QFT is solved if the rules of calculation of the S-matrix elements are formulated.

Let the total Lagrangian L or Hamiltonian H describing a quantum field system be given, i.e. its operator structure and parameters (masses, coupling constants and so on) are defined. The first step that should be made is to divide the Lagrangian (Hamiltonian) into free and interaction parts:

$$L = L_0 + L_I, \qquad \text{or} \qquad H = H_0 + H_I.$$

This division has the mathematical and physical sense if we can realize the system described by the free Hamiltonian H_0 (in- and out-fields) and formulate the rules of the perturbation calculation of the S-matrix using the interaction Hamiltonian H_I together with an effective coupling constant that is small enough. An extraction of H_0 from the total Hamiltonian means that definite representation of the CR is fixed and describes the in- and out-states.[3] In order to overcome the UV-divergences appearing in the perturbation expansion of the S-matrix one has to define the calculation rules including regularization and appropriate counter-terms, or, in other words, to define the renormalization scheme.

As soon as the total Hamiltonian H is divided into H_0 and H_I and the R-scheme is fixed, one has a representation of the canonical relations that is physically acceptable in the region of physical parameters where the effective (perturbation) coupling constants are small. We shall call this representation the "initial phase" of the field system under consideration.

Now let us consider the behaviour of a quantum field system when its parameters are changed and go out of the region where the effective coupling constants are small. We shall use the canonical transformations of field variables in order to find other representations of the same system. We should proceed from the initial representation where the theory is defined and keep the R-scheme invariable. Only in this case other representations can be considered as new phases of the system. In other words, the initial representation plays the role of "the initial conditions".

Thus we should start with the definition of "initial conditions". It is quite natural to suppose that *in the weak coupling regime at zero temperature* the self-interacting scalar field can be described by *the standard formalism of canonical quantization*, and the main theoretical aim is *the construction of the S-matrix according to the standard perturbation theory* (e.g., see [43]).

For definiteness we shall consider the model (2.1) in four dimensions. Let us write down the Lagrangian of the system in the standard form:

$$L(x) = \frac{1}{2}\varphi(x)\left(\Box - m^2(\mu)\right)\varphi(x) - \frac{g(\mu)}{4}\varphi^4(x) \qquad (2.6)$$

$$+\frac{1}{2}(Z_2 - 1)\varphi(x)\Box\varphi(x) - \frac{1}{2}\delta m^2(\mu)\varphi^2(x) - \frac{1}{4}(Z_1 - 1)g(\mu)\varphi^4(x) - \delta E.$$

We use the standard notation for the renormalized field φ, mass $m(\mu)$ and coupling constant $g(\mu)$. The parameter μ characterizes a renormalization scale. The renormalization is performed by the counter-terms written in the second line of (2.6). We will consider renormalization problems specially below.

The renormalized field φ and canonically conjugated momenta

$$\pi(\mathbf{x}, x_0) = \frac{\delta}{\delta\dot{\varphi}(\mathbf{x}, x_0)} \int d\mathbf{y} L(\mathbf{y}, x_0) = Z_2\dot{\varphi}(\mathbf{x}, x_0)$$

are considered as the canonical variables. The canonical commutation relations

---

[3] Usually the Fock representation corresponding to the free particles with physical masses is used.

$$[\varphi(x), \varphi(y)]_{x_0=y_0} = [\pi(x), \pi(y)]_{x_0=y_0} = 0,$$
$$[\varphi(x), \pi(y)]_{x_0=y_0} = i\delta(\mathbf{x} - \mathbf{y}) \tag{2.7}$$

are postulated. The Hamiltonian density has the form

$$H \qquad\qquad H = \pi\dot\varphi - L = \quad_0 + H_I + H_{ct},$$

where

$$H_0 = \frac{1}{2}\left[\pi^2 + (\nabla\varphi)^2 + m^2(\mu)\varphi^2\right], \quad H_I = \frac{1}{4}g(\mu)\varphi^4,$$

$$H_{ct} = \frac{1}{2}\left[\left(\frac{1}{Z_2} - 1\right)\pi^2 + (Z_2 - 1)(\nabla\varphi)^2 + \delta m^2(\mu)\varphi^2\right]$$

$$+ \frac{1}{4}(Z_1 - 1)g(\mu)\varphi^4.$$

In the interaction representation picture the total Hamiltonian H is divided into the free Hamiltonian H_0 (which determines the space of states) and the interaction Hamiltonian. The operators φ and π read

$$\varphi(x) = \int \frac{dk}{2\pi} \frac{1}{\sqrt{2\omega}} \left[a(k)\exp\{ikx - i\omega x_0\} + a^+(k)\exp\{-ikx + i\omega x_0\}\right],$$

$$\pi(x) = \frac{1}{i}\int \frac{dk}{2\pi} \sqrt{\frac{\omega}{2}} \left[a(k)\exp\{ikx - i\omega x_0\} - a^+(k)\exp\{-ikx + i\omega x_0\}\right],$$

$$\omega(k) = \sqrt{k^2 + m^2(\mu)}, \quad [a(k), a^+(k')] = \delta(k - k'). \tag{2.8}$$

The creation $a^+(k)$ and annihilation $a(k)$ operators are defined on the Fock space of free particles with the mass $m(\mu)$. The vacuum vector $|0\rangle$ of this space obeys the conditions

$$a(k)|0\rangle = 0 \quad \forall k, \quad \langle 0|0\rangle = 1. \tag{2.9}$$

The S-matrix is constructed by using iterations over degrees of the Hamiltonian $(H_I + H_{ct})$ and takes the well-known general form (e.g., see [44])

$$S = \lim_{\Lambda\to\infty} T_\Lambda \exp\left\{-i\int d^d x\, [H_I(x) + H_{ct}(x)]\right\}, \tag{2.10}$$

where operators H_I and H_{ct} are written in the interaction representation picture. The symbol T_Λ means the time ordering plus definite UV regularization.[4] The symbol $\lim_{\Lambda\to\infty}$ means the procedure of switching off the UV regularization. The operator H_{ct} is chosen in such a way that each term of the perturbation expansion of the S-matrix is finite in the limit $\Lambda \to \infty$.

---

[4] The regularization can be introduced in the perturbation series directly. One can use dimensional regularization [46, 50], nonlocal formfactors [45, 47, 60], etc.

The formulated procedure reflects the fact that in order to construct the S-matrix it is not enough to know only the Lagrangian, but it is necessary to define the calculation rules of the coefficient functions in perturbation series as well [43]. These rules relate to the presence of UV divergences and can be reduced to a redefinition of the chronological product of field operators. When the operator H_{ct} is introduced into the S-matrix (2.10) an ambiguity in the definition of the T-product is replaced by an ambiguity in the renormalization scheme (R-scheme), *which should be fixed.*

If on-shell renormalization is used, i.e., the renormalized mass coincides with the physical (pole) mass and the residue of the two-point Green function in the pole is equal to unity by a construction, then the Fock space induced by the free Hamiltonian describes the physical asymptotic *in*- and *out*-states.

Besides that, the reliability of perturbation theory is implied, since the S-matrix (2.10) is given by an iteration procedure only. Therefore the renormalized coupling constant is assumed to be small enough: $g(\mu) \ll 1$.

This quantization scheme provides a suitable description of most quantum field systems in the weak coupling regime $g(\mu) \ll 1$.

It should be noted that the formalism of the thermo field dynamics (TFD) [61, 67, 68] provides a natural way to consider the temperature effects within the canonical quantization. The TFD approach will be discussed in detail in the last section of the Part I.

2.3 The Renormalization Group

2.3.1 Renormalization Schemes

As has been already mentioned, the definition of the S-matrix according to (2.10) implies fixing the renormalization scheme. Prescriptions for removing the UV divergences used in specific calculations can be very different. All of these prescriptions can be divided into two classes (e.g., see [49]):

– mass-independent R-schemes;
– momentum subtraction R-schemes (e.g., the so-called canonical μ-scheme, subtractions at zero external momenta and so on).

As a representative example of mass-independent R-schemes one can recall the well-known minimal subtraction (MS) scheme [46, 50, 51]). In this case the renormalization procedure is actually reduced to the dimensional regularization of divergent diagrams and subtraction of the poles over the parameter $\varepsilon = d_{ph} - d$, where d_{ph} is the dimension of the physical space-time $R^{d_{ph}}$. At the regularization step one should introduce an arbitrary parameter μ having the dimension of mass and defining a renormalization scale. The value of this parameter is not determined by the MS-scheme itself. Some additional condition has to be introduced.

The canonical μ-scheme is defined by the following conditions on the renormalized two-point Green function:

$$D(p^2) \longrightarrow \frac{i}{p^2 - m^2(\mu) + i0} \quad \text{for } p^2 \to \mu^2 \qquad (2.11)$$

and on the four-point vertex function

$$\Gamma^{(4)}(s, t, u) = g(\mu) \quad \text{for } s = t = u = \frac{4}{3}\mu^2, \qquad (2.12)$$

where $m(\mu)$ and $g(\mu)$ are the renormalized mass and coupling constant. The Mandelstam variables s, t, u are defined in the usual way:

$$s = (p_1 + p_2)^2, \ t = (p_1 - p_3)^2, \ u = (p_1 - p_4)^2.$$

In the φ^4 theory the subtraction point μ must obey the inequality

$$\mu^2 \leq 9m^2(\mu),$$

but it is arbitrary in other respects. The renormalization scale μ should be fixed by an additional condition in analogy with the case of the MS-scheme.

These examples show that we can consider R-scheme to be fixed if one-parametric R_μ-class of renormalization prescriptions is chosen and renormalization scale μ is fixed by an additional condition. It is natural to express this condition in terms of internal parameters of a theory. For example, one can fix the ratio $m(\mu)/\mu$ $(m(\mu) \neq 0)$.

We shall say that the renormalization scheme is fixed if

– the one-parameter R_μ-class of renormalization prescriptions is chosen,
– the renormalization scale μ is fixed by the relation

$$\frac{m(\mu)}{\mu} = C,$$

where $m(\mu)$ is the renormalized mass, C is a constant.

It should be noted that the on-shell renormalization scheme corresponds to the canonical μ-scheme when $\mu = m(\mu)$ $(C = 1)$ (see (2.11)). In this case the renormalized mass coincides with the physical (pole) one. We shall use this simple fact for investigation of the model (2.6) in R^{3+1}.

2.3.2 Renormalization Group Equations

Let us rewrite the Lagrangian (2.6) in terms of the bare field φ_0, mass m_0 and coupling constant g_0:

$$L(x) = \frac{1}{2}\varphi_0(x)\left(\Box - m_0^2\right)\varphi_0(x) - \frac{g_0}{4}\varphi_0^4(x). \qquad (2.13)$$

The bare and renormalized quantities are connected by the relations

$$\varphi_0 = \sqrt{Z_2} \cdot \varphi, \quad m_0^2 = Z_m \cdot m^2, \quad g_0 = Z_g \cdot g.$$

The conception of renormalization invariance originates from the idea that a changing of the R-scheme leading to changing of the renormalization constants Z_i should be compensated by a corresponding transformation of the renormalized field, mass and coupling constant in such a way that the bare quantities are kept invariable. Such a demand provides an invariance of the S-matrix under changing of the R-scheme. This idea gets a constructive sense if the relations connected renormalized quantities in different R-schemes are known. Such relations are defined by the renormalization group (RG) equations.[5]

The changing of the R-scheme within the same R_μ-class, i.e. a variation of the scale parameter $\mu \rightarrow \nu$, leads to dependence of the renormalization constants Z_i on the scale ν as follows[6]:

$$Z_i = Z_i \left(g(\nu), \frac{m(\nu)}{\nu}, \frac{\Lambda}{\nu} \right), \qquad (i = 2, m, g).$$

Let us introduce the functions $\bar{\beta}$, $\bar{\gamma}_m$ and $\bar{\gamma}$, which are defined by the equations

$$\bar{\beta} \left(g(\nu), \frac{m(\nu)}{\nu}, \frac{\Lambda}{\nu} \right) = -\nu \frac{d}{d\nu} \ln Z_g,$$

$$\bar{\gamma}_m \left(g(\nu), \frac{m(\nu)}{\nu}, \frac{\Lambda}{nu} \right) = \nu \frac{d}{d\nu} \ln Z_m, \qquad (2.14)$$

$$\bar{\gamma} \left(g(\nu), \frac{m(\nu)}{\nu}, \frac{\Lambda}{\nu} \right) = \nu \frac{d}{d\nu} \ln Z_2,$$

The RG functions β, γ_m and γ are defined by the solutions of (2.14) when the regularization is turned off:

$$\beta = \lim_{\Lambda \rightarrow \infty} \bar{\beta}, \quad \gamma_m = \lim_{\Lambda \rightarrow \infty} \bar{\gamma}_m, \quad \gamma = \lim_{\Lambda \rightarrow \infty} \bar{\gamma}.$$

The RG equations following from the invariance of the bare quantities under RG transformations have the form [46]

$$\nu \frac{dg(\nu)}{d\nu} = \beta \left(g(\nu), \frac{m(\nu)}{\nu} \right),$$

$$\frac{\nu}{m^2(\nu)} \frac{dm^2(\nu)}{d\nu} = -\gamma_m \left(g(\nu), \frac{m(\nu)}{\nu} \right), \qquad (2.15)$$

$$\nu \frac{d\zeta(\nu)}{d\nu} = -\frac{1}{2} \gamma \left(g(\nu), \frac{m(\nu)}{\nu} \right),$$

---

[5] Different aspects of the RG are discussed in many books, e.g., see [43, 46]
[6] If dimensional regularization is used then the renormalization constants are $Z_i \left(g(\nu), \frac{m(\nu)}{\nu}, \varepsilon \right)$.

where ζ relates to a finite field renormalization. The initial conditions are as follows:

$$g(\nu) = g, \quad m(\nu) = m, \quad \zeta(\nu) = 1, \quad \text{if } \nu = \mu. \tag{2.16}$$

For mass-independent R-schemes the RG functions depend on the coupling constant only and (2.15) can be solved in the general form. Taking into account the conditions (2.16) one gets

$$\ln \frac{\nu}{\mu} = \int\limits_{g}^{g(\nu)} \frac{dg'}{\beta(g')},$$

$$\ln \frac{m(\nu)}{m} = -\frac{1}{2} \int\limits_{g}^{g(\nu)} dg' \frac{\gamma_m(g')}{\beta(g')}, \tag{2.17}$$

$$\ln \zeta = -\frac{1}{2} \int\limits_{g}^{g(\nu)} dg' \frac{\gamma(g')}{\beta(g')}.$$

The RG functions β, γ_m and γ contain complete information about the renormalization structure of the model.

Renormalization group transformations can be considered as a redistribution of the finite terms between the principal Lagrangian (the upper line of (2.6)) and counter-term Lagrangian (the lower line of (2.6)). Solution of the RG equations is equivalent to partial summations of perturbation series and can be used for many specific applications.

Usually the RG formalism is applied to analysis of asymptotic behaviour of the Green functions over energy variables. Therefore most of interest is the dependence of the renormalized masses, coupling constants and Green functions on the renormalization scale that is described by (2.15).

We will investigate a problem of another kind. Namely, we are interested in the dependence of the ground state of the system on the parameters g and m, contained in the initial conditions (2.16) for a fixed renormalization scheme. The RG formalism turns out to be useful for the solution of this problem.

2.4 Unitary Nonequivalent Representations of the Canonical Relations

The problem of unitary nonequivalent representations was first encountered in the Van Hove model [52]. Friedrichs was the first to give the mathematical analysis of this problem [53]. Very important results were obtained by Haag [55, 56], Wightman and others [54, 59]. Detailed analysis of all the problems concerning the nonequivalent representations can be found in monographs [48, 60, 61].

In this section we give a brief review of these problems in order to clarify how nonequivalent representations can be used for investigation of the phase structure of quantum field systems.

2.4.1 An Infinite Number of Degrees of Freedom

Quantum field systems have an infinite nondenumerable number of degrees of freedom. This circumstance leads to very specific consequences. Let us consider a neutral scalar field and put it into a box with periodic boundary conditions. The field can be represented as a superposition of independent oscillations of the plane-wave type:

$$\varphi(x) = \frac{1}{\sqrt{V}} \sum_{\mathbf{k}} \frac{1}{\sqrt{\omega(\mathbf{k})}} \left[a_{\mathbf{k}} \exp\{ikx\} + a_{\mathbf{k}}^{+} \exp\{-ikx\} \right],$$

where $\omega = \sqrt{\mathbf{k}^2 + m^2}$. Canonical commutation relations for $a_{\mathbf{k}}, a_{\mathbf{k}}^{+}$ have the form

$$\left[a_{\mathbf{k}}, a_{\mathbf{k}'}^{+} \right] = \delta_{\mathbf{k}\mathbf{k}'}.$$

One can see that each oscillation has the structure of the harmonic oscillator. Each oscillator has its own Hilbert space $\mathcal{H}_{\mathbf{k}}$, which contains the ground state $|0; \mathbf{k}\rangle$, obeying the conditions

$$a_{\mathbf{k}} |0; \mathbf{k}\rangle = 0, \quad \langle 0; \mathbf{k} | 0; \mathbf{k}\rangle = 1,$$

and an infinite set of eigenvectors of the particle number operator $N_{\mathbf{k}} = a_{\mathbf{k}}^{+} a_{\mathbf{k}}$. The total Hilbert space is the direct product of the oscillator Hilbert spaces:

$$\mathcal{H} = \prod_{\mathbf{k}} \mathcal{H}_{\mathbf{k}}.$$

Let us denote a state vector of the total field system as $|n_1, n_2, ..., n_i, ...\rangle$, where n_i denotes the number of quanta with the momentum $\mathbf{k}_i$. Each oscillator Hilbert space $\mathcal{H}_{\mathbf{k}}$ contains a countable set $\mathcal{M}$ of states and the total Hilbert space $\mathcal{H}$ contains a countable set $\mathcal{N}$ of the oscillator Hilbert spaces $\mathcal{H}_{\mathbf{k}}$. The dimension of the total Hilbert space of the field equals $\mathcal{M}^{\mathcal{N}}$, i.e. this space has the continuum power. The set $\{|n_1, n_2, ..., n_i, ...\rangle\}$ is uncountable and, hence, can not be a basis of the separable Hilbert space. To construct a separable Hilbert space one has to choose a countable subset and use it as a basis. There is an infinite number of such subsets. If two subsets provide representations of the operators $(a_i, a_i^{+} ; i = 1, 2, ...)$, then these representations are *unitary nonequivalent to each other* in the sense that a vector of one representation is not a superposition of the vectors of another.

If any interaction is absent in the system, the number of particles is conserved and the set $\{|n_1, ..., n_i, ...\rangle\}$ turns out to be unnecessarily large for a description of the free field. In this case the subset

$$\{|n_1, ..., n_i, ...\rangle \; ; \; \sum_i n_i \; = \; \text{finite number}\}$$

is usually taken as a basis. This subset is countable (e.g., see [61]). It generates the separable Hilbert space called the Fock space.

For interacting fields the situation is more complicated, which is reflected in the well-known Haag theorem [54]-[59]. In the most general form this theorem is formulated as follows.

Within the framework of the following requirements:

- *Poincare covariance of the theory;*
- *positive definiteness of the norm;*
- *locality;*
- *positive definiteness of the spectrum;*
- *equal time canonical relations and completeness of a set of the canonical variables;*
- *existence of an invariant and normalizable vacuum state;*
- *uniqueness of the vacuum*

there is no field theory differing from the free field theory. The proof can be found, for example, in [48].

One of the essential consequences of this theorem is the statement that the interaction representation in QFT is not defined mathematically. Firstly, let us prove the following theorem.

In the Euclidean invariant field theory based on the Fock representation of the canonical commutation relations the unique vacuum state is Euclidean invariant.

A proof of this statement is quite simple [60]. Let the canonical variables $\varphi(\mathbf{x}, t)$ and $\pi(\mathbf{x}, t)$ satisfy the standard commutation relations. If the theory is Euclidean invariant, then the unitary operator $U(\mathbf{a}, \omega)$ corresponding to some rotation ω and some shift $\mathbf{a}$ gives affects the canonical variables in the following way:

$$U(\mathbf{a}, \omega)\varphi(\mathbf{x}, t)U^+(\mathbf{a}, \omega) = \varphi(\omega\mathbf{x} + \mathbf{a}, t),$$
$$U(\mathbf{a}, \omega)\pi(\mathbf{x}, t)U^+(\mathbf{a}, \omega) = \pi(\omega\mathbf{x} + \mathbf{a}, t).$$

If $|0\rangle$ is a vacuum state, then for all $\mathbf{x}$ it satisfies the condition

$$\varphi^{(-)}(\mathbf{x}, t)|0\rangle = 0,$$

where

$$\varphi^{(-)}(\mathbf{x}, t) = \frac{1}{2}\left[\varphi(\mathbf{x}, t) + \frac{i}{\sqrt{m^2 - \nabla^2}}\pi(\mathbf{x}, t)\right].$$

Therefore, according to the transformation rule the vector $U(\mathbf{a}, \omega)|0\rangle$ satisfies the same condition as well. Owing to the uniqueness of the vacuum state this means that

$$U(\mathbf{a}, \omega)|0\rangle = e^{i\alpha(\mathbf{a}, \omega)}|0\rangle = |0\rangle, \qquad (2.18)$$

because a possible transformation phase $\alpha(\mathbf{a}, \omega) = 0$ since this transformation is the continuous one-dimensional representation of the Euclidean group. Thus the theorem is proven.

In QFT this theorem leads to quite remarkable consequences. Let

$$\mathbf{H} = \int d\mathbf{x} H(\mathbf{x}),$$

where $H(\mathbf{x})$ is the Hamiltonian density. Then for any translation-invariant state Ψ the matrix element $(\Psi, H(\mathbf{x})H(\mathbf{y})\Psi)$ depends only on $(\mathbf{x}\text{-}\mathbf{y})$, so that

$$\|\mathbf{H}\Psi\|^2 = \int d\mathbf{x} \int d\mathbf{y}(\Psi, H(\mathbf{x})H(\mathbf{y})\Psi) = \begin{cases} 0, & \text{if } \mathbf{H}\Psi = 0, \\ \infty, & \text{if } \mathbf{H}\Psi \neq 0. \end{cases} \qquad (2.19)$$

The Hamiltonian $\mathbf{H}$ is a well-defined operator if it annihilates any translation-invariant state. In the Fock representation the only invariant state is the vacuum $|0\rangle$, so that the Hamiltonian $\mathbf{H}$ is well-defined only if it annihilates $|0\rangle$. Therefore the vacuum has to be eigenstate of the operators $\mathbf{H}$ and $\mathbf{H}_0$ simultaneously:

$$\mathbf{H}|0\rangle = 0 \quad \text{and} \quad \mathbf{H}_0|0\rangle = 0. \qquad (2.20)$$

Relations (2.20) can not be satisfied simultaneously for all Hamiltonians of the form

$$\mathbf{H} = \mathbf{H}_0 + g\mathbf{H}_I,$$

where $\mathbf{H}_I$ is a function of the field operator φ and therefore contains at least one term with a product of the creation operators only. Thus one can see that the operator

$$V(t) = \exp\{i\mathbf{H}t\} \cdot \exp\{-i\mathbf{H}_0 t\}$$

linking the interaction and Heisenberg representations is not defined on the Fock space.

A crucial point in the proof of the Haag theorem is a condition of the uniqueness of the vacuum state, which implies an absence of nonequivalent representations and the suitability of the Fock space for a description of interacting fields [48]. Therefore a possible radical way to correct the situation could consist in the nonequivalent representations being included into the structure of quantum field models and interacting fields being defined as operators on a nonseparable Hilbert space. Discussion of the axiomatic basis of such a theory can be found, for example, in the paper [62]. However, constructive realization of this idea is damped by the fact that no suitable classification of the nonequivalent representations has yet been formulated.[7]

---

[7] Haag gave a very clear review of different approaches to the classification of nonequivalent representations [56]. A more general and detailed consideration can be found in [57].

We will undertake a less radical attempt to correct the description of the interacting fields. We suppose that the formalizm described in the previous sections permits a basis for the calculation of the S-matrix elements in the weak coupling regime $g(\mu) \ll 1$. It is well-known that this approach is successful in the case of electromagnetic and weak interactions where the coupling constants are small enough. Of course, all principal problems associated with the Haag theorem remain valid.

However, let us weaken the condition of uniqueness of the vacuum. We suppose that the state space has a set of normalized and invariant vectors. Necessarily these vectors are unitary nonequivalent (in the above mentioned sense). In this case the usual identification of the vacuum and ground state of a field system is no longer correct. The true ground state of a system should be defined by some physical principles and it has to be determined by an internal dynamics of the system. To realize this idea it is necessary to introduce a set of vacuum vectors (the Fock spaces) taking into account the dynamics and to choose among them the vacuum, which is an actual ground state for given dynamical and external parameters of a system.

With this possibility one does not need to construct of the complete (nonseparable) state space. It is enough to suppose that such a space exists and all necessary Fock spaces are its subspaces. The standard canonical quantization formalism is valid in these subspaces.

Our realization of this idea, which will be formulated in the next section, consists in the following.

- A set of trial vacuum vectors (nonequivalent representations) are introduced by means of the canonical transformations.
- The dynamics is taken into account by means of the standard renormalization procedures and the RG equations.
- The actual ground state is the vacuum for which the free energy and effective coupling constants are the smallest ones at given values of the dynamical and external parameters of a system.

Before the formulation of this method let us consider the most important nonequivalent representations. At the same time we shall introduce the formulas to be used below.

2.4.2 Canonical Transformations

Let us consider two pairs of the Boson operators $[a(\mathbf{k}), a^+(\mathbf{k})]$ and $[b(\mathbf{k}), b^+(\mathbf{k})]$ and denote the Fock spaces where these operators are defined as $\mathcal{H}[a, b]$. The vacuum state $|0\rangle$ of this Fock space obeys the conditions

$$a(\mathbf{k})|0\rangle = 0, \quad b(\mathbf{k})|0\rangle = 0 \quad \forall \mathbf{k}.$$

Algebraic properties of the operators are determined by the commutation relations

$$[a(\mathbf{k}), a^+(\mathbf{k}')] = \delta(\mathbf{k} - \mathbf{k}'), \qquad (2.21)$$
$$[b(\mathbf{k}), b^+(\mathbf{k}')] = \delta(\mathbf{k} - \mathbf{k}'),$$

and other commutators are equal to zero.

Let us introduce the operators $\alpha(\mathbf{k})$, $\beta(\mathbf{k})$ by the relations

$$\alpha(\mathbf{k}) = c(\mathbf{k})a(\mathbf{k}) - d(\mathbf{k})b^+(-\mathbf{k}),$$
$$\beta(\mathbf{k}) = c(\mathbf{k})b(\mathbf{k}) - d(\mathbf{k})a^+(-\mathbf{k}). \qquad (2.22)$$

Here the numerical coefficients c and d are real functions of $\mathbf{k}^2$ and satisfy the condition

$$c(\mathbf{k})^2 - d(\mathbf{k})^2 = 1.$$

The last relation warrants that α and β have the same algebraic properties as a and b. In other words, the transformation (2.22) is the canonical one. It is called the Bogoliubov transformation [63, 64, 65]. If the coefficient c is positive then one can put

$$c(\mathbf{k}) = \cosh \xi(\mathbf{k}), \qquad d(\mathbf{k}) = \sinh \xi(\mathbf{k}).$$

Let us introduce the operator

$$O[\xi] = \exp\{A[\xi]\}, \qquad (2.23)$$
$$A[\xi] = \int d\mathbf{k}\xi(\mathbf{k})[a(\mathbf{k})b(-\mathbf{k}) - b^+(\mathbf{k})a^+(-\mathbf{k})],$$

so that

$$[a(\mathbf{k}), A(\xi)] = -\xi(\mathbf{k})b^+(-\mathbf{k}),$$
$$[b^+(-\mathbf{k}), A(\xi)] = -\xi(\mathbf{k})a(\mathbf{k}).$$

Reiterating these relations we get

$$O^{-1}(\xi)a(\mathbf{k})O(\xi) = a(\mathbf{k})\cosh \xi - b^+(-\mathbf{k})\sinh \xi = \alpha(\mathbf{k}), \quad (2.24)$$
$$O^{-1}(\xi)b(\mathbf{k})O(\xi) = c(\mathbf{k})b(\mathbf{k}) - d(\mathbf{k})a^+(-\mathbf{k}) = \beta(\mathbf{k}).$$

These transformations look like the unitary ones. Actually they are not defined as operators on the Fock space. To verify this, let us consider the matrix element

$$o_0(\xi) = \langle 0|0; \xi\rangle, \quad \text{where} \quad |0; \xi\rangle = O^{-1}[\xi]|0\rangle.$$

After some calculations one can show [61] that this is equal to

$$o_0(\xi) = \exp\left\{-\delta^3(0)\int d^3k \ln[\cosh \xi(\mathbf{k})]\right\}, \quad \left(\delta^{(3)}(0) = \frac{V}{(2\pi)^3}\right).$$

In the limit of infinite volume $V \to \infty$ we get $o_0(\xi) \to 0$ independently of the convergence of the integral. Moreover, each coefficient of decomposition of $|0;\xi\rangle$ over the basis vectors of the space $\mathcal{H}[a,b]$ is equal to zero in this limit [61]:

$$o_n(\xi;\mathbf{q}) = n![\delta^3(0)]^n[\tanh(\mathbf{q})]^n \exp\left\{-\delta^3(0)\int d^3k \ln[\cosh\xi(\mathbf{k})]\right\} \to 0.$$

This means that the vector $|0;\xi\rangle$ does not belong to the space $\mathcal{H}[a,b]$. In other words, the transformation $O^{-1}[\xi]$ is not an operator in $\mathcal{H}[a,b]$.

On the other hand, according to (2.24) we have

$$\alpha(\mathbf{k})|0;\xi\rangle = 0, \quad \beta(\mathbf{k})|0;\xi\rangle = 0.$$

Hence $|0;\xi\rangle$ is the vacuum with respect to operators α, β and the corresponding Fock space $\mathcal{H}[\alpha,\beta]$ can be constructed. This consideration illustrates that the spaces $\mathcal{H}[a,b]$ and $\mathcal{H}[\alpha,\beta]$ correspond to two unitary nonequivalent CR representations in the sense that there is a vector in $\mathcal{H}[\alpha,\beta]$ which can not be represented as a superposition of the basis vectors of the space $\mathcal{H}[a,b]$. Moreover, this statement is true for any vector of the space $H[\alpha,\beta]$. One can say that the spaces $\mathcal{H}[a,b]$ and $\mathcal{H}[\alpha,\beta]$ are orthogonal to each other.

At the same time, the above consideration does not mean that it is impossible to define the action of the operators α and β on the vectors of $\mathcal{H}[a,b]$. *Such a definition is determined by relations (2.22). It was shown above that the canonical transformations (2.22) can not be realized as a unitary transformation and the operators α and β are not annihilation operators on the space $\mathcal{H}[a,b]$, where there is no vacuum related to α and β.*

Let us consider another canonical transformation generating nonequivalent representations. Let the operator α be related to the boson annihilation operator a:

$$\alpha(\mathbf{k}) = a(\mathbf{k}) + c(\mathbf{k}), \tag{2.25}$$

where $c(\mathbf{k})$ is a c-number function. This transformation generates a constant shift of the field and in the operator form looks like

$$\alpha(\mathbf{k}) = O^{-1}[c]a(\mathbf{k})O[c], \tag{2.26}$$

$$O[c] = \exp\left\{-\int d^3k[c^*(\mathbf{k})a(\mathbf{k}) - c(\mathbf{k})a^+(\mathbf{k})]\right\}.$$

The vacuum state of the Fock space $\mathcal{H}[\alpha]$ is

$$|0;c\rangle = O^{-1}[c]|0\rangle = \exp\left\{-\frac{1}{2}\int d^3k|c(\mathbf{k})|^2 - \int d^3k c(\mathbf{k})a^+(\mathbf{k})\right\}|0\rangle.$$

The matrix element connecting the old and new vacua is

$$\langle 0|0;c\rangle = \exp\left\{-\frac{1}{2}\int d^3k|c(\mathbf{k})|^2\right\}. \tag{2.27}$$

If the integral is divergent (e.g., for $c(\mathbf{k}) = \text{const}$ or $c(\mathbf{k}) = c\delta(\mathbf{k})$) we get

$$\int d\mathbf{k}|c(\mathbf{k})|^2 = \infty, \quad \text{and} \quad \langle 0||0;c\rangle = 0.$$

This means that the representations $\mathcal{H}[a]$ and $\mathcal{H}[\alpha]$ are unitary nonequivalent.

2.4.3 The Van Hove Model

The Hamiltonian of this model has the form [52]

$$H = \int dk \omega(\mathbf{k}) a^+(\mathbf{k}) a(\mathbf{k}) + g \int dk J(\mathbf{k})[a(\mathbf{k}) + a^+(\mathbf{k})],$$

where $a(\mathbf{k})$ and $a^+(\mathbf{k})$ are the boson operators and $J(\mathbf{k})$ is a real function. The transformation (2.26) with

$$c(\mathbf{k}) = -\frac{g J(\mathbf{k})}{\omega(\mathbf{k})}$$

diagonalizes the Hamiltonian

$$H = \int dk \omega(\mathbf{k}) a^+(\mathbf{k}) \alpha(\mathbf{k}) - \delta^3(0) E,$$

where $\delta^3(0) = \frac{V}{(2\pi)^3}$ and

$$E = -\frac{g^2}{(2\pi)^3} \int dk \frac{J^2(\mathbf{k})}{\omega(\mathbf{k})}$$

is the density of the vacuum energy.

If the source J is point-like, i.e., $J(\mathbf{k}) = \text{const}$, then any eigenstate of the total Hamiltonian H with a given value of the coupling constant g is orthogonal to any eigenstate of H with another value of g.

2.4.4 Renormalization Group Transformations

There are two important special cases among transformations of the type (2.22): the transition to a field with a new mass and the scale transformation of a field.

Let $\{\varphi, \pi\}$ be the canonical variables describing the free scalar field with the mass m, so that they obey the CR (2.7) and are expressed in terms of the creation a^+ and annihilation a operators by means of (2.8) with $\omega = \sqrt{k^2 + m^2}$.

Let us introduce the operators (α, α^+)

$$\alpha(\mathbf{k}) = \cosh \xi(\mathbf{k}) a(\mathbf{k}) - \sinh \xi(\mathbf{k}) a^+(-\mathbf{k}) = O^{-1}[\xi] a(\mathbf{k}) O[\xi], \quad (2.28)$$
$$\alpha^+(\mathbf{k}) = \cosh \xi(\mathbf{k}) a^+(\mathbf{k}) - \sinh \xi(\mathbf{k}) a(-\mathbf{k}) = O^{-1}[\xi] a^+(\mathbf{k}) O[\xi],$$

where

$$O[\xi] = \exp\left\{ \int d^3 k \xi(\mathbf{k})[a(-\mathbf{k}) a(\mathbf{k}) - a^+(\mathbf{k}) a^+(-\mathbf{k})] \right\}. \quad (2.29)$$

If one chooses the parameter ξ in the form

$$\xi(\mathbf{k}) = \frac{1}{2} \ln \left(\zeta^2 \frac{\omega}{\Omega} \right), \tag{2.30}$$

where

$$\zeta = \text{const}, \quad \Omega(\mathbf{k}) = \sqrt{\mathbf{k}^2 + M^2},$$

then the transformation (2.28) corresponds to a transition to the field with the new mass M and the scale transformation of the field. In other words, (2.28) and (2.30) determine canonical transformation of the form

$$\{\varphi, \pi\} \longrightarrow \{\zeta \Phi, \zeta^{-1} \Pi\}, \tag{2.31}$$

where the fields Φ and Π look like

$$\Phi(x) = \int \frac{d\mathbf{k}}{(2\pi)^{3/2}} \frac{1}{\sqrt{2\Omega(\mathbf{k})}} \left[\alpha(\mathbf{k}) e^{i\mathbf{k}\mathbf{x}} + \alpha^+(\mathbf{k}) e^{-i\mathbf{k}\mathbf{x}} \right], \tag{2.32}$$

$$\Pi(x) = \frac{1}{i} \int \frac{d\mathbf{k}}{(2\pi)^{3/2}} \sqrt{\frac{\Omega(\mathbf{k})}{2}} \left[\alpha(\mathbf{k}) e^{i\mathbf{k}\mathbf{x}} - \alpha^+(\mathbf{k}) e^{-i\mathbf{k}\mathbf{x}} \right].$$

The transformation (2.29) or (2.31) generates the representation $\mathcal{H}[M, \zeta]$, which is unitary nonequivalent to the initial representation $\mathcal{H}[m, 1]$.

Transformation (2.28) in the models (2.7)–(2.10) corresponds to the RG transformation if M and ζ satisfy the RG equations (2.15) and a finite renormalization of the coupling constant is performed.

On the other hand, let us turn to quantized φ^4-theory, defined by (2.7)–(2.10). The renormalization scheme is inferred to be fixed, i.e. a definite R_μ-class is chosen and a renormalization scale is fixed by the relation $m/\mu = C$. Let us introduce by means of the canonical transformations the field with the new mass $M = t \cdot m(\mu)$, where t is a parameter of the canonical transformation. It is clear that this transformation changes the R-scheme because $M/\mu \neq C$. It means that *if we perform a canonical transformation changing the field mass and at the same time we want to keep the R-scheme unchanged, then a compensating RG transformation $\mu \rightarrow \nu = t \cdot \mu$ should be performed in order to provide the relation*

$$\frac{m(\mu)}{\mu} = \frac{M(\nu)}{\nu} = C.$$

The ratio of the mass to renormalization scale should be the same in different representations of the CR.

The compensating RG transformation includes the scale transformation of the field (2.28) and finite renormalization of the coupling constant.

This is precisely the situation that arises within investigation of the phase structure: a set of nonequivalent representations is introduced by the canonical transformations and the R-schemes in all of these representations should be equivalent.

2.5 The Oscillator Representation Method

In previous sections we reviewed the methods and problems of QFT related to different aspects of the phase structure problem and discussed some features of the oscillator representation method. Now we will formulate the method completely.

The starting points of the OR method in QFT are:

– different phases of a quantum field system are revealed as nonequivalent representations of the canonical (anti)commutation relations;
– the renormalization structure of a quantum field model contains the basic information about its phase structure.

For definiteness let us turn to the models (2.1)–(2.3). We assume that the standard canonical quantization procedure using the Fock representation for the fields with renormalized mass m can be applied to these models if the renormalized coupling constant G is small and the temperature θ is equal to zero. The S-matrix elements can be calculated within the perturbation theory. The calculation of the S-matrix implies fixing the renormalization scheme. Thus we define the "initial" representation of the CR.

The point of main interest is to know what these systems are for other values of G and θ, when perturbation calculation of the S-matrix elements fails. We formulate the problem as follows:

What CR representation is suitable for different values of G and θ and what physical picture corresponds to this representation?

Nonequivalent representations occurring in the theory for given values of G and θ are identified with the different phases of the system.

Ideally the solution of the problem involves finding all possible CR representations formulating the principle how to choose among them the physically acceptable one. However, in practice we are limited to the canonical transformations described above, so that our investigation turns out to be restricted by a quite narrow class of representations.[8]

The oscillator representation method in QFT can be reduced to the following algorithm.

1. Canonical quantization of the theory is performed in a representation having an appropriate physical interpretation in the weak coupling regime, i.e. for $G \ll 1$. The renormalization R-scheme is fixed. This means that

– the one-parameter R_μ-class of renormalization prescriptions is chosen;

---

[8] Probably we should work with a representation of the S-matrix in the form of a functional integral. In this case additional possibilities can be found for taking into account the highest perturbation contributions into formation of the phases and bound states which can arise as a result of quantum field interactions. Some of these problems will be considered in Part ??.

– the renormalization scale μ is fixed by the relation

$$\frac{m(\mu)}{\mu} = C,$$

where $m(\mu)$ is the renormalized mass and C is a constant.

2. All possible canonical transformations are performed in such a way that the total Hamiltonian has *the correct form* in each CR representation. This means that

$$H = H_0 + H_I + H_{ct} + VE.$$

Here H_0 is the standard free Hamiltonian. An interaction Hamiltonian H_I contains the field operators in degree more then two. The operator H_{ct} is determined by H_0 and H_I and corresponds to the equivalent R-schemes in all representations.

The R-schemes in two representations with different masses m and M are considered to be equivalent if

– the same R_μ-class is used in both representations;
– renormalization scales μ and ν in the first and second representations obey the relation

$$\frac{m(\mu)}{\mu} = \frac{M(\nu)}{\nu} = C. \tag{2.33}$$

Equivalence condition (2.33) is realized by using of the compensating RG transformation.

3. As a result we get a set of nonequivalent CR representations describing different phases of the system under consideration. The main physical characteristic of the ground state of any system is the free energy density F, which is related to the vacuum energy density E as follows:

$$F = E - TS,$$

where S is the entropy density. At zero temperature E and F coincide. The system should be in the phase for which the free energy is minimal. This is a general physical principle.

However, in QFT the functions E and F can be calculated in practice in the lowest approximation only. The theoretical accuracy of these calculations will be defined by the value of the effective coupling constant $G_{eff}(G, \theta)$, so that the effective coupling constant can be considered as a physical parameter characterizing each representation of the CR.

Thus the physical principle for choosing the representation that is actually realized for a given G and θ can be formulated as follows:

A representation with minimal free energy density and the smallest effective coupling constant $G_{eff}(G, \theta)$ is considered to be the actual phase of a quantum field system.

Besides that, G_{eff} can be used to control the accuracy of approximations. For $d < 3$ the effective coupling constant is defined as

$$G_{\text{eff}} = \frac{g}{2\pi M^{3-d}}.$$

In the four-dimensional case the definition is more complicated and will be given below.

The rest of Part I is devoted to application of this algorithm to concrete QFT systems.

In conclusion we would like to give some additional comments. Usually a criterion based on comparison of the free energy densities is used in the phase transition theory. Meanwhile, the demand of the weak effective coupling seems more suitable in the QFT. From the physical viewpoint the quantity F does not play any role, since it does not contribute to the S-matrix elements. The free energy density can not be calculated exactly or at least with similar accuracy in different phases, so that comparison of the free energies loses its meaning. At the same time, it is natural to suppose that a large coupling constant in the interaction Hamiltonian means that the representation determined by H_0 does not describe the real states and can not be considered as a suitable representation for the total Hamiltonian. Nevertheless, our calculations have shown [33]-[36] that both criteria give similar results.

The demand of the equivalence of the R-schemes in different representations is conditioned by the formulation of the problem. We are interested in *how the ground state of a field system depends on the parameters g and m contained in the initial condition (2.16) for a fixed R-scheme.*

The demand of the correct form of the total Hamiltonian and weak coupling criterion relate to the usual scattering picture in QFT. The free Hamiltonian H_0 describes asymptotic fields, while H_I describes an interaction of the particles and should not contain terms linear or quadratic in fields, since they do not correspond to any real interaction, but only redefine the parameters of the free Hamiltonian (as in the Van Hove model). The usual perturbation expansion of scattering amplitudes has a physical sense only if the effective (perturbation) coupling constant is small enough.

3. The Phase Structure of the $(\varphi^2)^2$ Field Theory in R^{1+1}

In this chapter we consider the phase structure of models (2.1) and (2.3) in the two-dimensional space-time. These are the simplest QFT models in which the dynamical symmetry reconstruction is displayed. Model (2.1) was investigated intensively both within constructive QFT [8]-[12] and the variational approach [19, 23, 18]. It was also studied by using the renormalization group technique [29] and by the OR method [32]. It should be noted that the variational approach is well-defined in this case, since only normally ordered divergences are present in the two-dimensional scalar theory.

The GEP-approximation [19, 18], Ghang's approach [29] and the OR [32] lead to the same results for model (2.1). Namely, the critical coupling constant G_c corresponding to the first-order phase transition with symmetry breaking and the masses of the field in the broken symmetry (BS) phase are equal to each other in all approaches.

The case $N > 1$ was investigated within the GEP-approximation [21], the $1/N$-expansion [42] and the OR method [32]. According to the GEP and OR methods [21, 32], there is a phase transition of the first order accompanied by symmetry breaking for all N. The $(N-1)$-multiplet of the massive fields exists in the broken symmetry phase.

The Goldstone theorem forbids spontaneous breaking of the continuous symmetries in two-dimensional space-time, since particles with zero mass do not exist in R^2. Nevertheless, the results of the papers [21, 32] indicate the dynamical breaking of the $O(N)$-symmetry, but the "goldstone particles" have a nonzero mass.

At the same time, the $1/N$-expansion does not show a phase transition in the model (2.3) [42]. This contradiction has a natural explanation given in [21]. Application of the $1/N$-expansion is restricted by the condition $NG < 1$. Although the limit $N \gg 1$ of the GEP coincides with the effective potential obtained by the $1/N$-expansion, the critical point G_c is not seen in the $1/N$-expansion, since $NG_c \gg 1$ for $N \gg 1$.

3.1 The One-component φ^4 Model

3.1.1 The Initial Representation for $G \ll 1$

Let us consider the theory of a one-component scalar field which is described by the Lagrangian density (2.1). The Hamiltonian for Lagrangian (2.1) is written as

$$H = H_0 + H_I,$$

$$H_0[\varphi, \pi] = \frac{1}{2} \int_V dx : \left[\pi^2(x) + (\nabla\varphi(x))^2 + m^2\varphi^2(x)\right] :,$$

$$H_I[\varphi, \pi] = \frac{g}{4} \int_V dx : \varphi^4(x) :, \tag{3.1}$$

where

$$\varphi(x) = \int \frac{dk}{2pi} \frac{1}{\sqrt{2\omega(k)}} \left[a(k)e^{ikx} + a^+(k)e^{-ikx}\right],$$

$$\pi(x) = \frac{1}{i} \int \frac{dk}{2\pi} \sqrt{\frac{\omega(k)}{2}} \left[a(k)e^{ikx} - a^+(k)e^{-ikx}\right], \tag{3.2}$$

$$\omega(k) = \sqrt{k^2 + m^2}, \qquad [a(k), a^+(k')] = \delta(k - k').$$

The operators φ, π are canonical variables and obey the canonical commutation relations. The creation and annihilation operators $a^+(k)$ and $a(k)$ are defined on the Fock space of scalar particles with a mass m. The vacuum vector $|0\rangle$ obeys the condition

$$a(k)|0\rangle = 0, \qquad \forall k.$$

The Hamiltonian (3.1) is written in the normal form with respect to the vacuum $|0\rangle$, which is enough to remove all divergences from the perturbation series.

3.1.2 The Canonical Transformation

Let us perform the following canonical transformation:

$$\pi(x) \to \Pi(x), \quad \varphi(x) \to \Phi(x) + B.$$

The fields Φ and Π have the form of (3.2) but with a new mass M:

$$M^2 = m^2 \cdot t.$$

New creation and annihilation operators $a_t^+(k)$ and $a_t(k)$ are defined on the Fock space with the vacuum vector $|0\rangle\rangle = |0(t, B)\rangle$ satisfying the relations

$$|0\rangle\rangle = |0(t, B)\rangle = U_2^{-1}(t)U_1^{-1}(B)|0\rangle, \quad a_t(k)|0\rangle\rangle = 0, \qquad \forall k,$$

where U_1 and U_2 have the form of (2.27, 2.30) and look like

$$U_1(B) = \exp\left\{-2\pi m B \int dk\delta(k)[a(k) - a^+(k)]\right\},$$

$$U_2(t) = \exp\left\{\frac{1}{2}\int dk\lambda(k;t)[a(-k)a(k) - a^+(k)a^+(-k)]\right\}, \quad (3.3)$$

$$\lambda(k;t) = \frac{1}{2}\ln\left(\frac{\omega(k)}{\omega(k;t)}\right), \quad \omega(k;t) = \sqrt{k^2 + m^2 t^2}.$$

Now we should express the Hamiltonian in the new canonical variables (Φ, Π) and go to the normal product with respect to the new vacuum $|0\rangle\rangle$ in order to provide an equivalence of R-schemes in the initial and new representations. The Hamiltonian takes the form

$$H = H_0' + H_I' + H_1 + VE,$$

$$H_0' = \frac{1}{2}\int_V dx : \left[\Pi^2(x) + (\nabla\Phi(x))^2 + M^2\Phi^2(x)\right] :,$$

$$H_I' = \int_V dx : \left[\frac{g}{4}\Phi^4(x) + gB\Phi^3(x)\right] :, \quad (3.4)$$

$$H_1 = \int_V dx : \left[\frac{1}{2}R(t,B)\Phi^2(x) + P(t,B)\Phi(x)\right] :,$$

$$R(t,B) = m^2 - M^2 + 3g\left(B^2 - D\right),$$

$$P(t,B) = m^2 B + g\left(B^3 - 3BD\right),$$

$$E = \frac{1}{2}m^2 B^2 + L(t) + \frac{g}{4}[b^4 - 6B^2 D(t) + 3D^2(t)], \quad (3.5)$$

$$D(t) = \frac{1}{4\pi}\ln t, \quad L(t) = \frac{m^2}{8\pi}[t - 1 - \ln t].$$

The total Hamiltonian should have *the correct form*, i.e. the free Hamiltonian has a standard form and the interaction Hamiltonian can only contain field operators in degrees more than two. This means that the coefficients of the operators $: \varphi_t^2 :$ and φ_t in H_1 should be equal to zero. As a result, we get equations defining parameters B and t:

$$m^2 - M^2 + 3g\left(B^2 - D(t)\right) = 0,$$
$$m^2 B + g\left(B^3 - 3BD(t)\right) = 0,$$

where $t = M^2/m^2$. It is convenient to introduce the notation

$$G = \frac{g}{2\pi m^2}, \quad b^2 = 4\pi B^2.$$

Then these equations and energy density (3.5) can be written in the form:

$$1 - t + \frac{3}{2}G\left(b^2 - \ln t\right) = 0,$$

$$b\left[1 + \frac{G}{2}\left(b^2 - 3\ln t\right)\right] = 0, \qquad (3.6)$$

$$E = \frac{m^2}{8\pi}\left\{2b^2 + t - 1 - \ln t + \frac{G}{4}[b^4 - 6b^2\ln t + 3\ln^2 t]\right\}, \qquad (3.7)$$

It can be easily checked that (3.6) minimise the energy density (3.7) with respect to the variables t and B and coincide with the minimum conditions of the GEP.

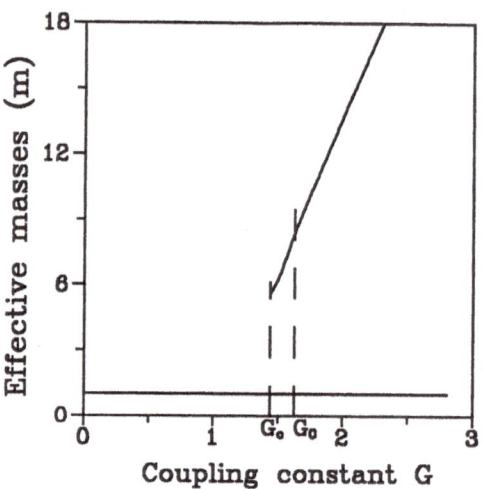

Fig. 3.1. The effective mass in two-dimensional model (2.1)

3.1.3 The Phase Structure

The second equation (3.6) has two solutions: $b = 0$ (symmetric) and $b \neq 0$ (spontaneously broken symmetry).

The Symmetric Representation.. Putting $b = 0$ in the first equation (3.6) we get

$$1 - t - \frac{3}{2}G\ln t = 0,$$

This equation has a unique real solution $t \equiv 1$, which leads to the initial representation (3.1) with energy density $E \equiv 0$.

The Broken Symmetry Representation.. Equations (3.6) and energy density in the broken symmetry phase for $b \neq 0$ can be rewritten after some transformations in the form

$$b^2 = \frac{t}{G}, \qquad G = \frac{t+2}{3\ln t},$$

$$E_{\mathrm{B}} = \frac{m^2}{8\pi}\left\{t - 1 - \frac{1}{2}\cdot\frac{t^2+2}{t+2}\cdot\ln t\right\}.$$

The equation for boson mass t has a real solution only for

$$G > G_{\mathrm{c}} = \min_t \frac{t+2}{3\ln t} \approx 1.44....$$

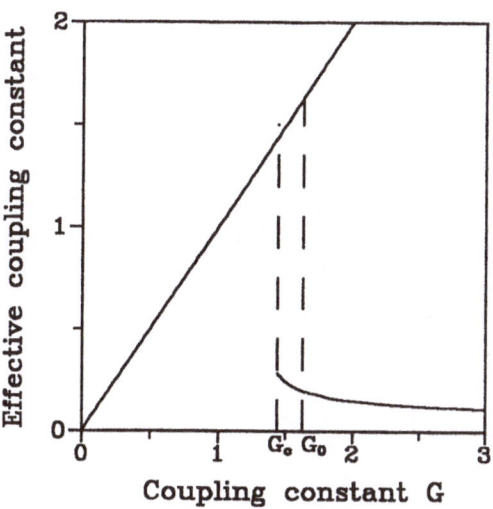

Fig. 3.2. The effective coupling constant for two-dimensional model (2.1)

The boson mass $t(G)$, the effective coupling constant $G_{\mathrm{eff}}(G)$ and the energy density $E_{\mathrm{B}}(G)$ as functions of G are shown in Figs. 3.1-3.3. The energy density is negative for all $G > G_0$, where the constant $G_0 \approx 1.625$ is determined by the equation $E_{\mathrm{B}}(G_0) = 0$. In the strong coupling regime $G \gg 1$ we obtain

$$E_{\mathrm{B}} \rightarrow -\frac{3m^2}{16\pi}G(\ln G)^2, \tag{3.8}$$

$$B^2 \rightarrow \frac{3}{4\pi}\ln G, \qquad t \rightarrow 3G\ln G,$$

$$G_{\mathrm{eff}} = \frac{g}{2\pi M^2} \rightarrow \frac{1}{3\ln G}.$$

According to the criteria of the smallest free energy density and effective coupling constant the asymptotic relations (3.8) show that in the strong coupling regime the phase with broken symmetry is preferable, since the energy density is negative and the effective coupling is small in this phase. Comparing the effective coupling constants $G_{\text{eff}} = G/t(G)$ in the BS-phase and $G_{\text{eff}} \equiv G$ in the S-phase, one finds that the phase transition takes place at $G = G_{\text{c}} \approx 1.44...$ (see Fig. 3.2), while a comparison of the energy densities leads to the critical coupling constant being equal to $G_0 \approx 1.625...$ (see Fig. 3.3). The value G_0 coincides with the critical coupling constant obtained within the GEP approximation, since, as has been mentioned above, (3.6) coincide with the minimum conditions for the GEP, and the energy density (3.7) is equal to the GEP at the minimum.

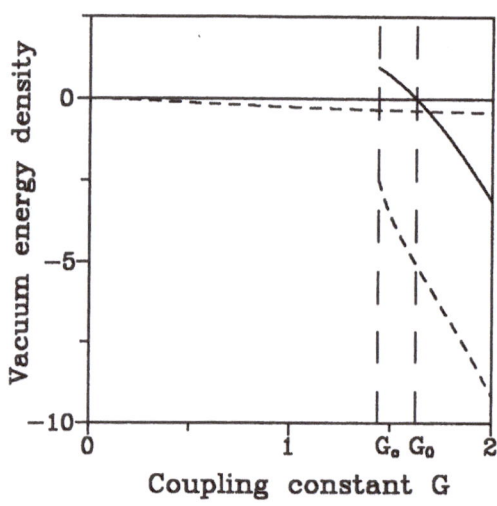

Fig. 3.3. The free energy density: the dashed line – F_S, F_B; the solid line – $F_S + \Delta F_S$, $F_B + \Delta F_B$

In the critical region the effective coupling constants are large enough in both phases (see Fig. 3.2), so that the perturbation corrections should be large and change the critical value of the coupling constant.

In order to estimate this change let us calculate the perturbation corrections to the free energy density. We shall calculate these corrections up to the order $O(G^3)$ for the S-phase and $O(G^2_{\text{eff}})$ for the BS-phase. The contributing diagrams are shown in Figs. 3.4 and 3.5. The result is

$$\Delta E_S = \frac{m^2}{8\pi}(-1.671G^2 + 4.039G^3 + O(G^4)),$$

$$\Delta E_B = \frac{M^2}{8\pi}(-1.758G_{\text{eff}} - 4.316G^2_{\text{eff}} - O(G^3_{\text{eff}})). \qquad (3.9)$$

One can see that the series for ΔE_B is nonalternating. This is a usual property of the systems with degenarate vacuum (e.g., see [69]). Doing the Borel summation in (3.9) we get

$$\Delta E_S = \frac{m^2}{8\pi} \left(\int_0^\infty dt\ e^{-t} \frac{1 + 0.806Gt}{1 + 0.806Gt + 0.836(Gt)^2} - 1 \right),$$

$$\Delta E_B = -\frac{M^2}{8\pi} \left(VP \int_0^\infty dt\ e^{-t} \frac{1 + 0.531G_{\mathrm{eff}}t}{1 - 1.228G_{\mathrm{eff}}t} - 1 \right).$$

The sign VP means the Gauchi mean value of the integral. The free energy density in the lowest approximation (solid line) and with account taken of the perturbative corrections (dashed line) is shown in Fig. 3.3. The corrections shift the critical point from $G_0 \approx 1.625...$ to $G_c \approx 1.44....$

Fig. 3.4. The diagrams $O(G^2)$ and $O(G^3)$ in the S-phase.

The order parameter $\sigma(G) = \pm b(G)$ is discontinuous at the critical point, so that the phase transition is of the first order. But we can not consider this result as well-established, since the effective coupling constants in both phases are large enough in the critical region.

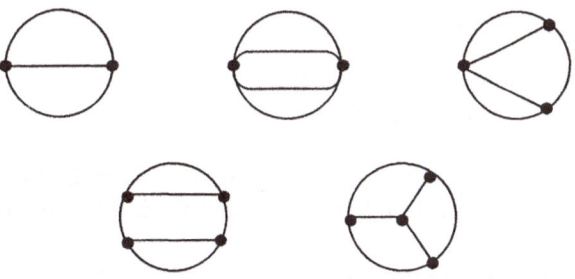

Fig. 3.5. The diagrams $O(G_{\mathrm{eff}})$ and $O(G_{\mathrm{eff}}^2)$ in the BS-phase.

3.2 The O(N)-Invariant Model $(\varphi^2)^2$ in R^{1+1}

In this section we will investigate the phase structure of model (2.3). The GEP-approximation [21] and the OR method [32] give the same results for model (2.3) in R^{1+1}.

3.2.1 Initial Representation

The Hamiltonian has the following form in the representation corresponding to weak coupling $G \ll 1$:

$$H = H_0 + H_I$$

$$H_0 = \frac{1}{2} \sum_i^N \int_V dx : [\pi_i^2 + (\nabla \varphi_i)^2 + m^2 \varphi_i^2] :,$$

$$H_I = \frac{g}{4} \int_V dx : \left[\sum_i^N \varphi_i^2\right]^2 : . \tag{3.10}$$

The fields φ_i and π_i have the form of (3.2) and satisfy the usual CR

$$[\phi_i(x), \pi_j(y)] = i\delta_{ij}\delta(x - y).$$

The states of the system are described by vectors in the Fock space with vacuum vector $|0\rangle$:

$$a_j(k)|0\rangle = 0, \quad \forall j, k, \quad \langle 0|0\rangle = 1.$$

The Hamiltonian (3.10) is taken in the normal form with respect to the vacuum $|0\rangle$, which is enough to remove all divergencies from the perturbation series (in R^{1+1}).

3.2.2 The Canonical Transformation

Let us perform the canonical transformation

$$(\varphi_i, \pi_i) \rightarrow (\Phi_i, \Pi_i) \quad (i = 1, ..., N - 1),$$
$$(\varphi_N, \pi_N) \rightarrow (\Phi + B, \Pi).$$

Here (Φ_i, Π_i) is the O(N-1)-multiplet of fields with a mass $M_0 = m \cdot t_0$, (Φ, Π) are the fields with a mass $M = m \cdot t$ and B is a constant field. The operator form of this transformation is analogous to (3.3).

The fields (Φ_i, Π_i) and (Φ, Π) are expressed in terms of creation and annihilation operators by (3.2), where new masses M_0 and M and new creation and annihilation operators (α_i^+, α_i) and (α^+, α) should be introduced. The operators are defined on the Fock space with vacuum vector $|0\rangle\rangle$:

$$\alpha_i(k)|0\rangle\rangle = 0, \quad \alpha(k)|0\rangle\rangle = 0, \quad \forall i, k.$$

The new Fock space is unitary nonequivalent to the initial one.

Now let us substitute the new canonical variables into the total Hamiltonian and go to the normal product with respect to the new vacuum $|0\rangle\rangle$. We get the following representation:

$$H = H_0' + H_I' + H_{ct}' + H_1 + E, \tag{3.11}$$

$$H_0' = \frac{1}{2} \int_V dx : \left[\sum_i^{N-1} \left(\Pi_i^2 + (\nabla \Phi_i)^2 + M_0^2 \Phi_i^2 \right) + \Pi^2 + (\nabla \Phi)^2 + M^2 \Phi^2 \right] :,$$

$$H_I' = \frac{g}{4} \int_V dx : \left[\Phi^4 + 4B\Phi^3 + 4B\Phi \sum_i^{N-1} \Phi_i^2 + 2\Phi^2 \sum_i^{N-1} \Phi_i^2 + \left(\sum_i^{N-1} \Phi^2 \right)^2 \right] :,$$

$$H_1 = \int_V dx : \left[\frac{1}{2} R(B, M, M_0)\Phi^2 + \frac{1}{2} P(B, M, M_0) \sum_i^{N-1} \Phi^2 + Q(B, M, M_0)\Phi \right] :,$$

$$R(B, M, M_0) = m^2 - M^2 - g\left(3D(t) + (N-1)D(t_0)\right) + 3gB^2,$$
$$P(B, M, M_0) = m^2 - M_0^2 - g\left(D(t) + (N+1)D(t_0)\right) + gB^2,$$
$$Q(B, M, M_0) = m^2 B + gB^3 - gB\left(3D(t) + (N-1)D(t_0)\right),$$
$$E = \frac{m^2}{2} B^2 + \frac{g}{4} B^4 + (N-1)L(t_0) + L(t)$$
$$+ \frac{g}{4} \Bigg\{ -2\left[3D(t) + (N-1)D(t_0)\right] B^2 + 3D^2(t) \tag{3.12}$$
$$+ (N-1)D^2(t_0) + 2(N-1)D(t)D(t_0) \Bigg\},$$

where $t = M^2/m^2$, $t_0 = M_0^2/m^2$. The functions D and L are the same as in the one-conponent case:

$$D(t) = \frac{1}{4\pi} \ln t, \quad L(t) = \frac{m^2}{8\pi}(t - 1 - \ln t).$$

In order to provide the correct form of the total Hamiltonian in the new representation one has to put $H_1 = 0$, so that one obtains the equations defining parameters of the canonical transformation:

$$m^2 - M^2 - g\left(3D(t) + (N-1)D(t_0)\right) + 3gB^2 = 0,$$
$$m^2 - M_0^2 - g\left(3D(t) + (N+1)D(t_0)\right) + gB^2 = 0, \tag{3.13}$$
$$m^2 B + gB^3 - gB\left(3D(t) - (N+1)D(t_0)\right) = 0.$$

Equations (3.13) define a minimum of the energy density (3.12) as a function of (t, t_0, B).

3.2.3 The Phase Structure

According to (3.13) we have two different cases: $B = 0$ (symmetric) and $B \neq 0$ (broken symmetry).

The Symmetric Representation.. The solution $B = 0$ of the third equation (3.13) leads to the initial representation with $M \equiv M_0 \equiv m$ and $E \equiv 0$. There are no other symmetric solutions.

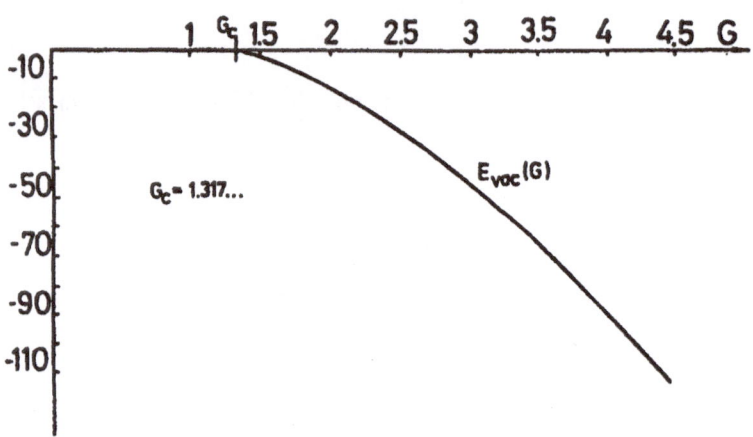

Fig. 3.6. The free energy density for the two-dimensional model (2.3).

Representations with Broken Symmetry.. The nonzero solution of the last equation (3.13) has the form

$$B^2 = 3D(t) + (N-1)D(t_0) - \frac{m^2}{g}.$$

Substitution of this solution into other equations and the energy density gives

$$t + 2 = G(3\ln t + (N-1)\ln t_0), \quad t_0 = G\ln\left(\frac{t}{t_0}\right),$$

$$E_{\mathrm{B}} = \frac{m^2}{8\pi}\left[t - 1 + 2\ln t + (N-1)t_0 - \frac{1}{G}\right.$$
$$\left. - \frac{G}{4}(9\ln^2 t + 2(N-1)\ln t \ln t_0 - (N-1)\ln^2 t_0)\right],$$

where $G = g/2\pi m^2$.

The critical coupling constant is determined by the condition $E_{\mathrm{B}}(G_{\mathrm{c}}) = 0$. For $N = 4$ it is equal to

$$G_{\mathrm{c}} = 1.317....$$

The energy density as a function of G is shown in Fig. 3.6. In the strong coupling regime $G \gg 1$ one obtains

$$t(G) \to (N+2)G \ln G + O(G \ln \ln G),$$
$$t_0(G) \to G \ln \ln G + O(G \ln \ln \ln G),$$
$$E_B(G) \to -\frac{m^2}{8\pi} \cdot \frac{3}{4} G \ln^2 G,$$
$$G_{\text{eff}}(G) = \frac{g}{2\pi M^2} \to \frac{1}{(N+2)\ln G},$$
$$G_{\text{eff}}^0(G) = \frac{g}{2\pi M_0^2} \to \frac{1}{\ln \ln G}.$$

The effective coupling constants as functions of G for $N = 4$ are represented in Fig. 3.7.

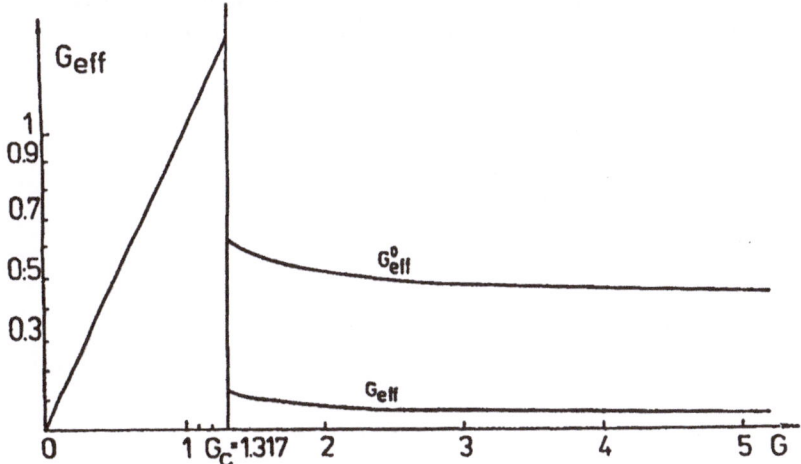

Fig. 3.7. The effective coupling constants for the two-dimensional model (2.3).

Figures 3.6 and 3.7 show that the point G_c is a critical one and corresponds to the phase transition accompanied by symmetry breaking. The phase transition is of the first order, since the order parameter B is discontinuous at the critical point G_c.

Now let us examine the Hamiltonian (3.11). The free Hamiltonian H_0' contains two kinds of scalar fields. The fields Φ_i $(i = 1, ..., N-1)$ describe scalar particles with a mass $M_0^2 = t_0 m^2$ and belong to the isotopic multiplet $O(N-1)$. The field Φ describes particles with a mass $M^2 = tm^2$. Moreover, we have

$$1 < \frac{t(G)}{t_0(G)} \xrightarrow{G \to \infty} (N+2)\frac{\ln G}{\ln \ln G}.$$

The fields Φ_i have nonzero masses, although from the point of view of the standard SSB picture these particles are "Goldstones" and should be massless. The mass of the field Φ is larger than the mass of the multiplet particles and this difference increases as $G \to \infty$. All these conclusions coincide with the results of the GEP approximation (see [21]).

In the present chapter we have considered the simplest models where the dynamical symmetry reconstruction occurs. The simplicity of these models is conditioned by the fact that in R^{1+1} all UV-divergencies can be removed by the normal ordering of the Hamiltonian. The ordered Hamiltonian is a well-defined operator on Fock space and variational methods, particularly the GEP-approximation, are applicable. We have seen that in this case the results of the OR and GEP coincide. Now we have to consider the case of the space-time R^{2+1}, when the GEP-approximation and other variational methods are invalid owing to the presence of UV-divergencies, which can not be removed by the normal ordering, but require additional mass renormalization. The OR method is valid for the three-dimentional case as well as for the two-dimensional one. No modifications of its algorithm are necessary.

4. The Phase Structure of the Three-Dimensional φ^4 Theory

4.1 The Hamiltonian φ^4 in R^{2+1}

In order to unify an investigation of both models (2.1) and (2.2) it is convenient to consider the model with the Lagrangian density:

$$L(x) = \frac{1}{2}\varphi(x)\left(\Box - m^2\right)\varphi(x) - g_3\varphi^3(x) - \frac{g_4}{4}\varphi^4(x), \qquad (4.1)$$

where g_3 and g_4 are the coupling constants, $x = (\mathbf{x}, t) = (x_1, x_2, t)$.

The Hamiltonian corresponding to the Lagrangian (4.1) has the form:

$$H = H_0 + H_I + H_{\text{ct}},$$

$$H_0 = \frac{1}{2}\int_V d\mathbf{x} : \left[\pi^2(\mathbf{x}) + (\nabla\varphi(\mathbf{x}))^2 + m^2\varphi^2(\mathbf{x})\right] :,$$

$$H_I = \int_V d\mathbf{x} : \left[\frac{1}{4}g_4\varphi^4(\mathbf{x}) + g_3\varphi^3(\mathbf{x})\right] :, \qquad (4.2)$$

where

$$\varphi(\mathbf{x}) = \int \frac{d\mathbf{k}}{2\pi}\frac{1}{\sqrt{2\omega}}\left[a(\mathbf{k})\exp\{i\mathbf{kx}\} + a^+(\mathbf{k})\exp\{-i\mathbf{kx}\}\right],$$

$$\pi(\mathbf{x}) = \frac{1}{i}\int \frac{d\mathbf{k}}{2\pi}\sqrt{\frac{\omega}{2}}\left[a(\mathbf{k})\exp\{i\mathbf{kx}\} - a^+(\mathbf{k})\exp\{-i\mathbf{kx}\}\right], \qquad (4.3)$$

$$\omega(\mathbf{k}) = \sqrt{\mathbf{k}^2 + m^2}, \quad [a(\mathbf{k}), a^+(\mathbf{k}')] = \delta(\mathbf{k} - \mathbf{k}').$$

The fields φ, π are the canonical variables and obey the standard CR:

$$[\varphi(\mathbf{x}), \pi(\mathbf{y})] = i\delta(\mathbf{x} - \mathbf{y}).$$

The creation and annihilation operators (a^+, a) act on the Fock space with vacuum vector satisfying the relations:

$$a(\mathbf{k})|0\rangle = 0 \quad \forall \mathbf{k} \in R^2. \qquad (4.4)$$

All operators in (4.2) are normally ordered with respect to the vacuum (4.4). The model is superrenormalizable. There are a finite number of divergent diagrams shown in Fig. 4.1. To remove the divergences one has to

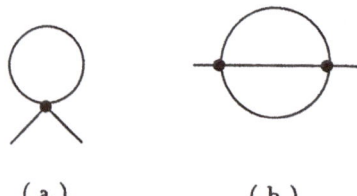

Fig. 4.1. Divergent diagrams in R^{2+1} contributing to mass renormalization.

(a)   (b)

introduce into the Hamiltonian the counter-term operator H_{ct}. The zero-momentum renormalization scheme determines H_{ct} in the form

$$H_{ct} = \int_V dx : \left[\frac{1}{2}A(m)\varphi^2(x) + C(m)\varphi(x) + \delta E(m) \right] :, \quad (4.5)$$

$$A(m) = 3!g_4^2 \Sigma(m),$$

$$C(m) = 3!g_3g_4 \Sigma(m),$$

$$\delta E(m) = \frac{3}{4}g_4^2 W_0(m) - \frac{9}{2}g_4^3 V(m) + \frac{1}{2}3!g_3^2 \Sigma(m).$$

The counter-terms can be calculated for instance in the Euclidean metrics where the propagator has the form

$$\Delta(x, m) = \int \frac{d^3k}{(2\pi)^3} \frac{e^{-ikx}}{k^2 + m^2} = \frac{1}{4\pi} \frac{e^{-mr}}{r},$$

where

$$x = (\mathbf{x}, x_3 = it), \quad r = \sqrt{\mathbf{x}^2 + x_3^2}, \quad k^2 = \mathbf{k}^2 + k_3^2.$$

After some calculations one gets

$$\Sigma(m) = \text{reg} \int d^3x \Delta^3(x, m) = \frac{1}{(4\pi)^2} \text{reg} \int_0^\infty \frac{ds}{s} \exp\{-3ms\},$$

$$W(m) = \text{reg} \int d^3x \Delta^4(x, m) = \frac{1}{(4\pi)^3} \text{reg} \int_0^\infty \frac{ds}{s^2} \exp\{-4ms\}, \quad (0.1)$$

$$V(m) = \text{reg} \int d^3k \Pi^3(k, m) = \frac{1}{4(2\pi)^5} \text{reg} \int_0^\infty \frac{ds}{s} \arctan^3 \left(\frac{s}{2m} \right),$$

$$\Pi(k, m) = \int d^3x \exp\{ikx\}\Delta^2(x, m) = \frac{1}{4\pi|k|} \arctan \left(\frac{|k|}{2m} \right).$$

An appropriate regularization is implied in these formulas.

4.2 The Canonical Transformations

Let us perform the following canonical transformation:

$$\pi(\mathbf{x}) \rightarrow \pi_t(\mathbf{x}), \quad \varphi(\mathbf{x}) \rightarrow \varphi_t(\mathbf{x}) + B, \tag{4.7}$$

where the fields φ_t and π_t have the form of (4.3), where the new mass of the field

$$M = m \cdot t$$

must be substituted. The constant B has the sense of the vacuum condensate. The transformation (4.7) can be expressed in terms of the creation and annihilation operators as

$$a(\mathbf{k}) \rightarrow a(\mathbf{k}, t) - 2\pi m B \delta(\mathbf{k}) = U_2^{-1}(t) U_1^{-1}(B) a(\mathbf{k}) U_1(B) U_2(t),$$

where

$$U_1(B) = \exp\left\{-2\pi m B \int d\mathbf{k} \delta(\mathbf{k})[a(\mathbf{k}) - a^+(\mathbf{k})]\right\},$$

$$U_2(t) = \exp\left\{\frac{1}{2} \int d\mathbf{k} \lambda(\mathbf{k}, t)[a(-\mathbf{k})a(\mathbf{k}) - a^+(\mathbf{k})a^+(-\mathbf{k})]\right\}. \tag{4.8}$$

The transformation U_1 shifts the field φ at the constant B. The transformation U_2 can be represented as

$$a(\mathbf{k}, t) = a(\mathbf{k}) \cosh(\lambda) - a^+(-\mathbf{k}) \sinh(\lambda),$$
$$a^+(\mathbf{k}, t) = a^+(\mathbf{k}) \cosh(\lambda) - a(-\mathbf{k}) \sinh(\lambda) \tag{4.9}$$

with the reversed transformation

$$a(\mathbf{k}) = a(\mathbf{k}, t) \cosh(\lambda) + a^+(-\mathbf{k}, t) \sinh(\lambda)$$
$$a^+(\mathbf{k}) = a^+(\mathbf{k}, t) \cosh(\lambda) + a(-\mathbf{k}, t) \sinh(\lambda). \tag{4.10}$$

Let the parameter λ be

$$\lambda(\mathbf{k}, t) = \frac{1}{2} \ln\left(\frac{\omega(\mathbf{k})}{\omega(\mathbf{k}, t)}\right), \quad \omega(\mathbf{k}, t) = \sqrt{\mathbf{k}^2 + m^2 t^2}.$$

Then using (4.10) one can express the fields φ_t, π_t with a mass M in terms of the operators $a(\mathbf{k}, t), a^+(\mathbf{k}, t)$. These operators act on the Fock space with the vacuum vector subjected to the conditions

$$|0(t, B)\rangle = U_2^{-1}(t) U_1^{-1}(B)|0\rangle, \quad a(\mathbf{k}, t)|0(t, B)\rangle = 0 \ \forall \mathbf{k}. \tag{4.11}$$

The CR representations defined by (4.4) and (4.11) are unitary nonequivalent for $B \neq 0$ or $t \neq 1$.

Now, one has to perform the canonical transformation in the Hamiltonian (4.2). This means that the Hamiltonian must be expressed in terms of new canonical variables and written in the form of the normal product with respect to the operators $(a(\mathbf{k}, t), a^+(\mathbf{k}, t))$. New counter-terms determined by the new representation of the free H'_0 and interaction H'_I Hamiltonians and by the zero-momentum R-scheme must be introduced. As a result one gets

$$H = H'_0 + H'_I + H'_{\text{ct}} + H_1 + E,$$

where

$$H'_0 = \frac{1}{2} : \left[\pi_t^2(\mathbf{x}) + (\nabla \varphi_t(\mathbf{x}))^2 + M^2 \varphi_t^2(\mathbf{x}) \right] :,$$

$$H'_I =: \left[\frac{1}{4} h_4 \varphi_t^4(\mathbf{x}) + h_3 \varphi_t^3(\mathbf{x}) \right] :, \qquad (4.12)$$

$$h_3 = g_3 + g_4 B, \quad h_4 = g_4.$$

The operator H'_{ct} has the structure of (4.5, 4.6), where the following substitution has to be done:

$$\varphi \to \varphi_t, \quad m \to M, \quad g_3 \to h_3, \quad g_4 \to h_4.$$

The operator H_1 looks like

$$H_1 =: \left[\frac{1}{2} R(t, B) \varphi_t^2(\mathbf{x}) + P(t, B) \varphi_t(\mathbf{x}) \right] :,$$

$$R = m^2 - M^2 + 3g_4 \left(B^2 - D \right) + 6g_3 B + 6g_4^2 \left(\Sigma(m) - \Sigma(M) \right),$$

$$P = m^2 B + g_4 \left(B^3 - 3BD \right) + 3g_3 \left(B^2 - D \right) \qquad (4.13)$$

$$+ 6g_4 \left(g_3 + g_4 B \right) \left(\Sigma(m) - \Sigma(M) \right).$$

The energy density E can be written in the form

$$E = E_0 + E_1 + E_2 + E_3, \qquad (4.14)$$

$$E_0 = \frac{1}{2} m^2 B^2 + L(t), \quad E_1 = \frac{1}{4} g_4 \left(B^4 - 6DB^2 + 3D^2 \right) + g_3 \left(B^3 - 3DB \right),$$

$$E_2 = \frac{3}{4} g_4^2 \left(W(m) - W(M) - 4D\Sigma(m) \right) + 3 \left(g_4 B + g_3 \right)^2 \left(\Sigma(m) - \Sigma(M) \right),$$

$$E_3 = -\frac{9}{2} g_4^3 \left(V(m) - V(M) \right),$$

$$D(t) = \Delta(0, m) - \Delta(0, M) = \int \frac{d\mathbf{k}}{(2\pi)^2} \left[\frac{1}{\omega(\mathbf{k})} - \frac{1}{\omega(\mathbf{k}, t)} \right] = \frac{m}{4\pi} (t - 1), (4.15)$$

$$L(t) = \frac{1}{2} \int \frac{d\mathbf{k}}{(2\pi)^2} \left[\omega(\mathbf{k}, t) - \omega(\mathbf{k}) - \frac{M^2 - m^2}{2\omega(\mathbf{k}, t)} \right] = \frac{m^3}{24\pi} (t - 1)^2 (2 + t), (4.16)$$

Using (4.6) we get

$$\Sigma(m) - \Sigma(M) = \frac{1}{(4\pi)^2} \ln t,$$

$$V(m) - V(M) = \frac{1}{64(4\pi)^2} \ln t, \qquad (4.17)$$

$$W(m) - W(M) - 4D\Sigma(m) = -\frac{4m}{(4\pi)^3} \left[(t-1)\left(\ln\left(\frac{4}{3}\right) - 1\right) + t\ln t\right].$$

The condition

$$H_1 = 0 \quad \Leftrightarrow \quad \begin{cases} R(t, B) = 0, \\ P(t, B) = 0 \end{cases} \qquad (4.18)$$

provides the correct form of the total Hamiltonian H and, at the same time, fixes the parameters t and B. Now, the Hamiltonian H describes scalar particles with a mass M depending on the coupling constant G. This dependence is regulated by (4.18) and (4.13). Introducing the dimensionless variables

$$G_4 = \frac{g_4}{2\pi m}, \quad G_3 = \frac{g_3}{m\sqrt{4\pi m}}, \quad b = B\sqrt{\frac{4\pi}{m}} \qquad (4.19)$$

and using (4.13) we represent (4.18) in the form

$$-\frac{1}{2}t^2 + \frac{1}{2} + \frac{3}{4}G_4\left(b^2 - t + 1\right) + 3G_3b + \frac{3}{4}G_4^2\ln t = 0 \qquad (4.20)$$

$$b + \frac{1}{2}G_4b\left(b^2 - 3t + 3\right) + 3G_3\left(b^2 - t + 1\right) + 3G_4\left(G_3 + \frac{G_4}{2}b\right)\ln t = 0.$$

In terms of the variables (4.19) the energy density (4.14) takes the form

$$E = \frac{m^3}{8\pi}(\varepsilon_0 + \varepsilon_1 + \varepsilon_2 + \varepsilon_3), \qquad (4.21)$$

$$\varepsilon_0 = b^2 + \frac{1}{3}(t-1)(2+t),$$

$$\varepsilon_1 = \frac{1}{4}G_4\left(b^4 - 6b^2(t-1) + 3(t-1)^2\right) + 2G_3b\left(b^2 - 3t + 3\right),$$

$$\varepsilon_2 = \frac{3}{2}(G_4b + 2G_3)^2\ln t - \frac{3}{2}G_4^2\left(t\ln t + (t-1)\left(\ln\frac{4}{3} - 1\right)\right),$$

$$\varepsilon_3 = -\frac{9}{32}\pi^2 G_4^3\ln t.$$

Different solutions of (4.20) describe the possible phases of the system under consideration. For a given value of the coupling constant G the system exists in the phase that corresponds to the smallest energy density and effective coupling constants

4.3 The Symmetric Model

Let us consider the phase structure of the model with symmetric Lagrangian (2.1). This case corresponds to the choice

$$G_4 = G, \quad G_3 = 0. \tag{4.22}$$

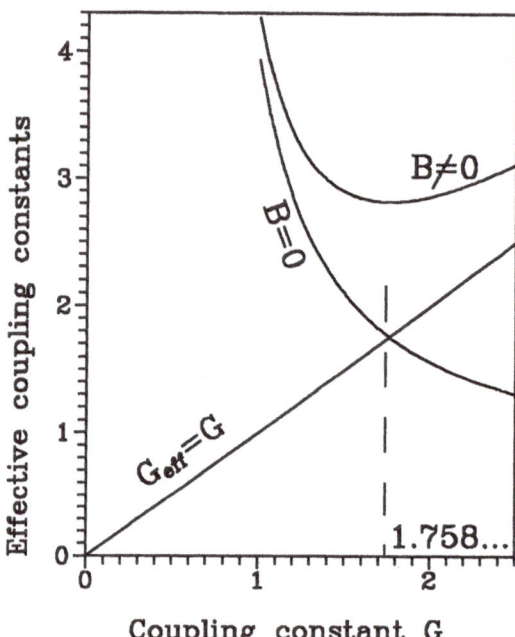

Fig. 4.2. The effective coupling constant in the model (2.1) in R^{2+1}.

Equations (4.20) take the form

$$t^2 - 1 - \frac{3}{2}G\left(b^2 - t + 1\right)) - \frac{3}{2}G^2 \ln t = 0, \tag{4.23}$$

$$b\left[1 + \frac{1}{2}G\left(b^2 - 3t + 3\right) + \frac{3}{2}G^2 \ln t\right] = 0.$$

Using the first equation, one can find two solutions for the second one:

$$b = 0 \text{ (symmetric)}, \quad b^2 = \frac{t^2}{G} \text{ (broken symmetry)}.$$

Let us first consider the phase with broken symmetry.

4.3.1 The BS-phase

We have in this case

$$b^2 = \frac{t^2}{G},$$

$$t^2 - 3Gt + 3G^2 \ln t + 2 + 3G = 0.$$

(4.24)

For all G the second equation (4.24) has the solution $t(G) < 1$, so that the effective coupling constant

$$G_{\text{eff}}(G) = \frac{G}{t(G)}$$

(4.25)

satisfies the inequality (see Fig. 4.2)

$$G_{\text{eff}}(G) > G \quad \forall G.$$

(4.26)

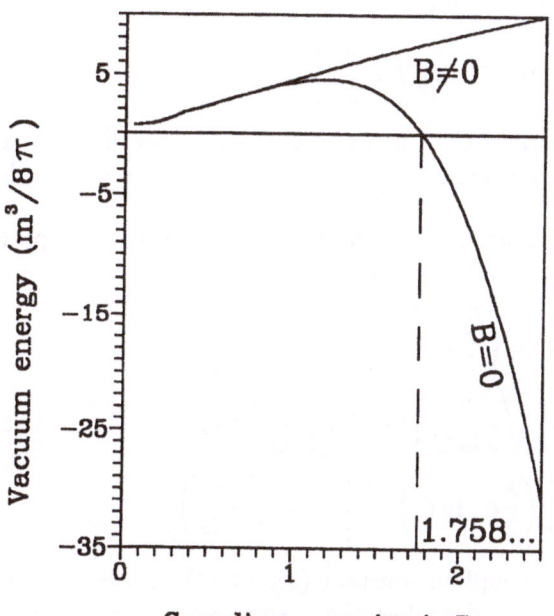

Fig. 4.3. The free energy density in the model (2.1) in R^{2+1}.

Substituting $t(G)$ into (4.21) ($G_4 = G$, $G_3 = 0$), one can check that the energy density is positive in the BS-phase for all G (see Fig. 4.3). Such a situation indicates that the BS-phase is not realized in the system. Consideration of the S-phase with $b = 0$ corroborates this conclusion.

4.3.2 The S-phases

Using (4.23) with $b = 0$, we get the following equation for the parameter t:

$$2t^2 + 3Gt - 3G^2 \ln t - 2 - 3G = 0. \tag{4.27}$$

First of all, one can see that (4.27) has two solutions: $t_1(G) \equiv 1$ (this is the initial representation (4.2) with zero energy density) and $t_2(G)$; moreover

$$t_2 < 1 \text{ if } G < G_c, \quad t_2 \geq 1 \text{ if } G \geq G_c,$$

$$G_c = \frac{1}{2}\left(1 + \sqrt{\frac{19}{3}}\right) \approx 1.758....$$

The critical coupling constant G_c is determined by the condition $t_2(G_c) = t_1 = 1$. The energy density ε as a function of G is shown in Fig. 4.3. One can see that the energy is negative for the solution t_2 for all $G > G_c$. The point G_c corresponds to phase transition from the initial phase with the mass $t_1 \equiv 1$ to the second S-phase with a mass $t_2(G)$. In the limit $G \to 0$ the function $t_2(G)$ to be

$$t_2(G) \xrightarrow{G \to 0} \exp\left\{-\frac{2}{3G}\right\}.$$

Nonanalyticity of t_2 at the origin $G = 0$ shows that difference between t_1 and t_2 can not be reduced to perturbation corrections, calculated within the initial representation (4.2).

In the strong coupling regime $G \gg 1$ the mass, the effective coupling constant and the energy density behave as follows:

$$M(G) = m \cdot G\sqrt{\frac{3}{2}\ln G}\left[1 + O\left(\frac{1}{\ln G}\right)\right],$$

$$G_{\text{eff}}(G) \longrightarrow \sqrt{\frac{2}{3\ln G}}\left[1 + O\left(\frac{1}{\ln G}\right)\right] \ll 1, \tag{4.28}$$

$$\varepsilon(G) = -\frac{2}{3}\left(\frac{3}{2}G^2 \ln G\right)^{3/2}\left[1 + O\left(\frac{1}{\ln G}\right)\right].$$

The smallness of the effective coupling constant (see (4.28)) shows that the second symmetric representation describes the system quite accurately in the strong coupling regime. At the same time, both representations cannot be considered suitable in the critical region $G \sim 1$, since $G_{\text{eff}}(1) \sim 1$.

Nevertheless the following conclusions are valid in any case:

- Symmetry breaking is absent in the three-dimensional model (2.1) for all G.
- There are two S-phases, a kind of phase transition without symmetry reconstruction takes plase at $G \approx 1.758$.

4.4 The Model with SSB in the Initial Representation

Now we will study the phase structure of model (2.2). Equations for t and b corresponding to this case are obtained from (4.20) by the substitution

$$G_4 = G, \quad G_3 = \frac{1}{2}\sqrt{G}$$

and looks like

$$t^2 - 1 - \frac{3}{2}G\left(b^2 - t + 1\right) - 3\sqrt{G}b - \frac{3}{2}G^2 \ln t = 0, \tag{4.29}$$

$$2b + Gb\left(b^2 - 3t + 1\right) + 3\sqrt{G}\left(b^2 - t + 1\right) + 3G\sqrt{G}\left(1 + \sqrt{G}b\right)\ln t = 0.$$

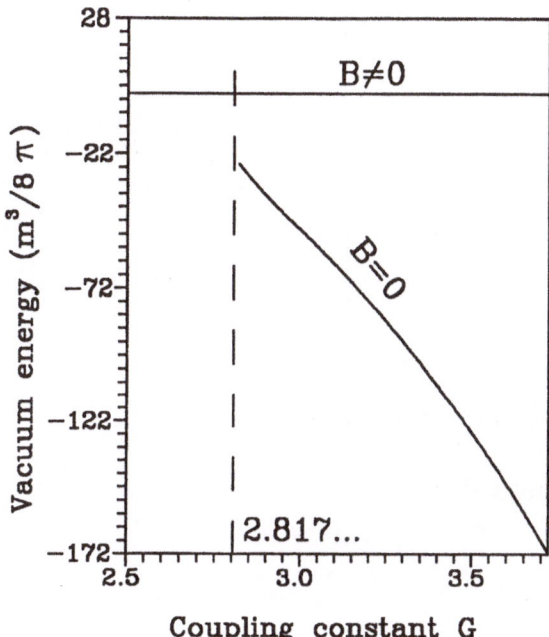

Fig. 4.4. The free energy density in the model (2.2) in R^{2+1}.

The same substitution has to be made in (4.21) for the energy density. Two solutions for b follow from (4.29):

$$b = -\frac{1}{\sqrt{G}} \text{ (symmetric)}, \quad b = -\frac{1}{\sqrt{G}} \pm \frac{t}{\sqrt{G}} \text{ (broken symmetry)}.$$

4.4.1 The BS-phase

Taking into account the solution with broken symmetry we rewrite the second equation (4.29) in the form

$$t^2 - 3Gt + 3G^2 \ln t - 1 + 3G = 0, \tag{4.30}$$

This equation has a unique solution $t \equiv 1$, which corresponds to the initial representation with SSB: $M = m$, $G_{\text{eff}} = G$, $\varepsilon = 0$.

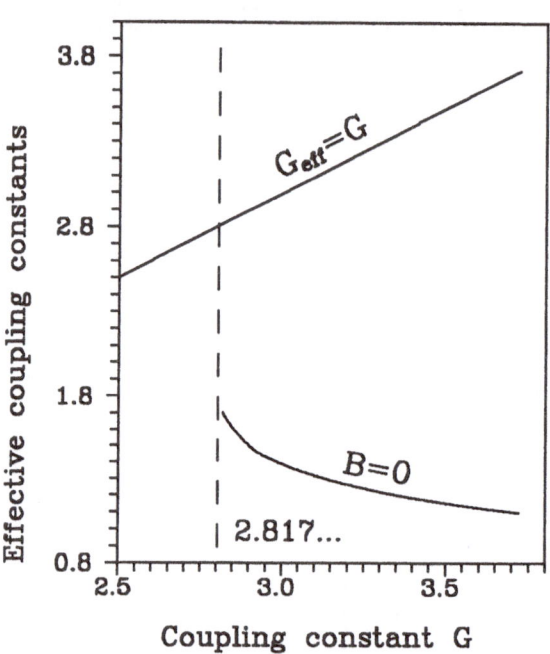

Fig. 4.5. The effective coupling constants in the model (2.2) in R^{2+1}.

4.4.2 The S-phase

For the symmetric solution $b = -1/\sqrt{G}$ we get the equation

$$2t^2 + 3Gt - 3G^2 \ln t + 1 - 3G = 0. \tag{4.31}$$

One can check that (4.31) has a real solution only for

$$G \geq G_c = 2.817....$$

Substitution of this solution into (4.21) gives the energy density as a function of G, which is plotted in Fig. 4.4. Dependence of the effective coupling constant on G is shown in Fig. 4.5. For $G \gg 1$ the mass, effective coupling constant and the energy density have the same asymptotic behaviour (4.28) as for the symmetric model. In this sense models (2.1) and (2.2) coincide in the strong coupling regime.

The critical value of coupling constant G_c is quite approximate, since $G_{\text{eff}}(G_c) \sim O(1)$ (see Fig. 4.5).

The order parameter

$$\sigma = \pm \frac{t(G)}{\sqrt{G}}$$

is discontinuous at the critical point G_c, so that the phase transition looks like the first order one. But this conclusion cannot be considered well-established, because none of the representations under consideration is suitable in the critical region.

Let us summarize the conclusions of this section.

- Symmetry is restored in the system (2.2) if the coupling constant is large enough.
- There is a phase transition between the BS- and S-phases near the point $G_c \approx 2.817$.

These conclusions agree with the results of [20, 29], but contrudict the result of the GEP approximation. Such a situation has a natural explanation. In contrast to the GEP, the OR and the method of papers [20, 29] take into account mass renormalization in the second perturbation order, which has a major influence on the phase structure formation.

4.5 The $O(N)$-Invariant Model in R^{2+1}

4.5.1 The Hamiltonian of the Model

Now we shall investigate how the UV divergences appearing in the highest perturbation orders influence the phase structure of the N-component model (2.3).

The Hamiltonian in the representation corresponding to the weak coupling regime $G = g/2\pi m \ll 1$ is written as

$$H = H_0 + H_I + H_{ct},$$

$$H_0 = \frac{1}{2} \sum_i^N \int_V d\mathbf{x} : [\pi_i^2 + (\nabla \varphi_i)^2 + m^2 \varphi_i^2] :,$$

$$H_I = \frac{g}{4} \int_V d\mathbf{x} : \left[\sum_i^N \varphi_i^2 \right]^2 :, \tag{4.32}$$

$$H_{ct} = \int_V d\mathbf{x} : \left[\frac{1}{2} A(m) \sum_i^N \varphi_i^2 + \delta E(m) \right] :,$$

where $A(m) = 2(N+2)g^2 \Sigma(m)$,

$$\delta E(m) = \frac{1}{4}N(N+2)g^2 W(m) - \frac{1}{2}N(N+2)^2 V(m).$$

The functions Σ, W, V are the same ones as in (4.6). The fields φ_i and π_i have the form of (4.3) and obey the usual CR:

$$[\phi_i(\mathbf{x}), \pi_j(\mathbf{y})] = i\delta_{ij}\delta(\mathbf{x} - \mathbf{y}).$$

The states of the system are described by vectors in Fock space with the vacuum vector $|0\rangle$:

$$a_j(\mathbf{k})|0> = 0, \quad \forall j, \, \mathbf{k}, \quad \langle 0|0\rangle = 1.$$

The Hamiltonian (4.32) is written in the form of a normal product with respect to the vacuum $|0\rangle$.

4.5.2 The Canonical Transformation

In complete analogy to the two-dimensional case let us perform the canonical transformation:

$$(\varphi_i, \pi_i) \to (\Phi_i, \Pi_i) \quad (i = 1, ..., N-1),$$
$$(\varphi_N, \pi_N) \to (\Phi + B, \Pi).$$

Here (Φ_i, Π_i) is a multiplet of the fields with mass $M_0 = m \cdot t_0$, (Φ, Π) are the fields with the mass $M = m \cdot t$ and B is a constant. The operator form of these transformations is analogous to (4.8, 4.10).

The fields (Φ_i, Π_i) and (Φ, Π) have the same form (4.3), but with new masses M_0 and M and new creation and annihilation operators (α_i^+, α_i) and (α^+, α), acting on the Fock space with the vacuum vector $|0\rangle\rangle$:

$$\alpha_i(\mathbf{k})|0\rangle\rangle = 0, \quad \alpha(\mathbf{k})|0\rangle\rangle = 0, \quad \forall i, \mathbf{k}.$$

The new Fock space is unitary nonequivalent to the initial one.

The Hamiltonian expressed in terms of the new canonical variables, normally ordered with respect to the vacuum $|0\rangle\rangle$ and corresponding to the zero-momentum R-scheme, has the form

$$H = H_0' + H_I' + H_{ct}' + H_1 + E,$$

$$H_0' = \frac{1}{2} : \int_V d\mathbf{x} \left[\sum_i^{N-1} (\Pi_i^2 + (\nabla\Phi_i)^2 + M_0^2\Phi_i^2) + \Pi^2 + (\nabla\Phi)^2 + M^2\Phi^2 \right] :,$$

$$H_I' = \frac{g}{4} \int_V d\mathbf{x} : \left[\Phi^4 + 4B\Phi^3 + 4B\Phi \sum_i^{N-1} \Phi_i^2 + 2\Phi^2 \sum_i^{N-1} \Phi_i^2 + \left(\sum_i^{N-1} \Phi^2 \right)^2 \right] :,$$

$$H'_{\text{ct}} =: \int_V d\mathbf{x} \left[\frac{1}{2} A_{\Phi}(M, M_0) \Phi^2 + \frac{1}{2} A_{\Phi_i}(M, M_0) \sum_i^{N-1} \Phi_i^2 \right.$$

$$\left. + C(M, M_0) \Phi + \delta E(M, M_0) \right] :,$$

$$H_1 = \int_V d\mathbf{x} : \left[\frac{1}{2} R(B, M, M_0) \Phi^2 + \frac{1}{2} P(B, M, M_0) \sum_i^{N-1} \Phi^2 \right.$$

$$\left. + Q(B, M, M_0) \Phi \right] :,$$

$$R(B, M, M_0) = m^2 - M^2 - g\left(3D(t) + (N-1)D(t_0)\right) + 3gB^2$$
$$+ A(m) - A_{\Phi}(M, M_0),$$
$$P(B, M, M_0) = m^2 - M_0^2 - g\left(D(t) + (N+1)D(t_0)\right) + gB^2$$
$$+ A(m) - A_{\Phi_i}(M, M_0),$$
$$Q(B, M, M_0) = m^2 B + gB^3 - gB\left(3D(t) + (N-1)D(t_0)\right)$$
$$+ BA(m) - C(M, M_0),$$

$$E = \frac{m^2}{2} B^2 + \frac{g}{4} B^4 + L(t_0)$$

$$+ \frac{g}{4} \left\{ -2\left[3D(t) + (N-1)D(t_0)\right] B^2 + 3D^2(t) \right.$$

$$\left. + (N-1)D^2(t_0) + 2(N-1)D(t)D(t_0) \right\}$$

$$+ \frac{1}{2} A(m)B^2 - \frac{1}{2} A(m)\left[D(t) + (N-1)D(t_0)\right] + \delta E(m) - \delta E(M, M_0),$$

where the following notation is used:

$$A_{\Phi} = 2g^2 \left[3\Sigma(M) + (N-1)\Sigma_1(M, M_0)\right],$$
$$A_{\Phi_i} = 2g^2 \left[3\Sigma_1(M_0, M) + (N+1)\Sigma_1(M, M_0)\right],$$
$$C = 2g^2 B \left[3\Sigma(M) + (N+1)\Sigma_1(M, M_0)\right],$$
$$\delta E(M, M_0) = \delta E_1 + \delta E_2 + \delta E_3,$$
$$\delta E_1 = g^2 B^2 \left[3\Sigma(M) + (N-1)\Sigma_1(M, M_0)\right],$$
$$\delta E_2 = \frac{1}{4} g^2 \left[3W(M) + (N^2-1)W(M_0) + 2(N-1)W_1(M, M_0)\right],$$
$$\delta E_3 = -\frac{1}{2} g^3 [9V(M) + (N-1)(N=1)^2 V(M_0) + 3(N-1)V_1(M_0, M) +$$
$$3(N-1)^2 V_1(M, M_0) + 8(N-1)V_2(M, M_0)].$$

The functions D, L, Σ, W and V have already been defined in (4.6, 4.15). Other divergent integrals look like

$$\Sigma_1(M, M_0) = \frac{1}{(4\pi)^2} \text{reg} \int_0^C n f t y \frac{ds}{s} \exp\left\{-s(M + 2M_0)\right\},$$

$$W_1(M, M_0) = \frac{1}{(4\pi)^3} \text{reg} \int_0^\infty \frac{ds}{s^2} \exp\left\{-2s(M + M_0)\right\},$$

$$V_1(M, M_0) = \frac{1}{4(2\pi)^5} \text{reg} \int_0^\infty \frac{ds}{s} \arctan\left(\frac{s}{2M_0}\right) \arctan^2\left(\frac{s}{2M}\right),$$

$$V_2(M, M_0) = \frac{1}{4(2\pi)^5} \text{reg} \int_0^\infty \frac{ds}{s} \arctan^3\left(\frac{s}{M + M_0}\right).$$

Correspondence between the divergent integrals and Feynman diagrams is illustrated in Fig. ??. In terms of the dimensionless variables (4.19) the energy density takes the form

$$E = \frac{m^3}{8\pi}(\varepsilon_1 + \varepsilon_2 + \varepsilon_3 + \varepsilon_4), \tag{4.33}$$

$$\varepsilon_1 = b^2 + \frac{G}{4}b^4 + f^2\left(1 + \frac{f}{3}\right) + (N-1)f_0^2\left(1 + \frac{f_0}{3}\right)$$
$$+ \frac{G}{4}[-(6f + 2(N-1)f_0)b^2 + 3f^2 + (N^2-1)f_0^2 + 2(N-1)ff_0],$$

$$\varepsilon_2 = \frac{1}{2}G^2 b^2\left[3\ln t + (N-1)\ln\left(\frac{1+t+t_0}{3}\right)\right],$$

$$\varepsilon_3 = -\frac{1}{2}G^2\left\{t\ln t - 3f\left(1 - \ln\left(\frac{4}{3}\right)\right)\right.$$
$$+ (N^2-1)\left[t_0\ln t_0 - f_0\left(1 - \ln\left(\frac{4}{3}\right)\right)\right]$$
$$\left. + (N-1)\left[(t+t_0)\ln\left(\frac{t+t_0}{2}\right) - (t-t_0)\left(1 - \ln\left(\frac{4}{3}\right)\right)\right]\right\},$$

$$\varepsilon_4 = -\frac{\pi^2}{32}G^3\left[9\ln t + (N+1)^2(N-1)\ln t_0 + 8(N-1)\ln\left(\frac{t+t_0}{2}\right)\right]$$
$$- \frac{G^3}{4\pi}3(N-1)J(t, t_0),$$

$$J(t, t_0) = \int_0^\infty \frac{ds}{s}\left\{N \arctan^3(s)\right.$$

$$- \arctan\left(\frac{s}{t}\right) \arctan\left(\frac{s}{t_0}\right) \left[\arctan\left(\frac{s}{t}\right) + (N-1)\arctan\left(\frac{s}{t_0}\right)\right]\Bigg\},$$

where $f = t - 1$, $f_0 = t_0 - 1$.

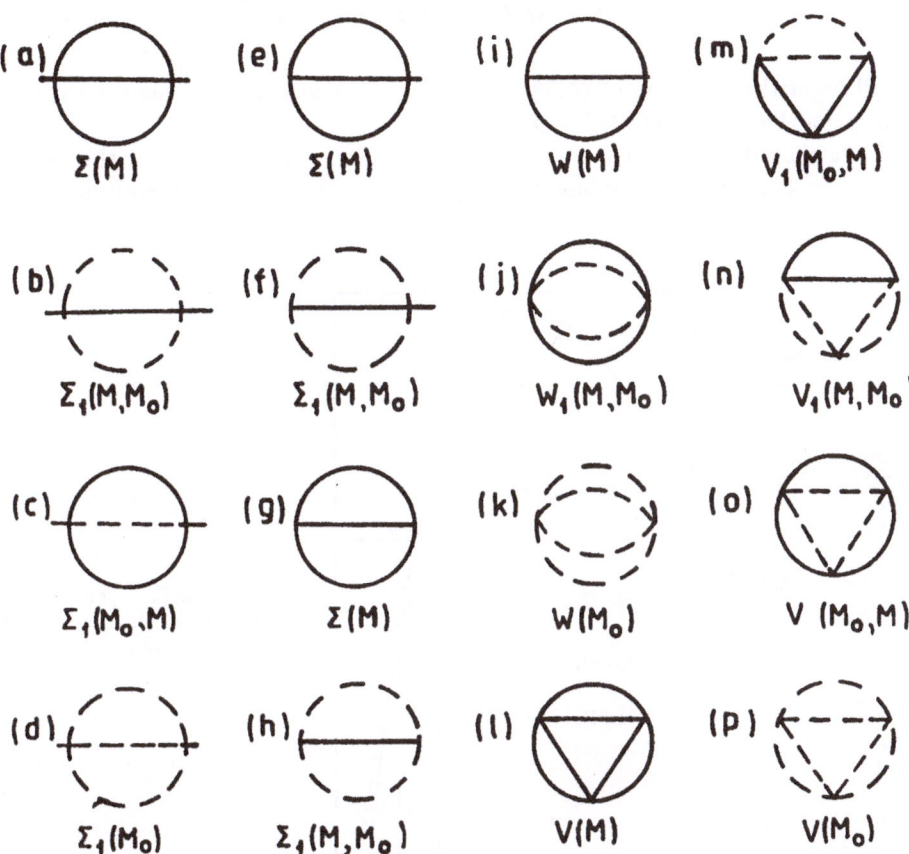

Fig. 4.6. Divergent vacuum diagrams: solid line – the field Φ, dashed – Φ_i; the diagrams (a,b) contribute into A_Φ; (c,d) – A_{Φ_i}; (e,f) – C; (g,h) – δE_1; (i-k) – δE_2; (l-p) – δE_3.

Now, we put the term H_1 equal to zero: $H_1 = 0$. This demand provides the correct form of the total Hamiltonian and gives equations for the parameters t, t_0, B. In dimensionless form these equations are

$$3Gb^2 - 2f(1 + 2f) - 3Gf$$
$$-(N-1)Gf_0 + 3G^2 \ln t + (N-1)G^2 \ln\left(\frac{t + 2t_0}{3}\right) = 0,$$

$$Gb^2 - 4f_0\left(1 + \frac{f_0}{2}\right) - Gf \tag{4.34}$$

$$-(N+1)Gf_0 + (N+1)G^2\ln t_0 + G^2\ln\left(\frac{t_0 + 2t}{3}\right) = 0,$$

$$b\left[2 + Gb^2 - 3Gf\right.$$

$$\left. -(N-1)Gf_0 + 3G^2\ln t + (N-1)G^2\ln\left(\frac{t + 2t_0}{3}\right)\right] = 0.$$

All logarithmic terms in (4.33) and (4.34) appear owing to renormalization in perturbation orders G^2 and G^3.

It should be noted that (4.34) do not define a minimum of the energy density E over the variables t, t_0 and b, as was so in the two-dimensional case.

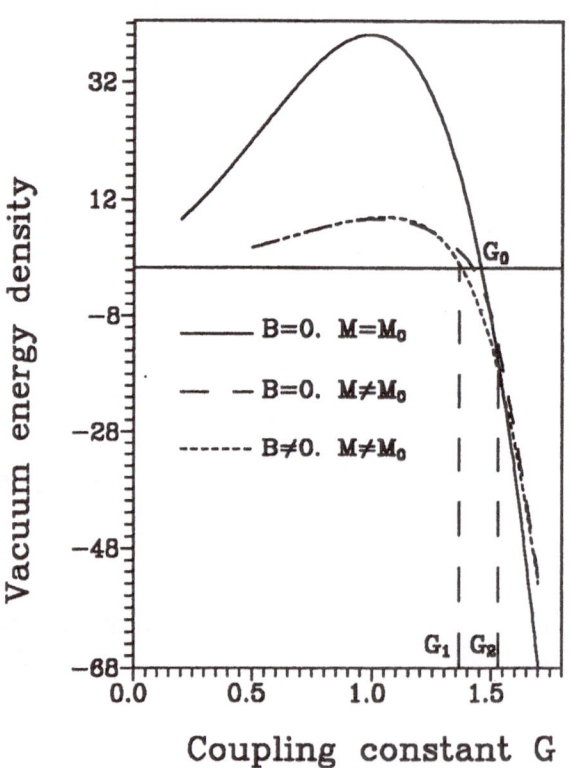

Fig. 4.7. Free energy density in the model (2.3) in R^{2+1} ($N = 4$).

4.5.3 The Phase Structure

Different phases correspond to different solutions of (4.34). We have the following possible phases:

- S_1 ($B = 0$, $M = M_0 = m$, $O(N)-$ and $\Phi \to -\Phi$ invariant)
- S_2 ($B = 0$, $M = M_0 \neq m$, $O(N)-$ and $\Phi \to -\Phi$ invariant)
- BS_1 ($B = 0$, $M \neq M_0$, $O(N-1)-$ and $\Phi \to -\Phi$ invariant)
- BS_2 ($B \neq 0$, $M \neq M_0$, $O(N-1)-$invariant).

We have to consider these phases and, by comparing the effective coupling constants and energy densities, determine the phases which are actually realized in the system.

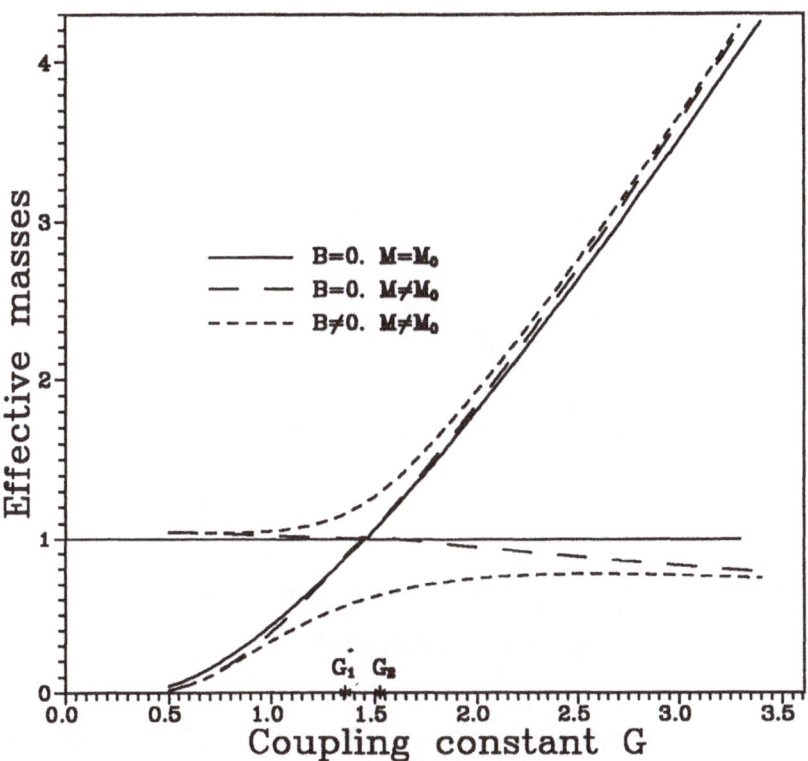

Fig. 4.8. The mass in different phases ($N = 4$): upper and lower dashed lines – M_0 and M in the BS-phases, correspondingly; solid line– M in the S-phase.

The $O(N)$-invariant phases correspond to the conditions:

$$b = 0, \quad t_0 = t.$$

Using these conditions in (4.34) we get the equation ($f = t - 1$)

$$2f(2 + f) + (N + 2)Gf - (N + 2)G^2 \ln(1 + f) = 0. \qquad (4.35)$$

The S_1-phase with $M_0 = M = m.$. Equation (4.35) has the solution $f \equiv 0$ $(t \equiv 1)$ for all G, which corresponds to the initial representation. The energy density, the mass and the effective coupling constant $(\equiv G)$ are plotted in Figs. 4.7–4.9 by solid lines.

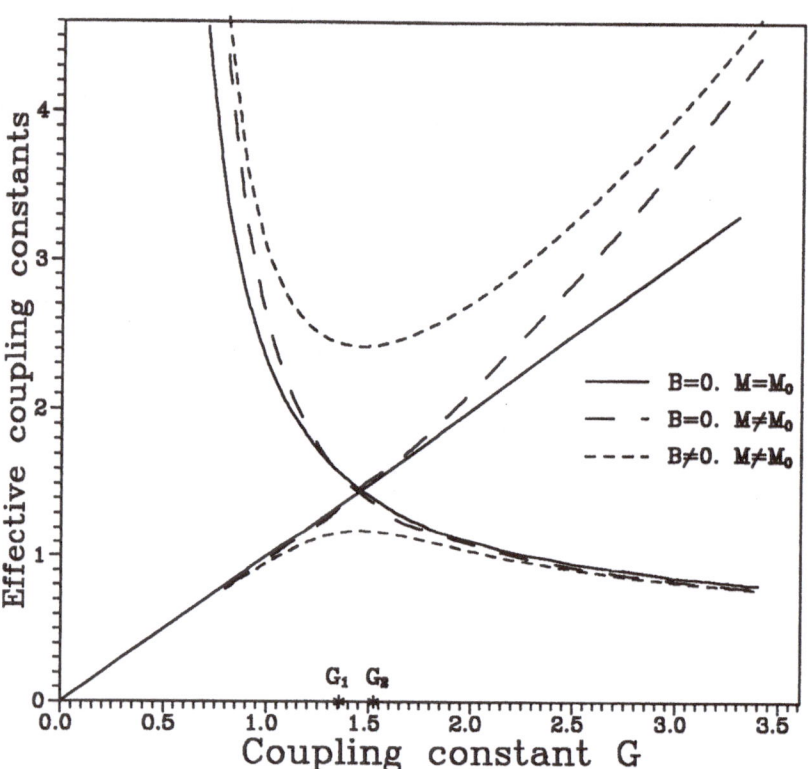

Fig. 4.9. The effective coupling constants in in different phases $(N = 4)$: upper and lower dashed lines – G_{eff} and $G_{\text{eff}}^{(0)}$ in the BS-phases, correspondingly; solid line– G_{eff} in the S-phase.

The S_2-phase with $M_0 = M \neq m.$. In order to exclude the solution $t \equiv 1$ we rewrite (4.35) in the form

$$2 + 2t + (N + 2)G = (N + 2)G^2 \frac{\ln t}{t - 1}. \tag{4.36}$$

This equation has a unique solution for all G. It can easily be checked that

$$M(G) \overset{G \to 0}{\longrightarrow} m \exp \left\{ -\frac{2}{(N+2)G^2} \right\}.$$

Nonanalyticity of the function $M(G)$ at the origin $G = 0$ means that difference between m and $M(G)$ can not be obtained by perturbation calculations. The critical point G_0, where the energy in the S_2-phase is equal to zero and a transition from one S-phase to another takes place, can be found from the condition $t(G_0) = 1$. Putting $t = 1$ in (4.36) one finds that

$$G_0 = \frac{1}{2} \left[1 + \sqrt{\frac{N+18}{N+2}} \right]. \tag{4.37}$$

Asymptotic behaviour in the strong coupling regime $G \gg 1$ looks like

$$t(G) \to \sqrt{\frac{N+2}{2}} G \sqrt{\ln G},$$

$$G_{\text{eff}}(G) \to \sqrt{\frac{2}{(N+2)\ln G}} \ll 1, \tag{4.38}$$

$$E(G) \to -\frac{m^3}{8\pi} N(N+2) \sqrt{\frac{N+2}{2}} G^3 \ln^{3/2}(G).$$

The functions $\varepsilon(G)$, $t(G)$ and $G_{\text{eff}}(G)$ for intermediate values of G are plotted in Figs. 4.7-4.9.

The BS_1-phase: mass splitting at zero condensate $b = 0$.. The breaking of the $O(N)$-symmetry owing to mass splitting in the multiplet $M \neq M_0$ corresponds to the case: $b = 0$, $t \neq t_0$. The Hamiltonian in this phase is invariant under the group $O(N-1)$ and discrete transformation $\Phi \to -\Phi$. This phase is described by (4.35) with $b = 0$. In order to exclude the solution $t = t_0$ from (4.35), it is convenient to introduce the variable $r = t_0/t$, subtract the second equation from the first one and devide the result by $1 - r$. Thus, the following equations are obtained:

$$2(r+1)t^2 + 2Gt - (N+1)G^2 \frac{\ln r}{1-r}$$

$$+(N-1)G^2 \frac{\ln\left(\frac{1+2r}{3}\right)}{1-r} - G^2 \frac{\ln\left(\frac{2+r}{3}\right)}{1-r} = 0,$$

$$2 + (N+2)G - rt(2rt + (N+1)G)$$

$$-Gt + (N+2)G^2 \ln(rt) + G^2 \ln\left(\frac{r+2}{3r}\right) = 0.$$

These equations and (4.33) define the asymptotic behaviour for $G \gg 1$:

$$t_0 \longrightarrow \sqrt{\frac{N+2}{2}} \cdot G \cdot \sqrt{\ln G},$$

$$t \longrightarrow \sqrt{\frac{N+2}{2}} \cdot G^{\frac{1-N}{3}} \cdot \sqrt{\ln G},$$

$$G_{\text{eff}}^0 = \frac{G}{t_0(G)} \longrightarrow \frac{\sqrt{2}}{\sqrt{(N+2)\ln G}} \ll 1, \tag{4.39}$$

$$G_{\text{eff}} = \frac{G}{t(G)} \longrightarrow \frac{\sqrt{2}}{\sqrt{(N+2)\ln G}} \cdot G^{\frac{N+2}{3}} \gg 1,$$

$$E \longrightarrow -\frac{m^3}{24\pi}(N^2 + 3N - 4)\sqrt{\frac{N+2}{2}} \cdot G^3 \sqrt{\ln^3 G}.$$

The energy, the masses and effective coupling constants as functions of G are shown in Figs. 4.7-4.9.

The BS_2-phase: mass splitting and nonzero condensate $b \neq 0$.. The last equation of the system (4.34) provides a nonzero solution for b:

$$Gb^2 = -2 + (N-2)G + 3Gt + (N-1)Gt_0 - 3G^2 \ln t$$
$$-(N-1)G^2 \ln\left(\frac{t+2t_0}{3}\right) \tag{4.40}$$

The phase is $O(N-1)$-symmetric and is characterized by the mass splitting $M \neq M_0$ and nonzero condensate $b \neq 0$. It is described by the first two equations (4.34) and (4.40). It is convenient to represent these equations in the form

$$b^2 = \frac{t^2}{G},$$

$$-2r^2t^2 + 2G(1-r)t$$
$$+3G^2 \ln r + G^2 \ln\left(\frac{2+r}{3r}\right) - (N-1)G^2 \ln\left(\frac{1+2r}{3r}\right) = 0,$$

$$(1-2r^2)t^2 - (N+1)Grt$$
$$-Gt + 2 + (N+2)G^2 \ln(rt) + G^2 \ln\left(\frac{2+r}{3r}\right) = 0,$$

where $r = t_0/t$.

Asymptotically the functions t, t_0, G_{eff}, G_{eff}^0 and E (see Figs. 4.7-4.9) are the same ones as in (4.39). For $G \gg 1$ the condensate $b(G)$ takes the form

$$b(G) \longrightarrow \sqrt{\frac{N+2}{2}} G^{-(2N+1)/6} \sqrt{\ln G}.$$

An essential point relating to both BS-phases is that the mass M_0 of the fields Φ_i belonging to the O(N-1)-multiplet is larger than the mass M of the field Φ (see Fig. 4.8). This contradicts the Goldstone theorem, which implies a zero mass for "Goldstone" bosons Φ_i. We should note that (4.34) has solutions $M_0 < M$ also, but the energy densities corresponding to these solutions are positive and grow as the coupling constant G grows.

The asymptotic relations (4.38, 4.39) show that in the strong coupling regime the S_2-phase is realized, since only in this phase does the effective coupling constant decrease when G increase. Comparison of the asymptotic behaviour of the energy densities guarantees this conclusion. For any N the energy of the S_2-phase is less than the energies of other phases. Transition between the first (initial) and the second symmetric phases takes place at the point G_0 (see (4.37)).

More detailed, but quite approximate (the effective coupling constants are of the order of unity), information follows from comparison of the energy densities at intermediate values of the coupling constant G. It follows from Fig. 4.9 that phase transitions with the following symmetry reconstruction are possible:

$$O(N) \longrightarrow O(N-1) \longrightarrow O(N).$$

At first, the system transits from the initial S_1-phase to the BS_2-phase with $B \neq 0$ and then to the S_2-phase with $M \neq m$. For instance, at $N = 4$ the critical coupling constants are equal to

$$G_1 = 1.357..., \quad G_2 = 1.525....$$

Qualitatively this picture does not depend on N:

$$G_1 < G_0 < G_2 \quad \forall N.$$

In any case, for $G \ll 1$ and $G \gg 1$ system (2.3) exists in the first and second symmetric phases respectively. The vacuum and the Hamiltonian are $O(N)$-invariant and the effective coupling is weak.

4.6 Stability Under Change of the R-scheme in the Initial Representation

Our previous calculations did not appeal directly to the RG formalizm. Now we wish to illustrate the role of the renormalization group more clearly. First of all, we have to investigate the dependence of the results obtained in previous sections on the choice of the R-scheme in the initial representation. In this section we shall reobtain the equations describing the phase structure of model (2.1) using the RG formalizm more actively and then consider these equations within different R-schemes.

Let us represent the problem in a form convenient for this purpose and write down the Lagrangian of model (2.1) in the form

$$L = \frac{1}{2}\varphi(x)(\Box - m_0^2)\varphi(x) - \frac{1}{4}g\varphi^4(x),$$

where $m_0^2 = m^2(\mu) + \delta m^2(\mu)$ is the bare mass, $m(\mu)$ and $\delta m(\mu)$ are the renormalized mass and the mass counter-term in some fixed R-scheme. The dimensionless coupling constant

$$G = \frac{g}{2\pi m(\mu)}$$

is a free parameter of the model. The Hamiltonian in the representation with the mass $m(\mu)$ looks like

$$H = H_0 + H_I + H_{ct}, \tag{4.41}$$

$$H_0 = \frac{1}{2} \int_V dx \left[\pi^2(x) + (\nabla\varphi(x))^2 + m^2(\mu)\varphi^2(x) \right],$$

$$H_I = \frac{1}{4}g \int_V dx\varphi^4(x), \quad H_{ct} = \frac{1}{2}\delta m^2(\mu) \int_V dx\varphi^2(x). \tag{4.42}$$

The operators φ and π satisfy the standard CR.

Then, as usual, we perform the canonical transformation to the representation with the new mass $M = m(\mu) \cdot t$ and condensate $B =$const:

$$(\varphi, \pi) \longrightarrow (\Phi + B, \Pi), \tag{4.43}$$

accompanied by the compensating renormalization group transformation

$$\mu \longrightarrow \nu = \mu \cdot t.$$

This RG transformation provides an equivalence of the R-schemes in different representations:

$$\frac{m(\mu)}{\mu} = \frac{M}{\nu}. \tag{4.44}$$

In the new representation the Hamiltonian takes the form

$$H = H_0' + H_I' + H_{ct}' + H_1, \tag{4.45}$$

$$H_0' = \frac{1}{2} \int_V dx \left[\Pi^2(x) + (\nabla\Phi(x))^2 + M^2\Phi^2(x) \right], \tag{4.46}$$

$$H_I' = \frac{1}{4}g \int_V dx \left[\Phi^4(x) + 4B\Phi^3(x) \right], \tag{4.47}$$

$$H_{ct}' = \frac{1}{2}\delta m^2(\mu \cdot t) \int_V dx \left[\Phi^2(x) + \delta m^2(\mu \cdot t)B\Phi(x) \right],$$

$$H_1 = \frac{1}{2} \int_V dx \left\{ \left[m^2(\mu \cdot t) + 3gB^2 - M^2 \right] \Phi^2(x) + \left[m^2(\mu \cdot t) + gB^2 \right] B\Phi(x) \right\}.$$

In order to provide the correct form of the total Hamiltonian let us put $H_1 = 0$. This condition leads to the equations for B and t:

$$m^2(\mu \cdot t) + 3gB^2 - m^2(\mu)t^2 = 0,$$
$$B\left[m^2(\mu \cdot t) + gB^2\right] = 0. \tag{4.48}$$

We have to investigate the R-scheme dependence of (4.48). In the case of the symmetric solution $B = 0$ the first equation (4.48) takes the form

$$\frac{m^2(\mu \cdot t)}{m^2(\mu)} = t^2. \tag{4.49}$$

Let us consider this equation calculating the running mass $m(\mu \cdot t)$ within different R-schemes.

4.6.1 Subtraction at Zero External Momentum and Arbitrary "Mass" in Propagators

In this R-scheme the mass counter-terms correspond to the diagrams shown in Figs. 4.1 with zero external momentum and arbitrary "mass" μ in the propagators. This is one of the possible ways to introduce the mass scale μ in the $(\varphi^4)_3$. Let us put

$$m_B^2 = \bar{m}^2(\mu) + \delta\bar{m}_a^2(\mu) + \delta\bar{m}_b^2(\mu), \tag{4.50}$$

where $\delta\bar{m}_a^2(\mu)$ and $\delta\bar{m}_b^2(\mu)$ relate to the diagrams a and b in Fig. 4.1. This counter-terms can be easily calculated:

$$\delta\bar{m}_a^2(\mu) = -3g\Delta_{\text{reg}}(\mu), \quad \delta\bar{m}_b^2(\mu) = 3!g^2\Sigma_{\text{reg}}(\mu),$$

$$\Delta_{\text{reg}} = \frac{1}{(2\pi)^2}\text{reg}\int_0^\infty \frac{du\,u^2}{u^2 + \mu^2},$$

$$\Sigma_{\text{reg}} = \frac{1}{(4\pi)^2}\text{reg}\int_0^\infty \frac{dt}{t}\exp\{-3\mu t\}. \tag{4.51}$$

Let the usual zero-momentum R-scheme be used in the initial representation. This corresponds to a particular choice of the parameter μ being equal to the renormalized mass m, i.e., the condition

$$\bar{m}(m) = m \tag{4.52}$$

fixes the standard zero-momentum R-scheme within the R_μ-class with arbitrary μ. Equation (4.49) takes the form

$$\frac{\bar{m}^2(m \cdot t)}{m^2} = t^2. \tag{4.53}$$

Using the RG-invariance of the bare mass m_0 and (4.52), one finds that

$$\bar{m}^2(\mu) = m^2 + 3g\left(\Delta_{\text{reg}}(\mu) - \Delta_{\text{reg}}(m)\right) - 6g^2\left(\Sigma_{\text{reg}}(\mu) - \Sigma_{\text{reg}}(m)\right).$$

Turning off the regularization, one gets the equation for running mass:

$$\bar{m}^2(\mu) = m^2 \left[1 + \frac{3}{2}G\left(1 - \frac{\mu}{m}\right) + \frac{3}{2}G^2 \ln\left(\frac{\mu}{m}\right) \right], \qquad (4.54)$$

where $G = g/2\pi m$. Using (4.54) with $\mu = m \cdot t$ in (4.53), one gets

$$t^2 - 1 + \frac{3}{2}G(t-1) - \frac{3}{2}G^2 \ln t = 0. \qquad (4.55)$$

This equation coincides with (4.27), obtained above.

4.6.2 Dimensional Regularization and the Minimal Subtraction Scheme

Let us introduce the following notation:

$$\varepsilon = 3 - d, \quad \alpha = \frac{g}{2\pi},$$
$$m_B^2 = m^2(\mu) + \delta m_a^2 + \delta m_b^2, \qquad (4.56)$$

where $m(\mu)$ is the running mass in the MS-scheme. By means of the standard calculations we find for the diagram (a) in Fig. 4.1 that

$$\delta m_a^2 = -3gm(\mu)\frac{1}{(2\pi)^d/2}\left(\frac{2\pi\mu}{m(\mu)}\right)^\varepsilon \Gamma\left(1 - \frac{d}{2}\right).$$

Putting $d = 3$ one gets the finite expression

$$\delta m_a^2 = \frac{3}{2}\alpha m(\mu), \qquad (4.57)$$

which is natural for dimensional regularization in the case of an odd physical dimension of the space-time.

Calculation of the diagram (b) in Fig. 4.1 at zero external momentum leads to the expression

$$\Sigma_{\text{reg}} = \frac{3}{4}\alpha^2\left[\frac{1}{\varepsilon} + \ln\left(\frac{4\pi\mu^2}{m^2(\mu)}\right) - \gamma_E + o(\varepsilon)\right]. \qquad (4.58)$$

In the MS-scheme only the divergent part of this expression is accounted for in counter-terms:

$$\delta m_b^2 = \frac{3}{4}\alpha^2\frac{1}{\varepsilon}. \qquad (4.59)$$

Using (4.57, 4.59) in (4.56), we get

$$m_B^2 = m^2(\mu) + \frac{3}{2}\alpha m(\mu) + \frac{3}{4}\alpha^2\frac{1}{\varepsilon}. \qquad (4.60)$$

Let us introduce in (4.60) a new scale ν: (e.g., see [46])

$$\alpha \to \alpha_{\text{new}} \left(\frac{\mu}{\nu}\right)^{\varepsilon} \quad \left(\alpha_{\text{new}} \xrightarrow{\varepsilon \to 0} \alpha\right).$$

Using this substitution, we get the expression

$$m_B^2 = m^2(\nu) + \frac{3}{2}\alpha_{\text{new}}m(\nu) + \frac{3}{4}\alpha_{\text{new}}^2\frac{1}{\varepsilon} + \frac{3}{2}\alpha_{\text{new}}^2 \ln\left(\frac{\mu}{\nu}\right) \tag{4.61}$$

with the obvious condition

$$m(\nu)|_{\nu=\mu} = m(\mu). \tag{4.62}$$

Taking the limit $\varepsilon \to 0$ in (4.60, 4.61) we get

$$m^2(\nu) + \frac{3}{2}\alpha m(\nu) - \frac{3}{2}\alpha^2 \ln\left(\frac{\nu}{\mu}\right) - m^2(\mu) - \frac{3}{2}\alpha m(\mu) = 0. \tag{4.63}$$

The solution of this equation satisfying condition (4.62) has the form

$$m(\nu) = -m(\mu) - \frac{3}{2}\alpha + \sqrt{\left[2m(\mu) + \frac{3}{2}\alpha\right]^2 + 6\alpha^2 \ln\left(\frac{\nu}{\mu}\right)}. \tag{4.64}$$

Putting $\nu = \mu \cdot t$ and $m(\nu) = m(\mu) \cdot t$ in (4.63) we get

$$t^2 - 1 + \frac{3}{2}G(t-1) - \frac{3}{2}G^2 \ln t = 0, \tag{4.65}$$

where $G = \alpha/m(\mu)$. Equation (4.65) coincides with (4.55). We see that the function $t(\cdot)$ turns out to be the same one in both cases, while the running masses $\tilde{m}(\nu)$ and $m(\nu)$ are completely different functions of ν (see (4.54, 4.64)).

4.6.3 Dimensional Regularization and Zero-Momentum Subtraction

Let us change the renormalization prescription for the diagram (b) in Fig. 4.1. Now we include in the counter-term δm_b^2 not only the pole part of (4.58), but its finite part as well. The bare mass takes the form

$$m_B^2 = \tilde{m}^2(\mu) + \frac{3}{2}\alpha\tilde{m}(\mu) + \frac{3}{4}\alpha^2 \left[\frac{1}{\varepsilon} - \gamma_E + \ln\left(\frac{4\pi\mu}{\tilde{m}^2(\mu)}\right) + o(\varepsilon)\right]. \tag{4.66}$$

Doing the standard transformation to the new scale ν in (4.66) we get the equation

$$\tilde{m}^2(\nu) - \tilde{m}^2(\mu) + \frac{3}{2}\alpha[\tilde{m}(\nu) - \tilde{m}(\mu)]$$
$$- \frac{3}{2}\alpha^2 \ln\left(\frac{\nu}{\mu}\right) + \frac{3}{4}\alpha^2 \ln\left(\frac{\tilde{m}^2(\mu)\nu^2}{\tilde{m}^2(\nu)\mu^2}\right) = 0. \tag{4.67}$$

One can see that the running mass $\tilde{m}$ defined by (4.67) differs from $\bar{m}$ and m (see (4.54, 4.64)). Nevertheless, the following substitution:

$$\nu = \mu \cdot t, \quad \tilde{m}(\nu) = \tilde{m}(\mu) \cdot t$$

in (4.67) leads to an equation that coincides with (4.55, 4.65). The difference in the dimensionless coupling constant is unimportant since this constant is a free parameter of the equations.

The above calculations means the equation defining the parameter t is quite stable under the changing of the R-scheme, while the running mass depends critically on the R-scheme.

5. The Four-Dimensional φ^4 Theory

5.1 The Hamiltonian $(\varphi^4)_4$ and Canonical Transformations

The models investigated in the previous chapters are superrenormalizable. Therefore we were able to calculate all necessary counter-terms. Now we would like to extend our consideration to the more complicated case of the renormalizable model and investigate the phase structure of the four-dimensional theory (2.1). The main point is that renormalization of the field and the coupling constant is necessary in the φ_4^4 case in contrast to the two- and three-dimensional cases.

The Hamiltonian for the Lagrangian (2.1) has in the weak coupling regime the form

$$H = H_0 + H_I + H_{ct},$$

$$H_0 = \frac{1}{2} \int_V d\mathbf{x} \left[\pi^2 + (\nabla\varphi)^2 + m^2(\mu)\varphi^2 \right], \quad H_I = \frac{1}{4} g(\mu) \int_V d\mathbf{x}\varphi^4,$$

$$H_{ct} = \int_V d\mathbf{x} \left\{ \frac{1}{2} \left[\left(\frac{1}{Z_2} - 1 \right) \pi^2 + (Z_2 - 1)(\nabla\varphi)^2 + \delta m^2(\mu)\varphi^2 \right] \right.$$
$$\left. + \frac{1}{4}(Z_1 - 1)g(\mu)\varphi^4 \right\}. \tag{5.1}$$

The operators φ and π are given by (2.8). These fields are the canonical variables and obey the standard canonical relations. The creation $a^+(\mathbf{k})$ and annihilation $a(\mathbf{k})$ operators are defined on the Fock space of the particles with a mass $m(\mu)$. The vacuum vector $|0\rangle$ satisfies the condition

$$a(\mathbf{k})|0> = 0 \ \forall \mathbf{k}, \quad <0|0> = 1.$$

The R-scheme in representation (5.1) is assumed to be fixed, i.e., a definite R_μ-class is chosen and the ratio $C = m(\mu)/\mu$ is fixed.

The representation (5.1) is quite appropriate if $g(\mu) \ll 1$. Bearing this in mind, we want to know, what the system is in the strong coupling regime $g(\mu) \gg 1$.

Let us perform the following canonical transformation:

$$(\varphi, \pi) \longrightarrow \left(\frac{1}{\zeta}\Phi + \frac{1}{\zeta}B, \zeta\Pi\right). \tag{5.2}$$

Here, (Φ, Π) are the fields with a mass $M = t \cdot m(\mu)$ and B is a constant condensate. According to the demand of the R-scheme equivalence in different representations the canonical transformation including the transition to the new mass M should be accompanied by the compensating scale transformation

$$\mu \longrightarrow \nu = t \cdot \mu,$$

which is the origin of the finite field renormalization ζ in (5.2).

The transformation (5.2) in terms of the creation and annihilation operators is given in (2.28, 2.29). The fields (Φ, Π) satisfy the CR. For $t \neq 1$ and (or) $B \neq 0$ they are defined on the Fock space unitary nonequivalent to the initial one.

The Hamiltonian in the new representation looks like

$$H = H'_0 + H'_I + H'_{ct} + H_1, \tag{5.3}$$

$$H'_0 = \frac{1}{2}\int_V dx \left[\Pi^2 + (\nabla\Phi)^2 + M^2\Phi^2\right], \quad H'_I = \frac{1}{4}g(\nu)\int_V dx[\Phi^4 + 4B\Phi^3],$$

$$H'_{ct} = \int_V dx \left\{\frac{1}{2}\left[\left(\frac{1}{Z'_2} - 1\right)\Pi^2 + (Z'_2 - 1)(\nabla\Phi)^2\right]\right.$$

$$+ \frac{1}{2}\left[\delta m^2(\nu) + 3(Z'_1 - 1)g(\nu)B^2\right]\Phi^2$$

$$+ \frac{1}{4}(Z'_1 - 1)g(\nu)(\Phi^4 + 4B\Phi^3)$$

$$\left. + \left[\delta m^2(\nu) + (Z'_1 - 1)g(\nu)B^2\right]B\Phi\right\},$$

$$H_1 = \int_V dx \left\{\frac{1}{2}\left[m^2(\nu) + 3g(\nu)B^2 - M^2\right]\Phi^2 + \left[m^2(\nu) + g(\nu)B^2\right]B\Phi\right\}.$$

Here $\nu = t \cdot \mu$, $M = t \cdot m(\mu)$.

To preserve the correct form of the Hamiltonian we should put $H_1 = 0$, which leads to the equations

$$m^2(\mu \cdot t) + 3g(\mu \cdot t)B^2 - m^2(\mu)t^2 = 0,$$

$$B\left[m^2(\mu \cdot t) + g(\mu \cdot t)B^2\right] = 0. \tag{5.4}$$

The quantities $m(\mu \cdot t)$ and $g(\mu \cdot t)$ are linked with $m(\mu)$ and $g(\mu)$ by the RG-transformation and are defined by the RG equations

$$t\frac{dg(\mu \cdot t)}{dt} = \beta\left(g(\mu \cdot t), \frac{m(\mu \cdot t)}{\mu \cdot t}\right),$$

$$\frac{t}{m^2(\mu \cdot t)}\frac{dm^2(\mu \cdot t)}{dt} = -\gamma_m\left(g(\mu \cdot t), \frac{m(\mu \cdot t)}{\mu \cdot t}\right), \tag{5.5}$$

with the initial conditions

$$g(\mu \cdot t) = g(\mu) \quad \text{if} \quad t = 1,$$
$$m(\mu \cdot t) = m(\mu) \quad \text{if} \quad t = 1. \tag{5.6}$$

Equations (5.4–5.6) describe the phase structure of the $(\varphi^4)_4$ theory in general form for arbitrary renormalization scheme. These equations reduce the problem of the phase structure to the properties of the RG functions γ_m and β.

Now we will examine the existance of solutions of the system (5.4–5.6). According to the second equation (5.4) we have two cases: $B = 0$ (S-phase) and $B \neq 0$ (BS-phase).

5.2 The Symmetric Phases

Putting $B = 0$ we get the equation for t:

$$\frac{m^2(\mu \cdot t)}{m^2(\mu)} = t^2. \tag{5.7}$$

It should be noted that (5.7) has the same form as the corresponding equation (4.49) in R^{2+1}.

The system of equations (5.7) and (5.5) can be solved in general form (without specification of the γ_m- and β-function) within an arbitrary R-scheme, while for (5.5) separately this can be done in the case of the mass-independent renormalization schemes only. Really, this is follows from the second equation (5.5) and equation (5.7) that the parameter t is a function of the coupling constant, which can be considered in these equations as a free variable, so that the system (5.5, 5.7) can be rewritten in the form

$$\frac{1}{t}\frac{dt}{d\bar{g}} = \beta^{-1}\left(\bar{g}, \frac{m(\mu \cdot t)}{\mu \cdot t}\right),$$
$$\frac{2}{t}\frac{dt}{d\bar{g}} = -\gamma_m\left(\bar{g}, \frac{m(\mu \cdot t)}{\mu \cdot t}\right)\beta^{-1}\left(\bar{g}, \frac{m(\mu \cdot t)}{\mu \cdot t}\right), \tag{5.8}$$
$$m(\mu \cdot t) = t \cdot m(\mu),$$

where $\bar{g}$ denotes the renormalized coupling constant. According to the initial conditions (5.6) we have

$$t(\bar{g}) = 1 \text{ for } \bar{g} = g. \tag{5.9}$$

Taking account of the third equation (5.8) in the first and second ones gives

$$\frac{d\ln t}{d\bar{g}} = \frac{1}{\beta(\bar{g}, C)},$$
$$2\frac{d\ln t}{d\bar{g}} = -\frac{\gamma_m(\bar{g}, C)}{\beta(\bar{g}, C)}. \tag{5.10}$$

The constant C is assumed to be fixed in the initial representation (5.1).

We see that the only free parameter in system (5.10) with initial condition (5.9) is the initial value of the coupling constant g. Integration of (5.10) over $\bar{g}$ and taking (5.9) into account gives

$$\ln t = \int\limits_{g}^{\bar{g}} \frac{dx}{\beta(x,C)}, \quad 2\ln t = -\int\limits_{g}^{\bar{g}} dx \frac{\gamma_m(x,C)}{\beta(x,C)},$$

or in a slightly different form

$$\ln t = \int\limits_{g}^{G_{\text{eff}}} \frac{dx}{\beta(x,C)}, \quad \int\limits_{g}^{G_{\text{eff}}} dx \frac{2 + \gamma_m(x,C)}{\beta(x,C)} = 0, \qquad (5.11)$$

where $G_{\text{eff}} = \bar{g}(g)$ is an effective coupling constant. According to (5.9) these equations have the solution $t \equiv 1$, $G_{\text{eff}} \equiv g$. The existence of other solutions depends on the properties of the RG-functions, which are different in different R-schemes (in particular, for different C). In order to remove the problem of the R-scheme dependence let us consider the case where the canonical μ-scheme with $c = 1$ is used in the initial representation (5.1). It should be remembered that the two-point Green function is normalized in the μ-scheme by the condition

$$\tilde{G}(p^2) \xrightarrow{p^2 \to \mu^2} \frac{i}{p^2 - m^2(\mu)}. \qquad (5.12)$$

Hence for $C = m(\mu)/\mu = 1$ we get the on-shell renormalization scheme, i.e., the renormalized mass $m(\mu)$ coincides with the physical mass m_{ph} by a construction. System (5.11) takes the form

$$\ln t = \int\limits_{g}^{G_{\text{eff}}} \frac{dx}{\beta(x)}, \quad \int\limits_{g}^{G_{\text{eff}}} dx \frac{2 + \gamma_m(x)}{\beta(x)} = 0, \quad M = m_{ph}t, \qquad (5.13)$$

where

$$\gamma_m(g) \equiv \gamma_m(g,1), \quad \beta(g) \equiv \beta(g,1),$$

$\gamma_m\left(g, \frac{m(\mu)}{\mu}\right)$ and $\beta\left(g, \frac{m(\mu)}{\mu}\right)$ are implied to be calculated within the canonical μ-scheme.

It should be noted that (5.7) takes the form

$$\frac{m(m_{ph}t)}{m_{ph}t} \equiv \frac{m(M)}{M} = 1, \qquad (5.14)$$

so that M satisfies the on-shell condition as well as m_{ph}:

$$\tilde{G}(p^2) \xrightarrow{p^2 \to M^2} \frac{i}{p^2 - M^2}, \qquad (5.15)$$

and, as a result, has the meaning of the physical mass in the new representation (5.3).

Let us return to (5.13). As soon as exact γ_m- and β-functions are not known, we will restrict ourselves to consideration of different possibilities. The behaviour of $\gamma_m(x)$ and $\beta(x)$ at small x can be defined within the perturbation theory. An integrand in the second equation (5.13) behaves like

$$F(x) = \frac{2 + \gamma_m(x)}{\beta(x)} \xrightarrow{x \to 0} \frac{2}{\beta_1 \alpha^2}, \tag{5.16}$$

here $\alpha = 3!x/(4\pi)^2$, $\beta_1 = 3/2$. It is known that the β-function is positive for $x \in (0, g^*)$, where the ultraviolet stable point g^* may be finite or infinite. If the function $F(x)$ does not change its sign in the interval $(0, g^*)$, then (5.13) has the trivial solution ($G_{\text{eff}} \equiv g$, $t \equiv 1$) only.

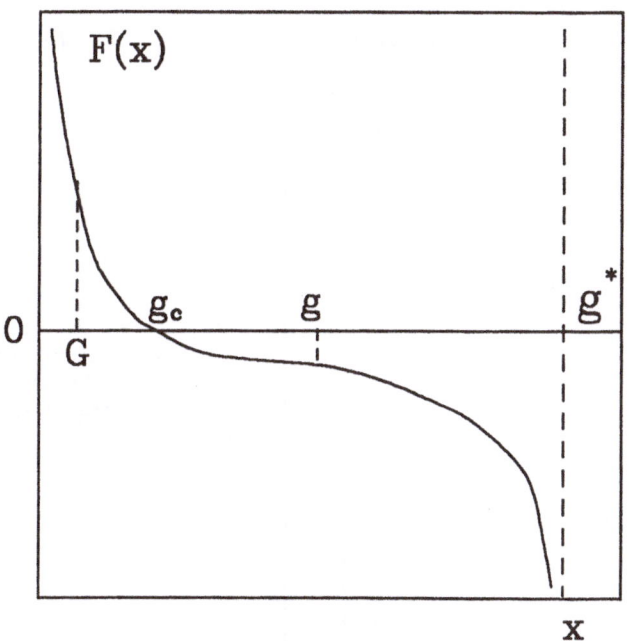

Fig. 5.1. Possible behaviour of the integrand.

Another possibility is illustrated in Fig. 5.1. The second solution of (5.13) exists if the function $F(x)$ changes its sign at some point $g_c \in (0, g^*)$.

For instance, let

$$\gamma_m = -ax, \quad \beta = bx^2 \ (g^* = \infty), \quad \text{where } a > 0, \quad b > 0.$$

After integration we get from (5.13)

$$-\frac{1}{G_{\text{eff}}} + \frac{1}{g} - \frac{a}{2} \ln \left(\frac{G_{\text{eff}}}{g} \right) = 0,$$

$$b \ln t = \frac{1}{g} - \frac{1}{G_{\text{eff}}}.$$

Asymptotic behaviour in the strong coupling regime looks like

$$t(g) \xrightarrow{g \to \infty} g^{-\frac{a}{2b}} \ll 1,$$

$$G_{\text{eff}}(g) \xrightarrow{g \to \infty} \frac{2}{a \ln g} \ll 1. \qquad (5.17)$$

This example illustrates a general picture, which is represented in Figs. 5.2, 5.3 (qualitatively of course). The effective coupling constant depends only on g, moreover

$$G_{\text{eff}}(g) \xrightarrow{g \to 0} g^*, \quad G_{\text{eff}}(g_c) = g_c, \quad G_{\text{eff}}(g) \xrightarrow{g \to g^*} 0, \qquad (5.18)$$

and, since $\beta(x) > 0$,

$$t(g) \xrightarrow{g \to 0} \infty, \quad t(g_c) = 1, \quad t(g) \xrightarrow{g \to g^*} 0. \qquad (5.19)$$

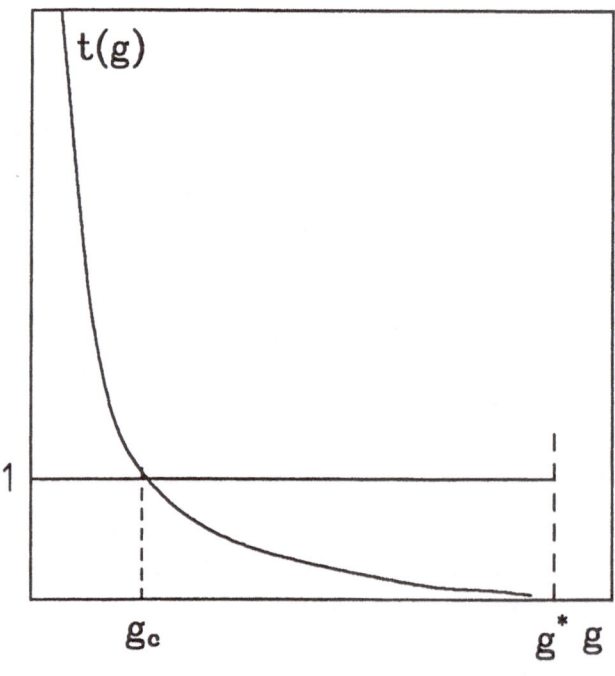

Fig. 5.2. Ratio of masses in the symmetric phases of the $(\varphi^4)_4$ model.

Comparing the asymptotic effective coupling constants we conclude that the system exists in the phase with a mass m_{ph} in the weak coupling regime, but another symmetric phase with a physical mass $M \ll m_{\text{ph}}$ (see (5.15)) and effective coupling constant $G_{\text{eff}} \ll 1$ can be realized in the strong coupling limit ($g \to g^*$). *A kind of phase transition occurs at a value g_c of the coupling* *constant g such that the anomalous dimension of the operator φ^2 compensates its canonical dimension:*

$$2 + \gamma_m(g_c) = 0.$$

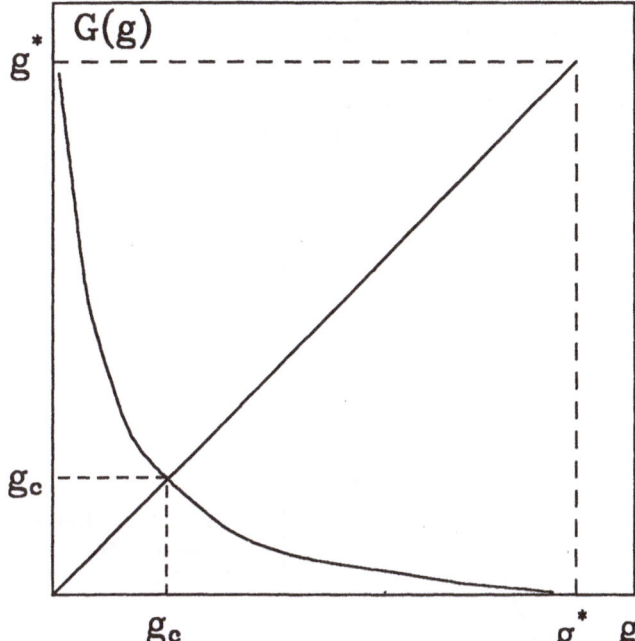

Fig. 5.3. The effective coupling constant in the $(\varphi^4)_4$ model.

It should be noted that the equation coinciding with the right-hand side of (5.14) (after the sign $' \equiv '$) has been used in [70, 71] applied to quantum chromodynamics. Namely, in [70] an equation of the type (5.14) links a current quark mass $m(\mu)$ ($\mu \gg m(\mu)$) with the pole mass M, which is interpretated as the constituent quark mass.

5.3 Dynamical Symmetry Breaking

The Hamiltonian H'_{ct} reflects the well-known fact that counter-terms for $(\varphi^4)_4$ with SSB are determined completely by the counter-trms for the symmetric $(\varphi^4)_4$ (e.g., see [46]), so that $m(\mu \cdot t)$ and $g(\mu \cdot t)$ in (5.4) are defined by the same equations (5.5) for $B = 0$ and $B \neq 0$.

Equations (5.4) can be rewritten in the form

$$B^2 = -\frac{m^2(\mu \cdot t)}{g(\mu \cdot t)}, \quad t^2 = -2\frac{m^2(\mu \cdot t)}{m^2(\mu)}. \tag{5.20}$$

Equations (5.20) do not have real solutions if $m^2(\mu \cdot t) > 0 \ \forall g(\mu), t$. Such a situation occurs for any mass-independent R-scheme. Thus, at least in this case, the system has no representations with $B \neq 0$ and, as a consequence, dynamical symmetry breaking does not occur. In the general case this conclusion can be invalid.

5.4 The Asymptotically Free Model

In order to clarify slightly a situation in the asymptotically free case let us consider the (of course unphysical, but representative) model $(\varphi^4)_4$ with negative coupling constant $g \to -g$. Such a model turns out to be asymptotically free [43].

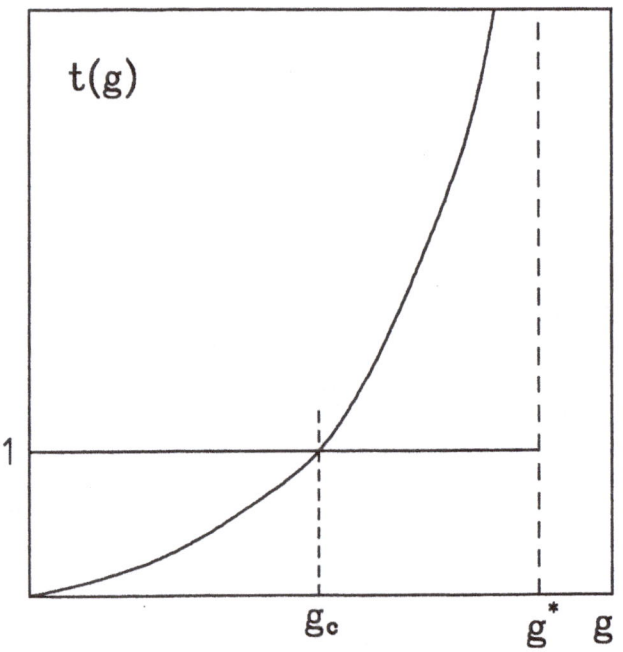

Fig. 5.4. Ratio of masses in the asymptotically free case.

The second equation (5.13) does not feel the sign of the β-function. Hence the asymptotic relations (5.18) do not change their form, and the qualitative behaviour of $G_{\text{eff}}(g)$ is the same as in Fig. 5.3. At the same time the function $t(g)$ is changed cardinally, as follows from the first equation (5.13). We get a situation opposite to (5.19) (see Figs. 5.2 and 5.4):

$$t(g) \xrightarrow{g \to 0} 0, \quad G_{\text{eff}}(g) \xrightarrow{g \to 0} g^*,$$
$$t(g) \xrightarrow{g \to g^*} \infty, \quad G_{\text{eff}}(g) \xrightarrow{g \to g^*} 0. \tag{5.21}$$

Such behaviour of the mass in the strong coupling regime $g \to g^*$ (g^* is an infrared stable point) can be interpreted as follows. The system exists in the phase with a mass $M = t(g)m$; moreover, a coupling between the particles is weak: $G_{eff} \ll 1$. Asymptotic relations (5.21) mean that the two-point Green function becomes an entire function in the limit $g \to g^*$. In other words, an interaction is effectively weak and decreases when $g \to g^*$. Nevertheless, the particles have infinite mass and can not be born as free ones. Such a situation illustrates scenario of the so-called analytical confinement, which can be displayed in realistic asymptotically free theories.

Table 5.1. The phases in the strong coupling regime.

	$G \ll 1$	$G \gg 1$
R^{1+1}	$\frac{1}{2}m^2\varphi^2 + \frac{1}{4}g\varphi^4$	$\frac{1}{2}M^2\Phi^2 + \frac{1}{4}g\Phi^4 + gB(g)\Phi^3$ (BS)
R^{2+1}		$\frac{1}{2}M^2\Phi^2 + \frac{1}{4}g\Phi^4$ (S)
R^{3+1}	$\frac{1}{2}m^2\varphi^2 + \frac{1}{4}g\varphi^4$	$\frac{1}{2}M^2\Phi^2 + \frac{1}{4}G_{eff}\Phi^4$, (S)
		if $\exists g_c \in (0, g^*)$: $2 + \gamma_m(g_c) = 0$
		?, if $\forall g \in (0, g^*)$ $2 + \gamma_m(g) > 0$
R^{1+1}	$\frac{1}{2}m^2\varphi^2 + \frac{1}{4}g\varphi^4 + m\sqrt{\frac{g}{2}}\varphi^3$	$\frac{1}{2}M^2\Phi^2 + \frac{1}{4}g\Phi^4 + gB(g)\Phi^3$ (BS)
R^{2+1}		$\frac{1}{2}M^2\Phi^2 + \frac{1}{4}g\Phi^4$ (S)
R^{1+1}	$\frac{1}{2}m^2 \sum_i^N \varphi_i^2 + \frac{1}{4}g\left[\sum_i^N \varphi_i^2\right]^2$	$\frac{1}{2}M^2\Phi^2 + \frac{1}{2}M_0^2 \sum_i^{N-1} \Phi_i^2$ (BS)
		$+\frac{1}{4}g\left[\sum_i^{N-1}\Phi_i^2 + \Phi^2\right]$
		$+gB(g)\Phi\left[\sum_i^{N-1}\Phi_i^2 + \Phi^2\right]$
R^{2+1}		$\frac{1}{2}M^2 \sum_i^N \Phi_i^2 + \frac{1}{4}g\left[\sum_i^N \Phi_i^2\right]^2$ (S)

5.5 A Correlation Between the Phase Structure and UV Divergences

Now we can compare the phase structure of models (2.1), (2.2), (2.3) in R^{d+1} with $d = 1, 2, 3$. It is seen from Table 5.1 that the behaviour of the systems for different d is completely different. Independent of the symmetry of the initial representation the BS-phase is realized in R^{1+1} for $G \gg 1$, while in R^{2+1} we have the S-phase. We can conclude that different UV-behaviour leads to completely different phase structure. The following heuristic consideration (e.g., see [8]) explains in some sense a correlation between UV divergences of a theory and its phase structure. An intuitively clear reason for dynamical symmetry breaking in the $(\varphi^4)_2$ consists in the normal ordering of the Hamiltonian. In other words, symmetry breaking arises in this case from the contribution of the diagram (a) in Fig. 4.1 to mass renormalization. This contribution changes the sign of the bare mass m_0 in the strong coupling regime. The opposite situation occurs in the $(\varphi^4)_3$, since two diagrams contribute to m_0 with different signs. The bare mass turns out to be positive for large g, and symmetry breaking does not occur. In the case of the $(\varphi^4)_4$ the picture is much more complicated, since the bare mass is represented by an alternating series, which can be positive for all g. Hence any reason for the apperance of the BS-phase is then completely absent.

6. The φ^4 Theory at Finite Temperatures

The problem of the dynamical reconstruction of a ground state under change of the coupling constant (like the fine structure constant) seems quite artificial, although it is important from a conceptual point of view. An investigation of the behaviour of field systems with respect to the external parameters such as external classical fields and temperature is much closer to real physics. In this chapter we will look into the phase structure of the φ^4 theory at nonzero temperatures. This problem relates directly to many aspects of the solid state physics [6, 61], to systems such as quark-gluon plasma or to the problem of the evolution of the Universe [3].

Intensive investigation of dynamical temperature effects in QFT began with the paper of Kirghnits [2]. The majority of papers are devoted to the study of four-dimensional theories within the one- or two-loop approximation of the effective potential. The phase structure of the two-dimensional (φ^4) model was investigated in papers [4, 5, 6] within the high temperature expansion.

Below, we shall obtain the phase diagrams in the (G, θ)-plane for models (2.1, 2.2). It should be remembered that $G = g/2\pi m^{3-d}$ and $\theta = T/m$ are dimensionless parameters of the models. As has been shown in previous chapters, behaviour of the systems when the variable G is changed is quite different in R^{1+1} and R^{2+1} (see Table 5.1). On the other hand, the dependence of the ground state on temperature is qualitatively the same both for two- and three-dimensional space-time: the systems exist in the symmetric phase if the temperature is high enough. This is the most general conclusion from the form of the phase diagrams.

6.1 The Two-Dimensional Model

6.1.1 The Hamiltonian at Zero Temperature

At finite temperatures, any representation of the canonical relations is sufficient to give a physical picture only in the case, where the masses of the particles depend on the temperature. This dependence is not known even in the weak coupling limit. That is why we must construct finite temperature field theory starting with the representation at zero temperature. Then, using the thermal canonical transformation we will introduce the temperature dependence. We shall deal with the Lagrangian (4.1) that unifies both the symmetric and the SSB models (2.1, 2.2).

The Initial Representation at $G \ll 1$, $T = 0$.. The Hamiltonian written in the normal form is

$$H = H_0 + H_I,$$

$$H_0 = \frac{1}{2} \int_V dx : \left[\pi^2(x) + (\nabla\varphi(x))^2 + m^2\varphi^2(x) \right] :,$$

$$H_I = \int_V dx : \left[\frac{1}{4} g_4 \varphi^4(x) + g_3 \varphi^3(x) \right] :, \qquad (6.1)$$

where

$$\varphi(x) = \int \frac{dk}{2\pi} \frac{1}{\sqrt{2\omega}} \left[a(k) \exp\{ikx\} + a^+(k) \exp\{-ikx\} \right],$$

$$\pi(x) = \frac{1}{i} \int_V \frac{dk}{2\pi} \sqrt{\frac{\omega}{2}} \left[a(k) \exp\{ikx\} - a^+(k) \exp\{-ikx\} \right], \qquad (6.2)$$

$$\omega(k) = \sqrt{k^2 + m^2}, \quad [a(k), a^+(k')] = \delta(k - k').$$

The operators φ, π obey the standard commutation relations. Creation and annihilation operators $a^+(k)$ and $a(k)$ act on the Fock space of the particles with a mass m. The vacuum vector $|0\rangle$ is subjected to the condition

$$a(k)|0\rangle = 0 \quad \forall k.$$

The Canonical Transformation.. Let us perform the canonical transformation

$$\pi(x) \to \pi_t(x), \quad \varphi(x) \to \varphi_t(x) + B.$$

The fields φ_t and π_t have the form of (6.2), but with a mass

$$M = m \cdot t$$

and new operators $a_t^+(k)$ and $a_t(k)$ defined on the Fock space with the vacuum vector

$$|0(t, B)\rangle = U_2^{-1}(t)U_1^{-1}(B)|0\rangle, \quad a_t(k)|0\rangle = 0 \quad \forall k.$$

The transformations U_1 and U_2 are given by (4.8).

The Hamiltonian expressed in the new canonical variables and normally ordered with respect to the vacuum $|0(t, B)\rangle$ takes the form

$$H = H_0' + H_I' + H_1 + E,$$

$$H_0' = \frac{1}{2} \int_V dx : \left[\pi_t^2(x) + (\nabla \varphi_t(x))^2 + M^2 \varphi_t^2(x) \right] :,$$

$$H_I' = \int_V dx : \left[\frac{1}{4} h_4 \varphi_t^4(x) + h_3 \varphi_t^3(x) \right] :,$$

$$h_3 = g_3 + g_4 B, \quad h_4 = g_4, \tag{6.3}$$

$$H_1 = \int_V dx : \left[\frac{1}{2} R(t, B) \varphi_t^2(x) + P(t, B) \varphi_t(x) \right] :,$$

$$R = m^2 - M^2 + 3g_4 \left(B^2 - D \right) + 6g_3 B,$$

$$P = m^2 B + g_4 \left(B^3 - 3BD \right) + 3g_3 \left(B^2 - D \right), \tag{6.4}$$

$$E = \frac{1}{2} m^2 B^2 + L + \frac{g_4}{4} [b^4 - 6B^2 D + D^2] + g_3 B[B^2 - 3D], \tag{6.5}$$

$$D(t) = \frac{1}{4\pi} \ln t, \quad L(t) = \frac{m^2}{8\pi} [t - 1 - \ln t].$$

In order to introduce the temperature dependence we have to perform an additional canonical transformation, which is a central idea of the thermo field dynamics (TFD).

6.1.2 The Thermo Field Dynamics

A detailed description of the TFD can be found in the review [67] or in the monograph [61]. We will restrict ourselves to a brief formulation of the main idea.

A principal feature of the TFD consists in doubling of the field variables at finite temperatures, $T > 0$. The intuitive motivation of such a doubling can be reduced to the following. The presence of a thermostat triggers a great number of excited quanta in a field system in thermodynamic equilibrium. Besides these, there are free energy levels (with positive energy), which are called holes. As a result, an absorption of energy by a system is achieved in two independent ways: either by means of an excitation of the additional quanta, or by the filling of free levels in the particle space, i.e., by annihilation of the holes with negative energy.[1] These two mechanisms of energy absorption lead to a doubling of the field variables.

In accordance with this motivation the following independent (commuting) operators have to be constructed:

$$\alpha(k) = a_t(k) \otimes 1, \quad \tilde{\alpha}(k) = 1 \otimes a_t(k). \tag{6.6}$$

They satisfy the usual CR:

---

[1] We would like to stress the difference between the above-mentioned holes and antiparticles in the usual sense. The antiparticles are the holes (free levels) in the space of states with negative energy.

$$[\alpha(k), \alpha^+(k')] = \delta(k - k'), \quad [\tilde\alpha(k), \tilde\alpha^+(k')] = \delta(k - k')$$

and act on the Fock space with the vacuum:

$$|0(t, B)\rangle\rangle = |0(t, B)\rangle \otimes |0(t, B)\rangle. \tag{6.7}$$

The operator $\alpha(k)$ describes the annihilation of the particles, while $\tilde\alpha(k)$ corresponds to annihilation of the holes. At nonzero temperatures the ground state of the system is not a state without particles. An averaged number of particles in the ground state is described by the standard boson distribution function.

The next step consists in a transition to the quasi-particle picture. The quasi-particles correspond to excitations of the physical ground state known as the thermal vacuum. Temprature dependent operators $\alpha_\beta(k)$, $\tilde\alpha_\beta(k)$ ($\beta = 1/T$) are introduced by means of the thermal Bogoliubov transformation

$$\alpha_\beta(k) = \alpha(k)\cosh(\xi) - \tilde\alpha^+(k)\sinh(\xi),$$
$$\tilde\alpha_\beta(k) = \tilde\alpha(k)\cosh(\xi) - \alpha^+(k)\sinh(\xi), \tag{6.8}$$

or the same thing in the operator form

$$\alpha_\beta(k) = U^{-1}(\beta)\alpha(k)U(\beta), \quad \tilde\alpha_\beta(k) = U^{-1}(\beta)\tilde\alpha(k)U(\beta),$$
$$U(\beta) = \exp\left\{\int dk\xi(k, \beta)[\tilde\alpha(k)\alpha(k) - \alpha^+(k)\tilde\alpha^+(k)]\right\}.$$

The operators $\alpha_\beta(k)$, $\tilde\alpha_\beta(k)$ are defined on the Fock space with the thermal vacuum

$$|0(\beta, t, B)\rangle = U^{-1}(\beta)|0(t, B)\rangle\rangle, \tag{6.9}$$

which satisfies the conditions

$$\alpha_\beta(k)|0(\beta, t, B)\rangle = \tilde\alpha_\beta(k)|0(\beta, t, B)\rangle = 0 \quad \forall k, \beta.$$

The parameter $\xi(k, \beta)$ is defined by the demand that an averaged number of particles in the state $|0(\beta, t, B)\rangle$ must be equal to the statistical dustribution function

$$\langle 0(\beta, t, B)|\alpha^+(k)\alpha(k)|0(\beta, t, B)\rangle = n(\omega(k, t)) = [\exp\{\beta\omega(k, t)\} - 1]^{-1}.$$

Using (6.6) and the reversed transformation (6.8) we find that

$$\sinh^2(\xi) = [\exp\{\beta\omega(k, t)\} - 1]^{-1}. \tag{6.10}$$

Condition (6.10) guarantees that the quantity $T = 1/\beta$ has the meaning of the statistical temperature. In accordance with (6.6) the fields (Φ, Π) and $(\tilde\Phi, \tilde\Pi)$ defined on the Fock space with vacuum (6.7) are introduced. The fields (Φ, Π) have the form of (6.2), where the substitution

$$\omega(k) \to \omega(k, t), \quad a(k) \to \alpha(k), \quad a^+(k) \to \alpha^+(k)$$

has to be made. The expressions for the tilde-conjugated variables $\tilde{\Phi}$, $\tilde{\Pi}$ are obtained from Φ, Π by means of the substitution $(\alpha, \alpha^+) \to (\tilde{\alpha}, \tilde{\alpha}^+)$.

To define the fields on the Fock space with the thermal vacuum (6.9) the reversed transformation (6.8) is used. By means of this transformation the fields Φ, Π, $\tilde{\Phi}$, $\tilde{\Pi}$ are expressed in terms of the operators $\alpha_\beta(k)$, $\alpha_\beta^+(k)$, $\tilde{\alpha}_\beta(k)$, $\tilde{\alpha}_\beta^+(k)$.

Within the TFD formalism the total Hamiltonian is defined by the formula

$$\hat{\mathcal{H}} = \mathcal{H} - \tilde{\mathcal{H}},$$

where

$$\mathcal{H} = H \otimes 1, \quad \tilde{\mathcal{H}} = 1 \otimes H. \tag{6.11}$$

The Hamiltonian $\mathcal{H}$ describes the particles, while $\tilde{\mathcal{H}}$ relates to the holes with negative energy, which is the reason for the sign " $-$ " before $\tilde{\mathcal{H}}$. The Hamiltonian $\mathcal{H}$ in the thermal vacuum representation (6.9) takes the form

$$\mathcal{H} = \mathcal{H}_0'' + \mathcal{H}_I'' + \mathcal{H}_1' + E',$$

$$\mathcal{H}_0'' = \frac{1}{2} : \left[\Pi^2(x) + (\nabla\Phi(x))^2 + M^2\Phi^2(x) \right] :,$$

$$\mathcal{H}_I'' =: \left[\frac{1}{4}h_4\Phi^4(x) + h_3\Phi^3(x) \right] :,$$

$$\mathcal{H}_1' =: \left[\frac{1}{2}\mathcal{R}(t, B; \beta)\Phi^2(x) + \mathcal{P}(t, B; \beta)\Phi(x) \right] :,$$

$$\mathcal{R} = m^2 - M^2 + 3g_4 \left(B^2 - D \right) + 6g_3 B,$$

$$\mathcal{P} = m^2 B + g_4 \left(B^3 - 3BD_\beta \right) + 3g_3 \left(B^2 - D_\beta \right), \tag{6.12}$$

$$D_\beta(t) = \frac{1}{4\pi} \ln t - \frac{1}{\pi}d(\theta/\sqrt{t}).$$

According to the demand of the R-scheme equivalence the normal ordering in these formulas relates to the operators $\alpha_\beta(k)$, $\alpha_\beta^+(k)$. The energy density E' can be obtained from E by means of the substitution $D \to D_\beta$, $L \to L_\beta$, where

$$L_\beta(t) = \frac{m^2}{8\pi}\{t - 1 - D_\beta(t) + 4t[2s(\theta/\sqrt{t}) + d(\theta/\sqrt{t})]\}, \tag{6.13}$$

$$d(z) = \int_0^\infty \frac{du}{\sqrt{1 + u^2}} \left(\exp\left\{ \frac{1}{z}\sqrt{1 + u^2} \right\} - 1 \right)^{-1},$$

$$s(z) = \int_0^\infty \frac{du\, u^2}{\sqrt{1 + u^2}} \left(\exp\left\{ \frac{1}{z}\sqrt{1 + u^2} \right\} - 1 \right)^{-1}.$$

The operator $\tilde{\mathcal{H}}$ is constructed according to the rule $\tilde{\mathcal{H}} = \mathcal{H}^*[\tilde{\Phi}, \tilde{\Pi}]$.

The energy density E' of the state $|0(t, B; \beta)\rangle$ is linked with the free energy density F by the equation

$$F = E' - TS, \qquad (6.14)$$

where S is the entropy density:

$$S = -\int \frac{dk}{\sqrt{2\pi}}[n(k, t) \ln n(k, t) - (1 - n(k, t)) \ln(1 - n(k, t))]$$

$$= \frac{m^3}{\pi}\frac{t}{T}[2s(\theta/\sqrt{t}) + d(\theta/\sqrt{t})], \qquad (6.15)$$

$$n(k, t) = [\exp(\beta\omega(k, t)) - 1]^{-1}.$$

Using (6.5), (6.13), (6.14) and (6.15) we get

$$F = \frac{1}{2}m^2B^2 + \frac{g_4}{4}[B^4 - 6B^2D_\beta(t) + 3D_\beta^2(t)] + g_3B[B^2 - 3D_\beta(t)]$$

$$+ \frac{m^2}{8\pi}\{t - 1 - 4\pi D_\beta(t) - 4t[2s(\theta/\sqrt{t}) + d(\theta/\sqrt{t})]\}. \qquad (6.16)$$

To provide the correct form of the total Hamiltonian we demand that

$$\mathcal{R}(t, B; \theta) = 0, \quad \mathcal{P}(t, B; \theta) = 0. \qquad (6.17)$$

It is easy to check the equivalence of (6.17) to the equations

$$\frac{\partial F(t, B)}{\partial B} = 0, \quad \frac{\partial^2 F(t, B)}{\partial B^2} = M^2 = m^2t,$$

which are analogous to the minimum and stability conditions for the effective potential [3]. At the same time, (6.17) define the minimum of the free energy density $F(t, B)$ as a function of two variables t and B for zero temperature $\theta = 0$ only. Thus only in this case must our results and the results of the GEP-approximation coincide.

6.1.3 The Symmetric Model

Putting $g_3 = 0$ and $g_4 = g$ we get from (6.17) and (6.12) the following equations for the parameters B and t $(G = g/2\pi m^2)$:

$$B[gB - 3gD_\beta(t) + m^2] = 0,$$

$$3gB^2 - 3gD_\beta(t) - m^2(t - 1) = 0. \qquad (6.18)$$

The S-phase.. Substituting the symmetric solution $B = 0$ of the first equation (6.18) into the second one we obtain

$$\frac{2}{3G}(t - 1) = -\ln t + 4d(\theta/\sqrt{t}). \qquad (6.19)$$

This equation has a unique solution for all G and θ; moreover $t(G, \theta) \equiv 1$ only at zero temperature $\theta = 0$. The free energy density in the S-phase looks like

$$F_S = \frac{m^2}{8\pi}\left\{\left(\frac{2}{3G} + 1\right)(t - 1) + \frac{(t - 1)^2}{3G} - 4t[2s(\theta/\sqrt{t}) + d(\theta/\sqrt{t})]\right\}. (6.20)$$

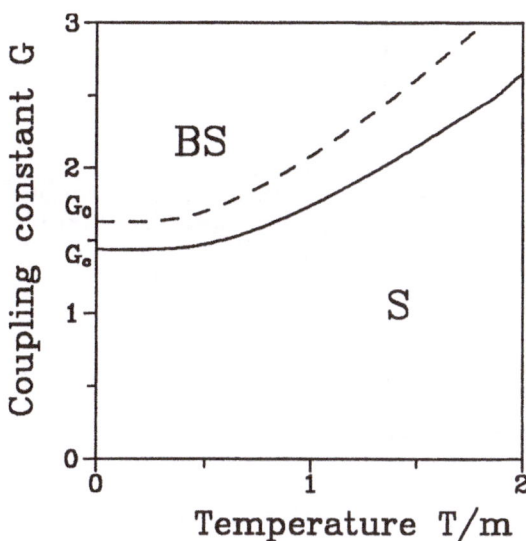

Fig. 6.1. The phase diagram for the symmetric $(\varphi^4)_2$.

The Phase with Broken Symmetry.. Using the nonzero solution for B we can rewrite (6.18) in the form

$$B^2 = \frac{t}{4\pi G}, \qquad \frac{t}{3G} + \frac{2}{3G} = \ln t - 4d(\theta/\sqrt{t}).$$

The second equation has solutions only for G and θ such that $G \geq G_c(\theta)$. The function $G_c(\theta)$ is plotted in Fig. 6.1. The free energy density in the BS-phase is given by the expression

$$F_B = \frac{m^2}{8\pi}\left\{-\frac{1}{2G} + \left(1 - \frac{1}{3G}\right)(t-1) - \frac{(t-1)^2}{6G} - 4t[2s(\theta/\sqrt{t}) + d(\theta/\sqrt{t})]\right\}.$$

$$(6.21)$$

Comparing the effective coupling constants $G_{\text{eff}} = G/t(G,\theta)$ in the S- and BS-phases we find that the phase boundary is given by the function $G_c(\theta)$ (solid line in Fig. 6.1). Comparison of the free energy densities F_S and F_B leads to the phase boundary plotted in Fig. 6.1 by the dashed line. One can see that these two boundaries do not contradict each other.

Near the critical region the effective coupling constants (see Fig. 6.2) are large enough in both phases, and the perturbation corrections will be large and can change the boundary shown by the dashed line. These corrections at zero temperature have been estimated in section 3.1.

6.1.4 The Model with SSB in the Initial Representation

Substituting into (6.17, 6.12) the coupling constants $g_4 = g$ and $g_3 = m\sqrt{g/2}$, corresponding to the Lagrangian (2.2), we get the equations

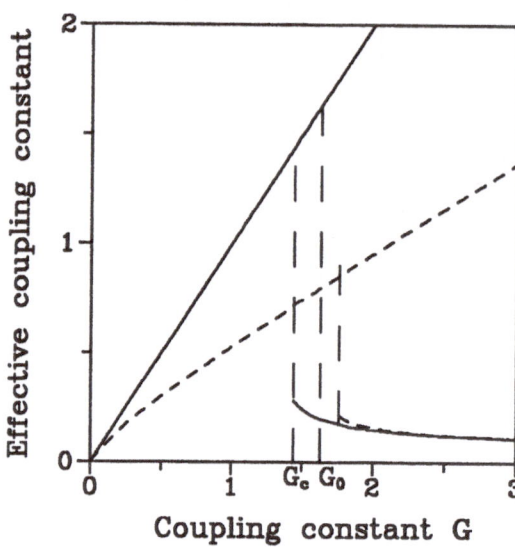

Fig. 6.2. Effective coupling constants for the symmetric $(\varphi^4)_2$: upper line – the S-phase; dashed line – $\theta = 1$, solid line – $\theta = 0$.

$$gB^3 + 3m\sqrt{g/2}B^2 + B[m^2 - 3gD_\beta(t)] - 3m\sqrt{g/2}D_\beta(t) = 0,$$
$$3gB^2 + 3m\sqrt{2g}B - 3gD_\beta(t) - M^2 + m^2 = 0. \tag{6.22}$$

The S-phase.. It is easy to check that the first equation (6.22) has the solution

$$B = -\frac{1}{\sqrt{4\pi G}}. \tag{6.23}$$

Substitution of (6.23) into the second equation (6.22) gives the equation

$$\frac{2}{3G}t + \frac{1}{3G} = -\ln t + 4d(\theta\sqrt{t}). \tag{6.24}$$

This equation has a unique solution for all G, θ. Taking into account relations (6.23) and (6.24) we get the following expression for free energy density in the S-phase:

$$F_S = \frac{m^2}{8\pi}\left\{\frac{1}{2G} + \left(1 + \frac{2}{3G}\right)(t-1) + \frac{(t-1)^2}{3G} - 4t[2s(\theta/\sqrt{t}) + d(\theta/\sqrt{t})]\right\}. \tag{6.25}$$

The Phase with Broken Symmetry.. The rest of the solutions of the first equation (6.22),

$$B = -\frac{1 \pm \sqrt{t}}{4\pi G},$$

substituted into the second one give

$$\frac{1}{3G}t - \frac{1}{3G} = \ln t - 4d(\theta/\sqrt{t}). \tag{6.26}$$

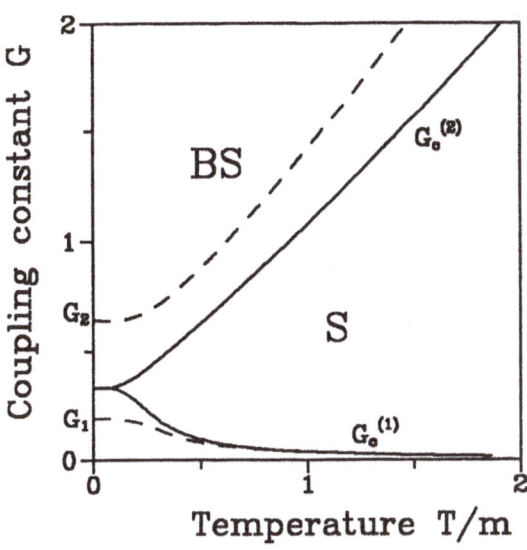

Fig. **6.3.** The phase diagram for the $(\varphi^4)_2$ model with initially broken symmetry.

This equation has solutions only for (G, θ) such that

$$G \leq G_c^{(1)}(\theta) \quad \text{or} \quad G \geq G_c^{(2)}(\theta).$$

The functions $G_c^{(1)}(\theta)$ and $G_c^{(1)}(\theta)$ are represented in Fig. 6.3. There are two solutions and they are equal to each other if

$$G = G_c^{(1)}(\theta) \quad \text{or} \quad G = G_c^{(2)}(\theta).$$

These solutions describe two different BS-phases. One can illustrate this easily at zero temperature. Since $d(0) = 0$, then (6.26) at zero temperature has the solution $t \equiv 1$ leading to the initial representation. To exclude this obvious solution one can rewrite (6.26) in the form

$$\frac{1}{3G} = \frac{\ln t}{t - 1}. \tag{6.27}$$

This equation has a unique solution for all G, which determines the BS-phase with the mass $M = m \cdot t(G, 0)$; moreover

$$M^2 \longrightarrow m^2 \exp\left\{-\frac{1}{3G}\right\} \quad \text{for } G \to 0.$$

Nonanalyticity at the origin means that a difference between the masses m and M can not be found within the perturbation expansion. One can see from Fig. 6.3 that $G_c^{(1)}(0) = G_c^{(2)}(0) = G_c$. Substituting $t = 1$ into (6.27) we find that $G_c = \frac{1}{3}$. The region below the line $G_c^{(1)}(\theta)$ corresponds to the first BS-phase, while the region above $G_c^{(2)}(\theta)$ represents the second BS-phase.

The free energy density looks like

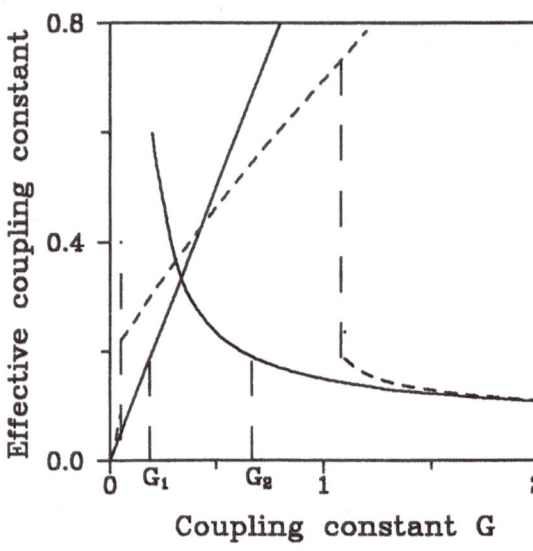

Fig. 6.4. The effective coupling constant for $(\varphi^4)_2$ with initially broken symmetry: upper line – the S-phase; dashed line – $\theta = 1$, solid line – $\theta = 0$.

$$F_B = \frac{m^2}{8\pi}\left\{\left(1-\frac{1}{3G}\right)(t-1) - \frac{(t-1)^2}{6G} - 4t[2s(\theta/\sqrt{t}) + d(\theta/\sqrt{t})]\right\}.$$

As seen from Fig. 6.3, the phase boundaries obtained by comparison of the effective coupling constants (see the solid lines in Fig. 6.4) and free energy densities (dashed lines) do not contradict each other. The points $G_1 = 0.19...$ and $G_2 = 0.64...$ in Fig. 6.3 coincide with the critical points obtained within the GEP approximation at zero temperature [19].

Thus we have two different BS-phases and one S-phase. At zero temperature the symmetry is violated for all G. There is a phase transition between the first and second BS-phases at $G = \frac{1}{3}$. At the same time, for any fixed G the symmetry is restored if the temperature is high enough. The phase transition at finite temperature is of the first order, since the mass $M = tm$ and the order parameter

$$\sigma = \pm\frac{\sqrt{t}}{4\pi G}$$

are discontinuous at the boundaries (see Fig. 6.5 for $\theta = 1$). The effective coupling constant is small everywhere besides the critical regions where $G_{\text{eff}} \sim O(1)$, so that our description is quite accurate only outside the regions of phase transitions.

The boundary $G_c^{(1)}(\theta)$ lies in the region of applicability of the high temperature expansion ($\theta \gg G$). Its form is in good quantitative agreement with the result of this method [4, 5, 6].

6.2 The Three-Dimensional Model φ^4 at Finite Temperatures

6.2.1 The Thermal Vacuum Representation

The canonical transformation to the fields with new mass and nonzero condensate has been performed in Chap. 3 (see (4.2–4.12)), and we do not reproduce these calculations here. Besides, the transition from the representation (4.12) to the formulas below is quite analogous to the two-dimensional case (see (6.6)–(6.11)). The only principal difference relates to additional counter-terms, which remove divergences in the diagrams of the second order (Fig. 4.1). That is why we start here with the expression for the total Hamiltonian $\mathcal{H}$ in the thermal vacuum $|0(\beta, t, B)\rangle$ (6.9) representation:

$$\mathcal{H} = \mathcal{H}_0'' + \mathcal{H}_I'' + \mathcal{H}_{ct}'' + \mathcal{H}_1' + VE,$$

$$\mathcal{H}_0'' = \frac{1}{2} \int_V d\mathbf{x} : \left[\Pi^2(\mathbf{x}) + (\nabla \Phi(\mathbf{x}))^2 + M^2 \Phi^2(\mathbf{x}) \right] :,$$

$$\mathcal{H}_I'' = \int_V d\mathbf{x} : \left[\frac{1}{4} h_4 \Phi^4(\mathbf{x}) + h_3 \Phi^3(\mathbf{x}) \right] : . \tag{6.28}$$

The normal ordering in these formulas relates to the temperature-dependent operators $\alpha(\mathbf{k}, \beta)$, $\alpha^+(\mathbf{k}, \beta)$. The Hamiltonian $\tilde{\mathcal{H}}$ is constructed according to the rule $\tilde{\mathcal{H}} = \mathcal{H}^*[\tilde{\Phi}, \tilde{\Pi}]$. The counter-term operator $\mathcal{H}_{ct}''$ takes the form in this representation

$$\mathcal{H}_{ct}''(M, \beta) = \int_V d\mathbf{x} : \left[\frac{1}{2} A(M) \Phi^2(\mathbf{x}) + C(M) \Phi(\mathbf{x}) \right] :, \tag{6.29}$$

with temperature-dependent functions A, C [37]:

$$A(M) = 3! g_4^2 \bar{\Sigma}(M), \quad C(M) = 3! g_3 g_4 \bar{\Sigma}(M),$$
$$\bar{\Sigma}(M) = \Sigma(M) + 3 \Sigma_\beta(M) + 3 \Sigma_{\beta\beta}(M),$$
$$\Sigma_\beta(M) = \frac{1}{2(2\pi)^2} \sigma_\beta(t, \theta), \quad \Sigma_{\beta\beta}(M) = \frac{1}{2(2\pi)^2} \sigma_{\beta\beta}(t, \theta),$$
$$\sigma_\beta(t, \theta) = -\ln 3 \cdot \frac{\theta}{t} \ln \left(1 - \exp \left\{ -\frac{t}{\theta} \right\} \right), \tag{6.30}$$

$$\sigma_{\beta\beta}(t, \theta) = \int_1^\infty \int_1^\infty dx\, dy \left[\exp \left\{ \frac{xt}{\theta} \right\} - 1 \right]^{-1} \left[\exp \left\{ \frac{yt}{\theta} \right\} - 1 \right]^{-1}$$

$$\times \left[\frac{1}{\sqrt{4(x^2 + y^2 + xy) - 3}} + \frac{1}{\sqrt{4(x^2 + y^2 - xy) - 3}} \right]. \tag{6.31}$$

Within the zero-momentum subtraction scheme the operator $\mathcal{H}_1'$ is

$$\mathcal{H}_1' = \int_V dx : \left[\frac{1}{2} \mathcal{R}(t, B, \beta) \Phi^2(x) + \mathcal{P}(t, B, \beta) \Phi(x) \right] :,$$
$$\mathcal{R} = m^2 - M^2 + 3g_4 \left(B^2 - D_\beta \right) + 6g_3 B + 6g_4^2 \left(\Sigma(m) - \bar{\Sigma}(M) \right),$$
$$\mathcal{P} = m^2 B + g_4 \left(B^3 - 3BD_\beta \right) + 3g_3 \left(B^2 - D_\beta \right) + \qquad (6.32)$$
$$6g_4 \left(g_3 + g_4 B \right) \left(\Sigma(m) - \bar{\Sigma}(M) \right),$$

where

$$D_\beta(t) = \frac{m}{4\pi} \left[t - 1 + 2\theta \ln \left(1 - \exp \left\{ -\frac{t}{\theta} \right\} \right) \right], \qquad (6.33)$$

The function Σ corresponds to $\bar{\Sigma}$ at zero temperature.

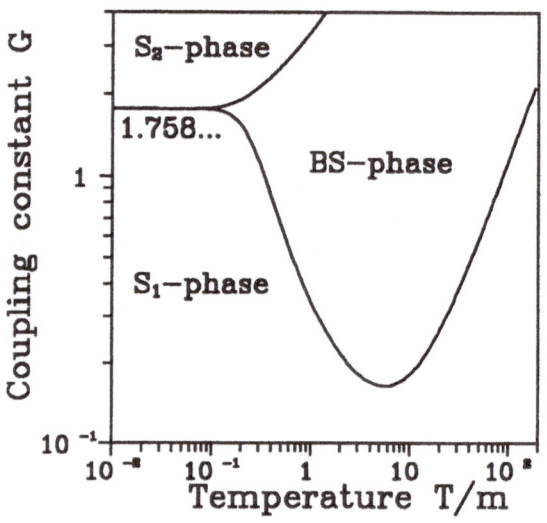

Fig. 6.5. The phase diagram for the symmetric $(\varphi^4)_3$.

The calculations throughout this part of the book show that the critical values of the parameters (G, θ) obtained by a comparison of the free energy densities and effective coupling constants gives similar results. We shall not calculate here the energy density E', which is quite complicated, and shall base our further analysis on comparison of the effective coupling constants only.

Let us demand that

$$\mathcal{H}_1' = 0 \iff \begin{cases} \mathcal{R}(t, B, \beta) = 0, \\ \mathcal{P}(t, B, \beta) = 0. \end{cases} \qquad (6.34)$$

This requirement provides the correct form of the Hamiltonian $\mathcal{H}$. It describes scalar particles with the mass M, which depends on the coupling constant G and temperature θ. This dependence is defined by (6.34).

It is convenient to introduce the following dimensionless quantities:

$$G_4 = \frac{g_4}{2\pi m}, \quad G_3 = \frac{g_3}{m\sqrt{4\pi m}}, \quad b = B\sqrt{\frac{4\pi}{m}}. \tag{6.35}$$

Using the definitions (6.35) we represent (6.32) in the form

$$-\frac{1}{2}t^2 + \frac{1}{2} + \frac{3}{4}G_4\left(b^2 - d(t,\theta)\right) + 3G_3 b$$
$$+\frac{3}{4}G_4^2\left(\ln t - 6\sigma_\beta(t,\theta) - 6\sigma_{\beta\beta}(t,\theta)\right) = 0, \tag{6.36}$$
$$b + \frac{1}{2}G_4 b\left(b^2 - 3d(t,\theta)\right) + 3G_3\left(b^2 - d(t,\theta)\right)$$
$$+3G_4\left(G_3 + \frac{G_4}{2}b\right)\left(\ln t - 6\sigma_\beta(t,\theta) - 6\sigma_{\beta\beta}(t,\theta)\right) = 0,$$

where

$$d(t,\theta) = t - 1 + 2\theta \ln\left(1 - \exp\left\{-\frac{t}{\theta}\right\}\right). \tag{6.37}$$

Different solutions of these equations describe possible phases of the system.

6.2.2 The Symmetric Model

Let us consider the model with the Lagrangian (2.1). Hence we choose

$$G_4 = G, \quad G_3 = 0,$$

and get from (6.36) the equations

$$t^2 - 1 - \frac{3}{2}G\left(b^2 - d(t,\theta)\right) - \frac{3}{2}G_4^2\left(\ln t - 6\sigma_\beta(t,\theta) - 6\sigma_{\beta\beta}(t,\theta)\right) = 0, \tag{6.38}$$
$$b\left[1 + \frac{1}{2}G\left(b^2 - 3d(t,\theta)\right) + \frac{3}{2}G^2\left(\ln t - 6\sigma_\beta(t,\theta) - 6\sigma_{\beta\beta}(t,\theta)\right)\right] = 0.$$

Using the first equation we get for the second one two solutions:

$$b = 0 \text{ (symmetric)}, \quad b^2 = \frac{t^2}{G} \text{ (nonsymmetric)}.$$

The S-phase.. The equation for t is obtained from (6.38), (6.37) and (6.30) in the form

$$2t^2 + 3Gt - 2 - 3G - 3G^2\ln t + 18G^2\sigma_{\beta\beta}(t,\theta) \tag{6.39}$$

$$+6G\left(\theta - 3\ln 3 \cdot G\frac{\theta}{t}\right)\ln\left(1 - \exp\left\{-\frac{t}{\theta}\right\}\right) = 0,$$

where the function $\sigma_{\beta\beta}$ is defined by (6.31). Equation (6.39) has two solutions if (G,θ) belongs to the regions S_1 or S_2 in Fig. 6.5, while in the region BS solutions are absent.

High temperature asymptotics $(\theta \gg 1, t \gg 1)$ are determined by the terms of (6.39), which are linear over the coupling constant G. This becomes obvious when one notices that (see (6.31))

$$\sigma_{\beta\beta}(t,\theta) \xrightarrow{\theta \gg t \gg 1} C \cdot \frac{\theta^2}{t^2} + O\left(\frac{\theta}{t}\ln\frac{\theta}{t}\right), \tag{6.40}$$

$$C = \int_1^\infty \int_1^\infty \frac{dx\, dy}{xy}\left[\frac{1}{\sqrt{4(x^2+y^2+xy)-3}} + \frac{1}{\sqrt{4(x^2+y^2-xy)-3}}\right]$$

and

$$\theta \gg \frac{\theta}{t}, \quad \theta\ln\theta \gg \frac{\theta^2}{t^2}.$$

The asymptotics of the mass $M = m \cdot t$ and effective coupling constant $G_{\text{eff}} = G/t$ have the form

$$t \xrightarrow{\theta \gg G} \sqrt{3G\theta\ln\theta}, \quad t \xrightarrow{G \gg \theta} \sqrt{\frac{3}{2}G^2\ln G},$$

$$G_{\text{eff}} \xrightarrow{\theta \gg G} \sqrt{\frac{G}{3\theta\ln\theta}} \ll 1, \quad G_{\text{eff}} \xrightarrow{G \gg \theta} \sqrt{\frac{2}{3\ln G}} \ll 1. \tag{6.41}$$

Let us return for a moment to the zero temperature case. For $\theta = 0$ we get from (6.39) the equation

$$2t^2 + 3Gt - 3G^2\ln t - 2 - 3G = 0,$$

which coincides with the equation obtained in the section 4.3. It should be remembered that there are two solutions: $t_1(G,0) \equiv 1$ corresponding to the initial representation (4.2) and $t_2(G,0)$, which satisfy the relations:

$$t_2 < 1 \text{ if } G < G_c, \quad t_2 \geq 1 \text{ if } G \geq G_c,$$

$$G_c = \frac{1}{2}\left(1 + \sqrt{\frac{19}{3}}\right) = 1.758....$$

The point G_c corresponds to the phase transition between two S-phases.

At finite temperatures $\theta > 0$ analysis of (6.39) shows that for

$$(\theta, G) \in S_1, S_2$$

there are two solutions

$$t_2(G, \theta) < t_1(G, \theta) \not\equiv 1 \quad \text{if} \quad (\theta, G) \in S_1$$

$$t_2(G, \theta) > t_1(G, \theta) \not\equiv 1 \quad \text{if} \quad (\theta, G) \in S_2.$$

Comparing the effective coupling constants we should put

$$t(G, \theta) = \begin{cases} t_1(G, \theta), & \text{if } (\theta, G) \in S_1 \\ t_2(G, \theta), & \text{if } (\theta, G) \in S_2. \end{cases} \tag{6.42}$$

We will stress that *at finite temperatures neither t_1 nor t_2 corresponds to the initial representation* (4.2).

The BS-phase.. In the case of the broken symmetry solution we have the equations

$$b^2 = \frac{t^2}{G},$$
$$t^2 - 3Gt + 3G^2 \ln t - 18G^2 \sigma_{\beta\beta}(t, \theta) \tag{6.43}$$
$$-6G\left(\theta - 3\ln 3 \cdot G\frac{\theta}{t}\right) \ln\left(1 - \exp\left\{-\frac{t}{\theta}\right\}\right) = 0,$$

where we have used (6.38), (6.37) and (6.30).

The second equation (6.43) for all (θ, G) has a unique solution. One can check that the solution with asymptotic behaviour of the type

$$1 \ll t \ll \theta \quad \text{or} \quad t \gg \theta \quad \text{for } \theta \gg 1$$

is absent. This means that at high temperatures the effective coupling constant $G_{\text{eff}}(G, \theta)$ in the S-phase is smaller then in the BS-phase. We conclude that the system is in the symmetric phase at high temperature. Numerical solution of (6.39) and (6.43) indicates that the same conclusion is valid for all $(\theta, G) \in S_1, S_2$.

It is convenient to represent the result in the form of the diagram given in Fig. 6.5. The phase boundaries correspond to transitions of the first order since the order parameter

$$\sigma = \pm\frac{t(G, \theta)}{\sqrt{G}}$$

is discontinuous at the critical points. The asymptotic relation (6.41) for effective coupling constants shows that our description is accurate enough outside the critical region. At the same time, the region of phase transitions has been investigated roughly, so that the phase boundaries in Fig. 6.5 are defined very approximately.

Let us summarize the results of this subsection.

- Symmetry breaking is absent in the three-dimensional model (2.1) for all
  θ, if $G \ll 1$.
- There are two S-phases and one BS-phase; the phase transition accompanied by symmetry reconstruction takes place at intermediate values of G, θ
  (see Fig. 6.5).
- The system is in the symmetric phase if the temperature θ or the coupling
  constant G is large enough.
- Outside the critical region the effective coupling constant is small $G_{\text{eff}} \ll 1$
  and we can do perturbation calculations using the Hamiltonian (6.28),
  (6.29).

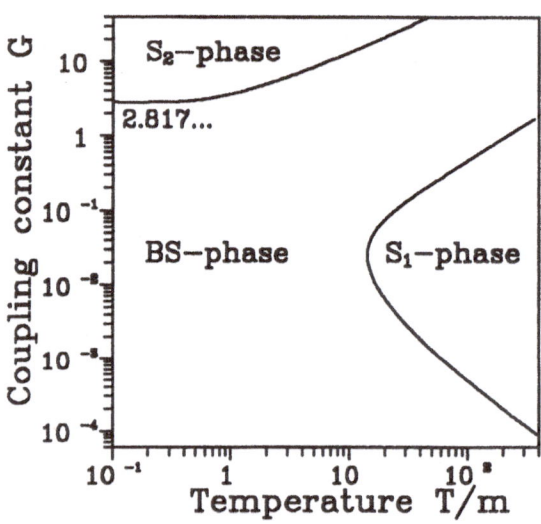

Fig. 6.6. The phase diagram for the $(\varphi^4)_3$ model with initially broken symmetry.

6.2.3 The Case of SSB in the Initial Representation

In this subsection we look at the phase structure of model (2.2). The equations
for the parameters t and b are obtained from (6.36) by the substitution

$$G_4 = G, \quad G_3 = \frac{1}{2}\sqrt{G}.$$

As a result we get

$$t^2 - 1 - \frac{3}{2}G\left(b^2 - d(t,\theta)\right) - 3\sqrt{G}b - \frac{3}{2}G^2\left(\ln t - 6\sigma_\beta(t,\theta) - 6\sigma_{\beta\beta}(t,\theta)\right) = 0,$$

$$2b + Gb\left(b^2 - 3d(t,\theta)\right) \tag{6.44}$$

$$+3\sqrt{G}\left(b^2 - d(t,\theta)\right) + 3G\sqrt{G}\left(1 + \sqrt{G}b\right)\left(\ln t - 6\sigma_\beta(t,\theta) - 6\sigma_{\beta\beta}(t,\theta)\right) = 0.$$

Two solutions for b follow from (6.44):

$$= - \qquad b \qquad \frac{1}{\sqrt{G}} \text{ (symmetric)}, \quad b = -\frac{1}{\sqrt{G}} \pm \frac{t}{\sqrt{G}} \text{ (nonsymmetric)}.$$

Using the solution with broken symmetry we get the equation

$$t^2 - 3Gt - 1 + 3G + 3G^2 \ln t - 18G^2 \sigma_{\beta\beta}(t,\theta)$$
$$-6G \left(\theta - 3\ln 3 \cdot G\frac{\theta}{t} \right) \ln \left(1 - \exp \left\{ -\frac{t}{\theta} \right\} \right) = 0, \quad (6.45)$$

while for the symmetric case $b = -1/\sqrt{G}$ we have:

$$2t^2 + 3Gt + 1 - 3G - 3G^2 \ln t + 18G^2 \sigma_{\beta\beta}(t,\theta)$$
$$+6G \left(\theta - 3\ln 3 \cdot G\frac{\theta}{t} \right) \ln \left(1 - \exp \left\{ -\frac{t}{\theta} \right\} \right) = 0. \quad (6.46)$$

The S-phase.. There are two solutions of (6.46) in the regions S_1, S_2 in Fig. 6.6, and there are no solutions in the region BS. The asymptotics of the mass $M = m \cdot t$ and effective coupling constant $G_{\text{eff}} = G/t$ are the same as in the symmetric model (see (6.41)).

The BS-phase.. Equation (6.45) has a unique solution for all (G, θ). Analysis of the asymptotic behaviour, numerical solution of (6.46) and (6.45) and comparison of the effective coupling constants indicate the phase diagram shown in Fig. 6.6. The phase boundary at $G \ll 1$ agrees with the boundary that can be expected from perturbation calculation of the effective potential [3]. The order parameter is discontinuous at the boundaries, so that we have first-order phase transitions. The description is quite accutate outside the critical regions (see (6.41)), but the phase boundaries are determined very approximately, since the effective coupling constant is large enough in the vicinity of phase transitions.

The conclusions of this subsection can be summarized as follows.

- The symmetry, is restored in the system (2.2) if the temperature or the coupling constant is large enough.
- There are phase transitions between BS- and S-phases; the phase boundaries are shown approximately in Fig. 6.6.
- The OR method provides the possibility of determining the temperature dependence of the mass. The procedure of this determination is quite correct outside the critical regions.
- In the strong coupling $g \gg \theta$ and in the high temperature $\theta \gg G$ regimes the effective coupling is weak $G_{\text{eff}}(G, \theta) \ll 1$ and one can do the standard perturbation calculations using the Hamiltonian (6.28), (6.29).

6.3 The Systems in R^{1+1} and R^{2+1}

Let us compare the phase structure of models (2.1), (2.2) in R^{2+1} and R^{1+1} at finite temperatures. The phase diagrams for the two-dimensional models are plotted in Figs. 6.1 and 6.3. The behaviour of the systems over the variable G is quite different in R^{1+1} and R^{2+1} (see also Table 5.1). We have the BS-phase in the space-time R^{1+1} and the S-phase in R^{2+1} for $G \gg 1$ independently of the symmetry of the classical Lagrangians (2.1), (2.2). At the same time, the behaviour over the variable θ is qualitatively the same in R^{1+1} and R^{2+1}. The systems are symmetric if the temperature is large enough.

7. The Two-Dimensional Yukawa Theory

Lattice calculations indicate a rich phase structure of the Yukawa-type theories [72]. Particularly, it has been argued in the paper [73] that in the two-dimensional Yukawa theory fermion has a dynamically generated mass even for a very weak Yukawa coupling.

In this chapter we apply the oscillator representation method to investigation of the phase structure of the Yukawa model supplemented with the self-interaction of the pseudoscalar field φ

$$L(x) = \bar{\psi}(x) i \hat{\partial} \psi(x) \tag{7.1}$$
$$+ \frac{1}{2} \varphi(x) \left(\Box - m_B^2 \right) \varphi(x) - y \varphi(x) \bar{\psi}(x) i \gamma_5 \psi(x) - \frac{g}{4} \varphi^4(x)$$

in space-time R^{1+1} ($x = (x_0, x_1)$). The fermion field ψ is massless. The parameters m_B^2, g and y are positive. The Dirac matrices are related to the Pauli matrices as

$$\gamma_0 = \sigma_3, \quad \gamma_1 = i\sigma_2, \quad \gamma_5 = \sigma_1.$$

The Lagrangian is invariant under the parity (P) transformation

$$\varphi(x_0, x_1) \rightarrow -\varphi(x_0, -x_1), \quad \psi(x_0, x_1) \rightarrow \gamma_0 \psi(x_0, -x_1).$$

Model (7.1) gives a simple example for studying the dynamical P violation and generation of the fermion mass [74]. If the dimensionless coupling constants

$$G = \frac{g}{2\pi m_B^2} \quad \text{and} \quad Y = \frac{y^2}{2\pi m_B^2} \tag{7.2}$$

are small enough, the Lagrangian (7.1) should describe in quantum theory the system symmetric under parity transformation. Is this statement really true and what happens in the strong coupling regime?

In order to get an answer to this question we shall find the boson $M_B(G, Y)$ and fermion $M_F(G, Y)$ masses, effective coupling constants

$$G_{\text{eff}}(G, Y) = \frac{g}{2\pi M_B^2(G, Y)}, \quad Y_{\text{eff}}(G, Y) = \frac{y^2}{2\pi M_B^2(G, Y)}, \tag{7.3}$$

order parameter and free energy density as functions of (G, Y) for different CR representations. The phase diagram in the (Y, G)-plane will be constructed. The Hamiltonians describing system (7.1) in each phase will be obtained. Two different symmetric phases and the phase with violated parity occur in the system. The parity breaking in the strong coupling regime $G \gg Y$ is conditioned by the boson self-interaction. This is in accordance with the vacuum structure of pure φ_2^4 theory [19, 32].

Another representation with the symmetry breaking coursed by the Yukawa coupling is not realized, since the vacuum energy in this representation is larger than the energy of symmetric phases for any Y, G. Therefore the Yukawa interaction does not lead to an instability of the symmetric phase. At the first glance this contradicts to the results of lattice calculations [73]. However, statement of the problem of the phase structure of a field system and investigation technique within the regularized (lattice) and renormalized (as in our case) formulations of quantum field theory are basically different. We analyze this point in the last section of this chapter and show that the results of [73] and ours neither agree nor contradict to each other.

7.1 The Hamiltonian and Renormalization

The quantized Hamiltonian corresponding to the Lagrangian (7.1) has the following form:

$$H = H_0 + H_I + H_{ct}, \tag{7.4}$$

$$H_0 = \int_V dx_1 \left\{ \frac{1}{2} : \left[\pi^2(x) + (\partial_1\varphi(x))^2 + m_B^2\varphi^2(x) \right] : \right.$$
$$\left. + : \bar{\psi}(x)i\gamma_1\partial_1\psi(x) : \right\},$$

$$H_I = \int_V dx_1 \left\{ y\varphi(x) : \bar{\psi}(x)i\gamma_5\psi(x) : + \frac{g}{4} : \varphi^4(x) : \right\},$$

$$H_{ct} = \int_V dx_1 \left\{ \frac{1}{2}\delta m_B^2 : \varphi^2(x) : + \delta E \right\}.$$

Standard equal time canonical relations are postulated:

$$[\pi(x_0, x_1), \varphi(x_0, x_1')]_- = i\delta(x_1 - x_1'),$$
$$[i\psi^+(x_0, x_1), \psi(x_0, x_1')]_+ = i\delta(x_1 - x_1'). \tag{7.5}$$

The Hamiltonian (7.4) is constructed in such a way that CR (7.5) are represented in the Fock space of bosons with the renormalized mass m_B and massless fermions. The Hamiltonian is normally ordered with respect to the vacuum vector $|0\rangle$ of this Fock space.

The model under consideration is superrenormalizable. Boson mass and vacuum energy renormalization comes from the divergent diagrams given in Fig. 7.1. It is convenient to fix the renormalization scheme by the following prescription:

– **mass renormalization**: external momentum in the diagram (a) in Fig. 7.1 is on the mass shell $(p^2 = m_B^2)$;

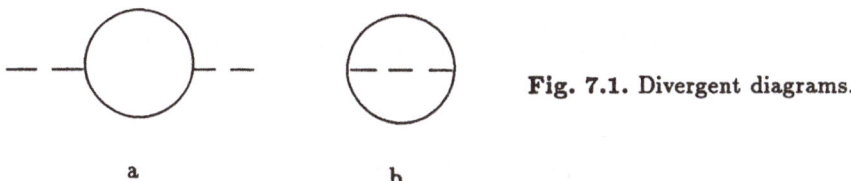

Fig. 7.1. Divergent diagrams.

– **vacuum energy renormalization**: diagram (b) in Fig. 7.1 is subtracted completely.

Simple calculation gives the following result for the counter-terms:

$$\delta m_B^2 = y^2 \tilde{\Pi}_{\text{reg}}^{\text{P}}(m_B^2|0), \quad \delta E = \frac{y^2}{8\pi}\text{reg}\int_0^\infty \frac{du}{u+m_B^2}\tilde{\Pi}_{\text{reg}}^{\text{P}}(-u|0),$$

$$\tilde{\Pi}_{\text{reg}}^{\text{P}}(p^2|0) = i\text{reg}\int \frac{d^2q}{(2\pi)^2}\text{Tr}\left\{i\gamma_5\tilde{S}(q-p|0)i\gamma_5\tilde{S}(q|0)\right\},$$

where $\tilde{S}$ is the fermion propagator

$$\tilde{S}(q|m_F) = 1/(m_F - \hat{q} - i\epsilon).$$

An appropriate regularization is implied in (7.6). Now the S-matrix is defined, all terms of the perturbation series over $Y \ll 1$ and $G \ll 1$ are ultraviolet finite and can be calculated. The strong coupling regime $Y \gg 1$ and (or) $G \gg 1$ and self-consistency of this construction in the weak coupling regime will be investigated by means of the canonical transformation method.

7.2 The Canonical Transformation

Let us transform the canonical variables as

$$\{i\psi^+, \psi\} \longrightarrow \left\{i\Psi^+\exp\left(-i\frac{\alpha}{2}\gamma_5\right), \exp\left(i\frac{\alpha}{2}\gamma_5\right)\Psi\right\},$$
$$\{\pi, \varphi\} \longrightarrow \{\Pi, \Phi + B\}. \tag{7.6}$$

Here Ψ is the fermion field with a new mass $M_F^2 = fm_B^2$, Φ is the boson field with the mass $M_B^2 = tm_B^2$, the angle α is a parameter of chiral transformations and B is a constant boson condensate. Such a transformation can be realized in terms of the creation and annihilation operators [48]. Transformation (7.6) is the canonical one, i. e., the fields (Π, Φ) and $(i\Psi^+, \Psi)$ obey the same canonical relations (7.5). New fields are defined on the Fock space with the vacuum state $|0\rangle\rangle$. This space is unitary nonequivalent to the initial one with the vacuum $|0\rangle$.

The Hamiltonian takes the following form in the new canonical variables

$$H = H_0' + H_I' + H_{\text{ct}}' + VE + H_1, \tag{7.7}$$

$$H_0' = \int_V dx_1 \left\{ \frac{1}{2} : [\Pi^2(x) + (\partial_1\Phi(x))^2 + M_B^2\Phi^2(x)] : \right.$$
$$+ : \bar{\Psi}(x)\,[i\gamma_1\partial_1 + M_F]\,\Psi(x) : \Big\},$$

$$H_I' = \int_V dx_1 \left\{ y\Phi(x) : \bar{\Psi}(x)\,[\sin\alpha - i\gamma_5\cos\alpha]\,\Psi(x) : \right.$$
$$+ \frac{g}{4} : [\Phi^4(x) + 4B\Phi^3(x)] : \Big\},$$

$$H_{\text{ct}}' = \int_V dx_1 \left\{ \frac{1}{2}\delta M_B^2 : \Phi^2(x) : + \delta E' \right\}.$$

Here the sign :: means normal ordering with respect to $|0\rangle\!\rangle$. The counter-terms H_{ct}' are determined by the new Hamiltonians H_0' and H_I' and correspond to the renormalization scheme which is equivalent to the initial one: the inner lines in the diagram (a) in Fig. 7.1 correspond to the new fermion propagator and external momentum is on the mass shell $p^2 = M_B^2$, the vacuum diagram in Fig. 7.1 with the new propagators is subtracted completely. We get

$$\delta M_B^2 = y^2\,\tilde{\Pi}_{\text{reg}}^{\text{PS}}(M_B^2|M_F), \tag{7.8}$$

$$\delta E = \frac{y^2}{8\pi}\text{reg}\int_0^\infty \frac{du}{u + M_B^2}\,\tilde{\Pi}_{\text{reg}}^{\text{PS}}(-u|M_F), \tag{7.9}$$

$$\tilde{\Pi}_{\text{reg}}^{\text{PS}}(p^2|M_F) = i\text{reg}\int \frac{dq}{(2\pi)^2}\text{Tr}\{(\sin\alpha - i\gamma_5\cos\alpha)\tilde{S}(q - p|M_F)$$
$$\times(\sin\alpha - i\gamma_5\cos\alpha)\tilde{S}(q|M_F)\}.$$

The quantity E in (7.7) is the vacuum energy density and looks like

$$E = E_0 + E_I + E_{\text{ct}}$$

$$E_0 = \frac{1}{2}m_B^2 B^2 + L(t) + \langle\!\langle 0|\bar{\Psi}(i\partial_1\gamma_1 + M_F)\Psi|0\rangle\!\rangle - \langle 0|\bar{\psi}i\partial_1\gamma_1\psi|0\rangle,$$

$$E_I = \frac{g}{4}\left[B^4 - 6B^2 D(t) + 3D^2(t)\right] + yB\sin\alpha\langle\!\langle 0|\bar{\Psi}\Psi|0\rangle\!\rangle, \tag{7.10}$$

$$E_{\text{ct}} = \delta E - \delta E' - \frac{1}{2}\delta m_B^2 D(t) + \frac{1}{2}\delta m_B^2 B^2,$$

$$D(t) = \int \frac{d^2k}{(2\pi)^2 i}\left[\frac{1}{m_B^2 - k^2 - i\varepsilon} - \frac{1}{M_B^2 - k^2 - i\varepsilon}\right] = \frac{1}{4\pi}\ln t. \tag{7.11}$$

The function $L(t)$

$$L(t) = \frac{1}{2}\langle\!\langle 0|\Pi^2 + (\partial_1\Phi)^2 + m_B^2\Phi^2|0\rangle\!\rangle - \frac{1}{2}\langle 0|\pi^2 + (\partial_1\varphi)^2 + m_B^2\varphi^2|0\rangle$$

$$= \frac{m_B^2}{8\pi}(t - 1 - \ln t)$$

comes from the normal reordering of the free Hamiltonian. The last term H_1 in (7.7) has the form

$$H_1 = \int_V dx_1 \left\{ \frac{1}{2} : \Phi^2 : [m_B^2 - M_B^2 - 3gD(t) + 3gB^2 + \delta m_B^2 - \delta M_B^2] \right.$$

$$+ \Phi \left[m_B^2 B - 3gBD(t) + gB^3 + \delta m_B^2 B - y \sin \alpha \operatorname{Tr} S(0|M_F) \right]$$

$$\left. + [yB \sin \alpha - M_F] : \bar{\Psi}\Psi : -yB \cos \alpha : \bar{\Psi} i\gamma_5 \Psi : \right\}.$$

To preserve the correct form of the total Hamiltonian in the new representation we demand that $H_1 \equiv 0$. This requirement leads to equations for the parameters M_F, M_B, B and α of the canonical transformation:

$$yB \sin \alpha - M_F = 0,$$
$$yB \cos \alpha = 0, \tag{7.12}$$
$$m_B^2 - M_B^2 - 3gD(t) + 3gB^2 + \delta m_B^2 - \delta M_B^2 = 0,$$
$$m_B^2 B - 3gBD(t) + gB^3 - y \sin \alpha \operatorname{Tr} S(0|M_F) + \delta m_B^2 B = 0.$$

Using ((7.6),(7.8)) and introducing the dimensionless quantities (7.2) and

$$f = \frac{M_F^2}{m_B^2}, \quad t = \frac{M_B^2}{m_B^2}, \quad b = \sqrt{\pi}B$$

one can rewrite (7.12) in the form

$$\sqrt{2Y}b \sin \alpha - \sqrt{f} = 0,$$
$$\sqrt{2Y}b \cos \alpha = 0, \tag{7.13}$$
$$1 - t - \frac{3}{2}G \ln t + Y \ln f + Y \left(1 - 4\frac{f}{t}\right) F \left(\frac{f}{t}\right) + 6Gb^2 = 0,$$
$$b \left[1 - \frac{3}{2}G \ln t + Y \ln f + 2Gb^2 \right] = 0.$$

where

$$F(z) = \int_0^1 \frac{dx}{x(1-x) - z} = \begin{cases} \frac{1}{\sqrt{1-4z}} \ln \left(\frac{1+\sqrt{1-4z}}{1-\sqrt{1-4z}} \right), & \text{if } z \leq \frac{1}{4} \\ \frac{2}{\sqrt{4z-1}} \operatorname{arctg} \sqrt{4z-1}, & \text{if } z \geq \frac{1}{4}. \end{cases} \tag{7.14}$$

Using these equations we can rewrite the energy density (7.10) as

$$E = \frac{m_B^2}{8\pi} \left\{ 4b^2 + t - 1 - \ln t \right. \tag{7.15}$$

$$\left. + 2f \ln f + G \left[4b^4 - 6b^2 \ln t + \frac{3}{4} \ln^2 t \right] - \frac{1}{2} Y \ln^2 t + YJ(t/f) \right\},$$

$$J(s) = 2 \int_0^1 \frac{dx(1-x^2)}{x((1-x)^2 + sx)} \left[\frac{x}{x-1} \ln x - \ln(1-x) \right].$$

Equations (7.13) do not minimize the energy density (7.15) in the variables t, f, b. These equations do not relate to any variational principle. They follow from the demand of the correct form of the total Hamiltonian. This demand, combined with the canonical transformations, provides a regular prescription for dealing with the highest ultraviolet divergencies (like the diagrams in Fig. 7.1). At the same time, the results of our and variational methods coincide in the case $Y = 0$ (the pure φ_2^4 theory) when the variational approach is well-defined [19, 32].

7.3 The Phase Structure

Different solutions of equations (7.13) define the nonequivalent representations of the CR or different phases of the model (7.1). The proper Hamiltonians in these phases are given by (7.7). It is convenient to formulate the following definitions. Let us suppose that (7.13) have N different solutions, which can be denoted as

$$S_j(Y, G) = \{t_j(Y, G), f_j(Y, G), b_j(Y, G), \alpha_j(Y, G)\} \quad (j = 1, ..., N).$$

The effective coupling constants (7.3) and energy density (7.15) corresponding to the j-th solution are denoted by

$$Y_{\text{eff}}^{(j)}(Y, G) = \frac{Y}{t_j(Y, G)}, \quad G_{\text{eff}}^{(j)}(Y, G) = \frac{G}{t_j(Y, G)},$$

$$E_j(Y, G) = E\left(t_j(Y, G), f_j(Y, G), b_j(Y, G), \alpha_j(Y, G), Y, G\right).$$

We shall say that in the region $\Gamma_k \subset R_+^2 = \{(Y, G) : Y \geq 0, G \geq 0\}$ the Yukawa system (7.1) exists in the phase described by the solution $S_k(Y, G)$ if for $(Y, G) \in \Gamma_k$

$$\min_j E_j(Y, G) = E_k(Y, G), \tag{7.16}$$

$$\min_j Y_{\text{eff}}^{(j)}(Y, G) = Y_{\text{eff}}^{(k)}(Y, G), \quad \min_j G_{\text{eff}}^{(j)}(Y, G) = G_{\text{eff}}^{(k)}(Y, G). \tag{7.17}$$

The regions Γ_k cover all the space R_+^2, i.e., $\cup \Gamma_k = R_+^2$. It is quite possible that some solutions are not realized as actual phases of the system, since they do not minimize the effective coupling constants and energy density for any Y and G.

7.3.1 The Pure Yukawa Interaction

First of all, let us study the case $G = 0$, i.e., the pure Yukawa model.

We will show that for any coupling constant Y the Yukawa interaction does not lead to dynamical generation of the fermion mass and parity violation.

For $G = 0$ equations (7.13) are reduced to the form

$$\sqrt{2Y}b\sin\alpha = \sqrt{f}, \quad \sqrt{Y}b\cos\alpha = 0,$$

$$1 - t + \frac{Y}{2}\ln f + Y\left(1 - 4\frac{f}{t}\right)F(f/t) = 0, \tag{7.18}$$

$$b\left[1 + Y\ln f\right] = 0.$$

Energy density (7.15) looks in this case like

$$E = \frac{m_B^2}{8\pi}\left\{4b^2 + t - 1 - \ln t - \frac{Y}{2}\ln^2 t + 2f\ln f + YJ(t/f)\right\}. \tag{7.19}$$

Equations (7.18) has three different solutions.

I. $b_1 \equiv 0$, $t_1 \equiv 1$, $f_1 \equiv 0$, $\sin\alpha_1 = 0$, $Y_{eff}^{(1)} \equiv Y$, $E_1 \equiv 0$.
This is the initial representation (7.4).

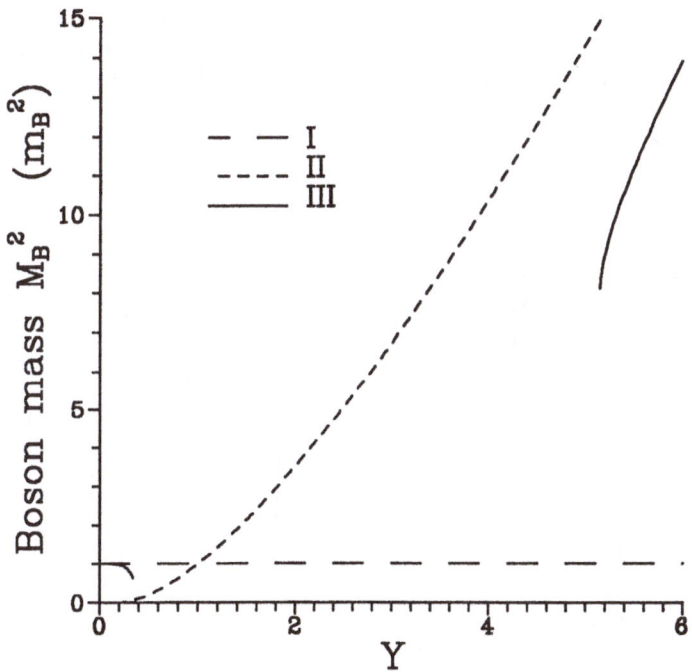

Fig. 7.2. Boson mass for different phases of the pure Yukawa model.

II. $b_2 \equiv 0$, $t_2(Y)$, $f_2 \equiv 0$, $\sin\alpha_2 = 0$, $Y_{eff}^{(2)}(Y)$, $E_2(Y)$.
Equation for the boson mass can be represented in the form:

$$\frac{t_2 - 1}{\ln t_2} = Y. \tag{7.20}$$

Using this equation one can represent energy density (7.19) as

$$E_2 = \frac{m_B^2}{8\pi}\left\{t_2 - 1 - \frac{1}{2}(t_2 + 1)\ln t_2\right\}. \tag{7.21}$$

The functions $t_2(Y)$ and $E_2(Y)$ are plotted in figures 7.2 and 7.3 by the short-dashed lines. In the strong coupling regime $Y \gg 1$ we get from (7.20) and (7.21)

$$t_2(Y) \to Y\ln Y, \quad Y_{\text{eff}}^{(2)}(Y) \to \frac{1}{\ln Y} \ll 1,$$

$$E_2(Y) \to -\frac{m_B^2}{8\pi}\frac{1}{2}Y\ln^2 Y. \tag{7.22}$$

Neither equation (7.20) for boson mass t nor the energy density (7.21) depends on the angle α. We have a family of degenerate (in the masses and energy density) vacua enumerated by the angle α. Representations with $\sin\alpha \neq 0$ correspond to the symmetry broken by the interaction of the *pseudoscalar field* Φ with the *scalar* fermion current (see the interaction Hamiltonian H_I' (7.7)). Below we will consider for definiteness only the symmetric representation with $\sin\alpha = 0$.

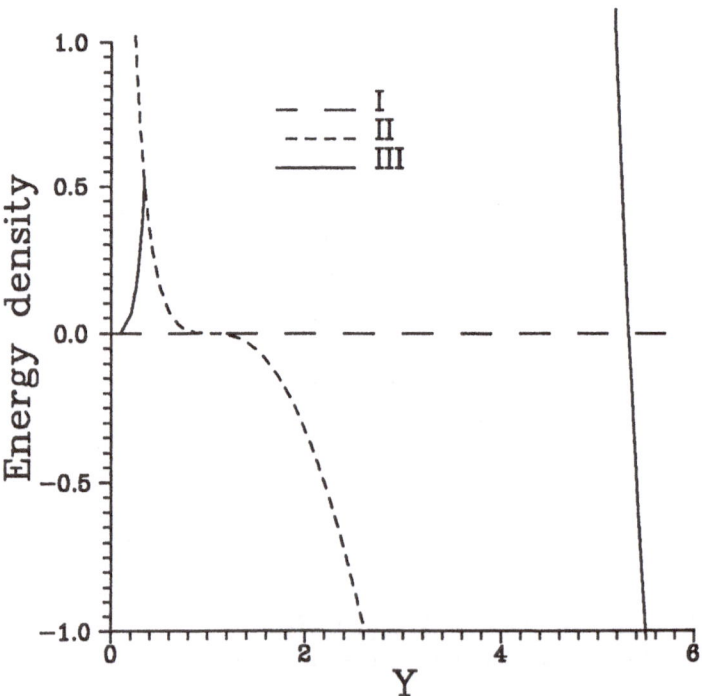

Fig. 7.3. Energy density for different phases of the pure Yukawa model.

III. $b_3 = \pm \frac{1}{\sqrt{2Y}} \exp\{-1/2Y\}$, $t_3(Y)$, $f_3 = \exp\{-1/Y\}$, $\sin \alpha_3 = \pm 1$.

The sign "±" corresponds to two degenerate vacua connected by parity transformation. In this case an equation for the boson mass has the form

$$t_3 - Y \left(1 - 4\frac{f_3}{t_3}\right) F \left(\frac{f_3(Y)}{t_3}\right) = 0. \tag{7.23}$$

The energy density (7.19) takes the form:

$$E_3 = \frac{m_B^2}{8\pi} \left\{ t_3 - 1 - \ln t_3 - \frac{Y}{2} \ln^2 t_3 + Y J(t_3/f_3) \right\}. \tag{7.24}$$

The function $t_3(Y)$ is plotted in Fig. 7.2 by the solid line. One can see that the following relation takes place

$$t_3(Y) < \max[1, t_2(Y)], \quad \forall Y > 0. \tag{7.25}$$

As a consequence of this relation, the effective coupling constant $Y_{\text{eff}}^{(3)}$ is larger in the broken symmetry representation III than in the symmetric ones

$$Y_{\text{eff}}^{(3)} > \min\left[Y, Y_{\text{eff}}^{(2)}(Y)\right], \quad \forall Y > 0. \tag{7.26}$$

The energy density $E_3(Y)$ (the solid line in Fig. 7.3) is positive or larger than $E_2(Y)$ owing to inequality (7.25) and presence of the positive term $YJ(t/f)$ in (7.24)

$$E_3(Y) > \min[0, E_2(Y)], \quad \forall Y > 0. \tag{7.27}$$

An asymptotic behavior of all functions in the weak ($Y \ll 1$) and strong ($Y \gg 1$) coupling regimes can be found from ((7.23),(7.24)).

For $Y \to 0$ we get:

$$t_3(Y) \to 1 - 2 \exp\left\{-\frac{1}{Y}\right\}, \tag{7.28}$$

$$Y_{\text{eff}}^{(3)}(Y) = \frac{Y}{t_3(Y)} \to Y \left(1 + 2 \exp\left\{-\frac{1}{Y}\right\}\right),$$

$$E_3(Y) \to \frac{m_B^2}{4\pi} \frac{1}{2Y} \exp\left\{-\frac{1}{Y}\right\}.$$

The asymptotic expression for the energy density originates from the term $YJ(t/f)$ in (7.24), i.e., it is conditioned by contribution of the diagram (b) in Fig. I.7.1. One can see that the energy density is non-analytic at $Y = 0$.

In the strong coupling regime $Y \gg 1$ one gets:

$$t_2(Y) - t_3(Y) \to \frac{1}{Y \ln Y} > 0,$$

$$Y_{\text{eff}}^{(2)}(Y) - Y_{\text{eff}}^{(3)}(Y) \to -\frac{1}{\ln Y} < 0, \tag{7.29}$$

$$E_2(Y) - E_3(Y) \to -\frac{m_B^2}{8\pi} \ln Y < 0.$$

Comparing the energy densities and effective coupling constants we get the following relations

$$\min\left[Y, Y_{\text{eff}}^{(2)}(Y), Y_{\text{eff}}^{(3)}(Y)\right] = \begin{cases} Y & \text{if } Y \le 1 \\ Y_{\text{eff}}^{(2)}(Y) & \text{if } Y \ge 1 \end{cases}, \qquad (7.30)$$

$$\min\left[0, E_2(Y), E_3(Y)\right] = \begin{cases} 0 & \text{if } Y \le 1 \\ E_2(Y) & \text{if } Y \ge 1 \end{cases}. \qquad (7.31)$$

Equations (7.30) and (7.31) show that according to both definitions (7.17) and (7.16) a kind of phase transition between the phases I and II occurs at $Y = 1$. The phase III with broken symmetry is not realized for any $Y > 0$.

Thus we conclude that parity is not violated dynamically in the two-dimensional Yukawa model. The fermion is massless for any values of the coupling constant Y. This conclusion differs from the results of the lattice calculations [73]. We discuss an origin of this difference in the last section of this chapter.

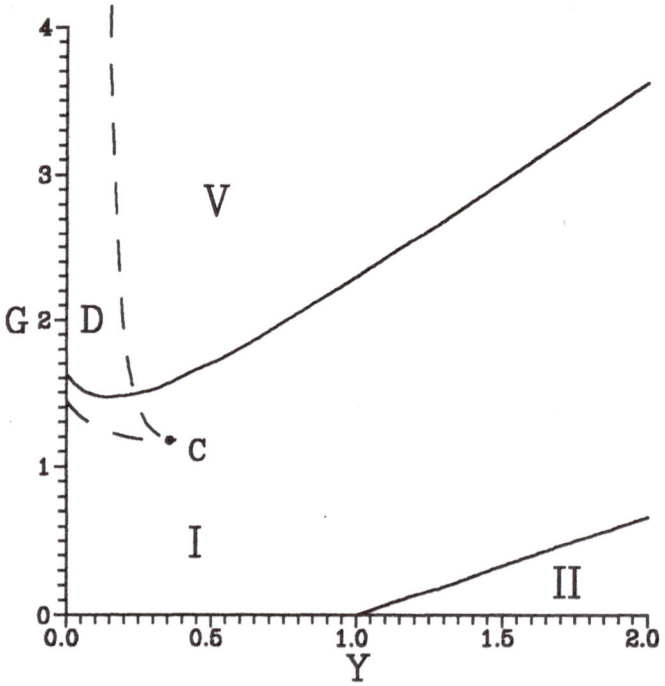

Fig. 7.4. Phase diagram in the plane (Y, G). The dashed lines restrict the region D where Eqs.(7.35) have three solutions.

The effects described in this subsection are determined by non-analyticity of the physical parameters of the system (like the masses and boson condensate) at $Y = 0$. Such a non-analytical behavior can be obtained neither in perturbation theory nor within the variational approach like the Gaussian effective potential which does not take into account a major contribution of the divergent diagrams given in Fig. 7.1.

7.3.2 The Yukawa Model with Boson Self-Interaction

The main effect of boson self-interaction is that the parity is dynamically violated and the fermion gets a nonzero mass in the strong coupling regime $G \gg Y$. This is illustrated by the phase diagram shown in Fig. 7.4. The solid lines correspond to the phase boundaries. In the regime $G \gg Y$ the broken symmetry phase conditioned by the boson self-interaction exists, while for $Y \gg G$ the nontrivial symmetric phase caused by the Yukawa coupling is realized.

In the general case equations (7.13) have five different solutions.

I. $b_1 \equiv 0$, $t_1 \equiv 1$, $f_1 \equiv 0$, $\sin \alpha_1 = 0$, $Y_{\text{eff}}^{(1)} \equiv Y$, $G_{\text{eff}}^{(1)} \equiv G$, $E_1 \equiv 0$.
This is the initial representation (7.4).

II. $b_2 \equiv 0$, $t_2(Y, G)$, $f_2 \equiv 0$, $\sin \alpha_2 = 0$, $Y_{\text{eff}}^{(2)}(Y, G)$, $G_{\text{eff}}^{(2)}(Y, G)$, $E_2(Y, G)$
For $b = 0$ and $f = 0$ the equation for the boson mass can be written in the following form (see the third equation (7.13))

$$\frac{t_2 - 1}{\ln t_2} = Y - \frac{3}{2}G. \tag{7.32}$$

This equation has a unique solution for all Y and G obeying the condition

$$Y - \frac{3}{2}G > 0$$

and does not have solutions for other values of (Y, G). Using (7.32) in (7.15) one can reduce the energy density to the form

$$E_2 = \frac{m_B^2}{8\pi} \left\{ t_2 - 1 - \frac{1}{2}(t_2 + 1) \ln t_2 \right\}, \tag{7.33}$$

which coincides with (7.21). In the strong coupling regime $Y \gg G$ one finds

$$t_2(Y, G) \to Y \ln Y,$$

$$Y_{\text{eff}}^{(2)}(Y, G) \to \frac{1}{\ln Y} \ll 1, \quad G_{\text{eff}}^{(2)}(Y, G) \to \frac{G}{Y \ln Y} \ll 1,$$

$$E_2(Y, G) \to -\frac{m_B^2}{8\pi} \frac{1}{2} Y \ln^2 Y. \tag{7.34}$$

Using the formulas (7.33) and (7.3) we get the inequalities

$$E_2 \leq 0, \quad G_{\text{eff}}^{(2)} \leq G, \quad Y_{\text{eff}}^{(2)} \leq Y \quad \text{for} \quad t_2 \geq 1.$$

Equation (7.32) indicates that $t_2 \geq 1$ if $Y - 3/2G \geq 1$. Thus, according to both criteria (7.17) and (7.16) the phase transition from the first symmetric phases I to the second symmetric phase II takes place on the curve $Y - 3/2G = 1$ shown in Fig 7.4 by the solid line starting at the point $(Y = 1, G = 0)$.

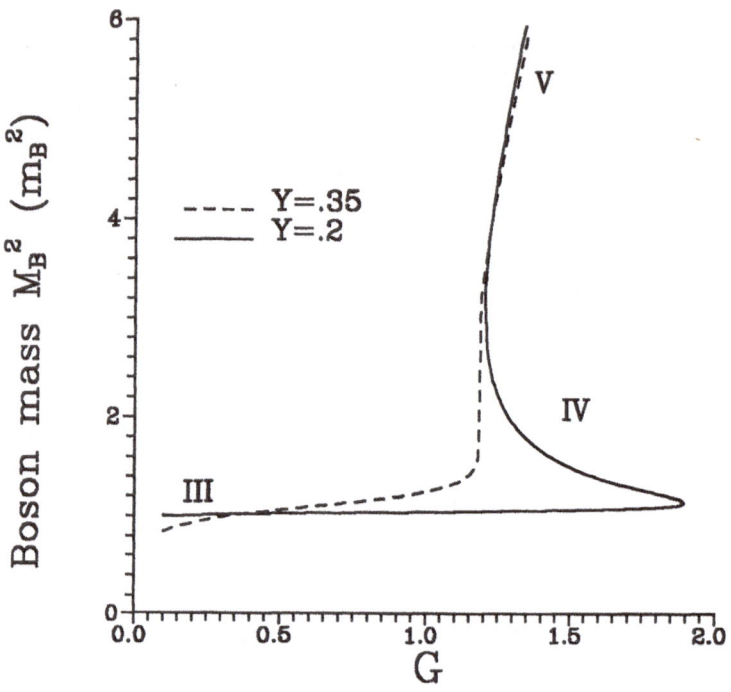

Fig. 7.5. Boson mass for the phases III, IV and V with broken symmetry.

Solutions III, IV and V with nonzero boson condensate:

$b_j(Y, G) = \pm\sqrt{\frac{f_j(Y,G)}{Y}}$, $t_j(Y, G)$, $f_j(Y, G)$, $\sin \alpha_j = \pm 1$ $(j = 3, 4, 5)$.

The sign "$\pm$" corresponds to two degenerate vacua connected by the P-transformation. The P-symmetry breaking is provided by two reasons. These are the terms : Φ^3 : and $\Phi : \bar{\Psi}\Psi$: in the interaction Hamiltonian H'_I (7.7). The energy density for the broken symmetry representations is defined in (7.15). For description of these solutions it is convenient to introduce the variable $s = f/t$, to subtract the last equation (7.13) from the third one and to rewrite (7.13) in the form $(f = st)$

$$t\left(1 - 2\frac{G}{Y}s\right) = Y(1 - 4s)F(s),$$

$$1 + (Y - \frac{3}{2}G)\ln t + Y \ln s + \frac{G}{Y}st = 0. \tag{7.35}$$

The function $F(s)$ is defined by (7.14).

The analysis of equations (7.35) shows that there are two qualitatively different regions in the (Y, G) plane. There are three solutions inside the region D restricted by the G-axis and dashed lines in Fig. 7.4, while outside this region only one solution exists. All solutions are equal to each other at the point C in Fig. 7.4 which corresponds to $Y_C = .341...$ and $G_C = 1.12....$ Comparing the limit $G \to 0$ of equations (7.35) and equation (7.23) for the pure Yukawa model we see that one of the three different solutions of equations (7.35) is a continuation of the pure Yukawa solution III (see subsection 7.3.1) on the (Y, G) plane. This solution describes the Yukawa-type phase with broken symmetry. The existence of this phase is conditioned by the divergent diagrams (Fig. 7.1) appearing due to the Yukawa coupling.

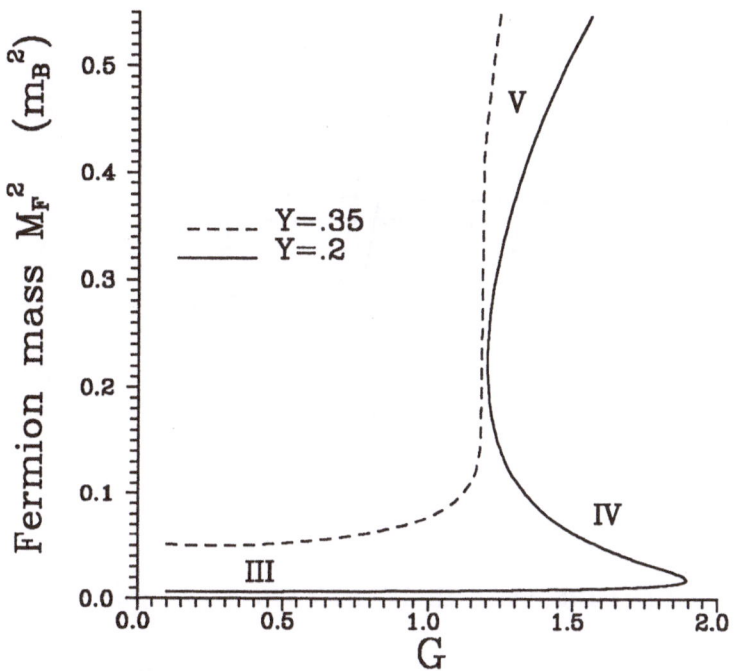

Fig. 7.6. Fermion mass for the phases III, IV and V.

In the strong coupling regime $Y \gg G$, we get from $((7.35),(7.13))$ the following asymptotic relations

$$t_2(Y, G) - t_3(Y, G) \to \frac{1}{Y \ln Y} > 0,$$

$$Y_{\text{eff}}^{(2)}(Y, G) - Y_{\text{eff}}^{(3)}(Y, G) \to -\frac{1}{\ln Y} < 0,$$

$$G_{\text{eff}}^{(2)}(Y, G) - G_{\text{eff}}^{(3)}(Y, G) \to -\frac{G}{Y \ln Y} < 0,$$

$$f_3(Y,G) \rightarrow \exp\left\{-\frac{1}{Y}\right\}, \quad b_3 \rightarrow \pm\frac{1}{\sqrt{2Y}},$$

$$E_2(Y,G) - E_3(Y,G) \rightarrow -\frac{m_B^2}{8\pi} \ln Y < 0,$$

which are exactly the same as (7.29). The boson mass t_3 approaches t_2 from below. At the same time, the divergence between energy densities E_3 and E_2 grows due to the contribution of the term $YJ(f/t)$ in (7.13).

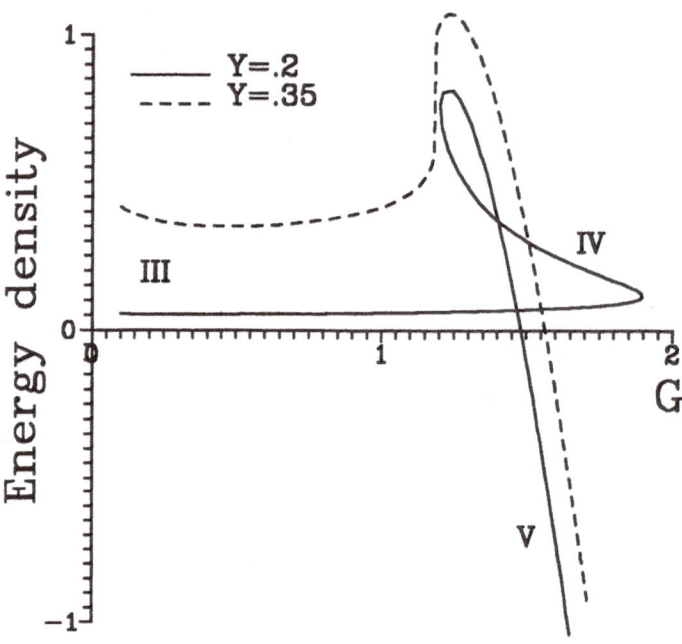

Fig. 7.7. Energy density for the phases III, IV and V with broken symmetry.

When $Y < Y_C$ and G grows two additional solutions of (7.35) appear at the lower dashed line restricting the region D in Fig. 7.4. This solutions are of the φ_2^4-type since they are a continuation of the pure φ_2^4 broken symmetry representations [32] on the (Y,G) plane. The φ_2^4-type phases originate from the divergences caused by the boson self-interaction (the bubble diagrams). On the upper dashed line in Fig. 7.4 solutions III and IV terminate and above this line we have only the φ_2^4 type phase V with broken symmetry.

The following asymptotic solutions can be obtained from (7.35) for $G \gg Y$

$$t_5(Y,G) \rightarrow 3G \ln G, \tag{7.36}$$

$$f_5(Y,G) \rightarrow \frac{3}{2}Y \ln G, \quad b_5 \rightarrow \pm\sqrt{\frac{3}{4} \ln G},$$

$$E_5(Y, G) \to -\frac{m_B^2}{8\pi}\frac{3}{2}G\ln^2 G$$

The asymptotic behavior of the boson mass and energy density is the same as in the pure φ_2^4 theory [32]. The point C in Fig. 7.4 is quite analogous to a critical point known in the classical thermodynamical systems like gas-liquid [1]. Different phases do not exist and the system is always homogeneous outside the region D. One can say that at the critical point (Y_C, G_C) the difference between phases disappears (solutions of (7.35) are equal to each other at the critical point C). As soon as the critical point exists, a continuous transition between the phases III and V is possible, in which the separation into phases does not occur at any point. To do this, the change of coupling constants must take place along some curve in the (Y, G) plane nowhere cutting the lower dashed line in Fig. 7.4. This curve may pass through the critical point C.

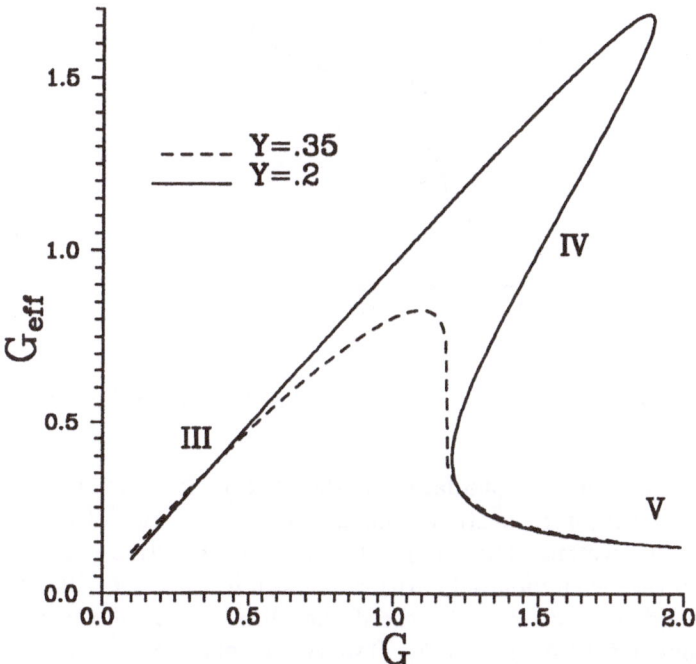

Fig. 7.8. Boson self-coupling effective constant for the phases III, IV and V.

Boson and fermion masses as the functions of G for a fixed value of Y are shown in figures 7.5 and 7.6 for two different paths in the (Y,G) plane. The solid line represents the case $Y < Y_C$: the path cuts the region D and we see the separation into the phases III, IV and V. The dashed line corresponds to $Y > Y_C$: the path does not cut the region D, the separation does not occur and a continuous transition from the Yukawa-type phase III to the φ_2^4-type phase V takes place. The difference between these two phases is purely quantitative. Strictly speaking, one can speak of two phases only in the case when they exist at the same time touching each other, i.e., for points (Y, G) situated inside the region D.

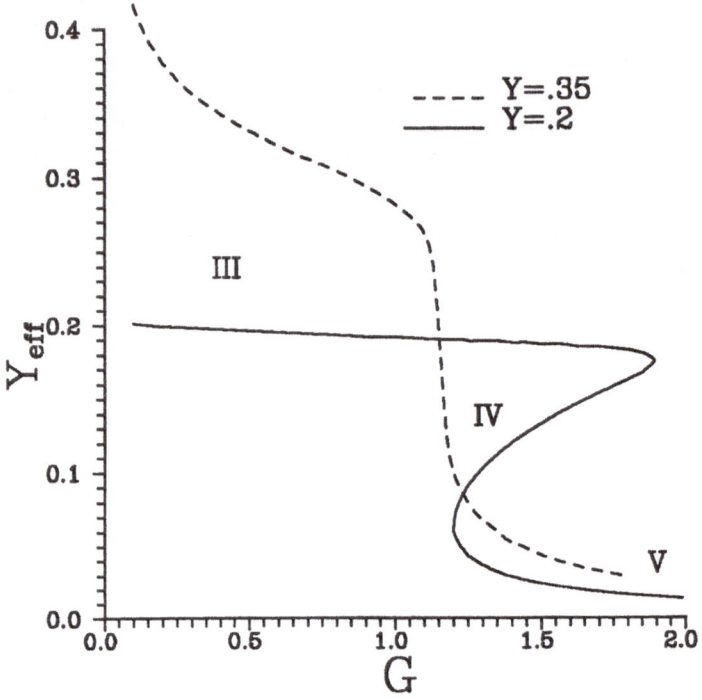

Fig. 7.9. The Yukawa effective coupling constant for the phases III, IV and V.

In order to find the phase boundaries in the (Y, G) plane we have to compare the effective coupling constants and the energy densities of all the possible phases of the system. The energy densities and effective coupling constants for the phases with violated parity are shown in Figs. 7.7–7.9. The solid lines correspond to $Y = .2 < Y_C$, the dashed lines represent the case $Y = .35 > Y_C$. Following the definition (7.17) we get the phase diagram given in Fig 7.4 by the solid lines. On the right hand side from the boundary starting at the point $(Y = 1, G = 0)$ the nontrivial symmetric phase II is realized, while the φ_2^4-type phase V with violated parity occurs above the line starting at the G-axis. The transition from the initial phase I to the phase V is of the first order since the order parameter (see Fig 7.10) has a jump at the boundary. Asymptotic relations (7.34) and (7.36) shows that the description of the phases is quite accurate outside the critical regions, since the effective coupling constants are small and tend to zero when the coupling constant G

or Y grows. At the same time our description of the phase boundaries and the region in Fig 7.4 where the phase I is realized is very approximate, the effective coupling constants are large enough as can be seen from figures 7.8 and 7.9.

In any case, we can conclude that parity is violated and the fermion has a dynamical mass in the strong coupling regime $G \gg Y$ owing to the self-interaction of the pseuc oscalar field. The Yukawa coupling does not lead to dynamical generation of the fermion mass and parity violation but only courses the phase transition I→II without symmetry rearrangement.

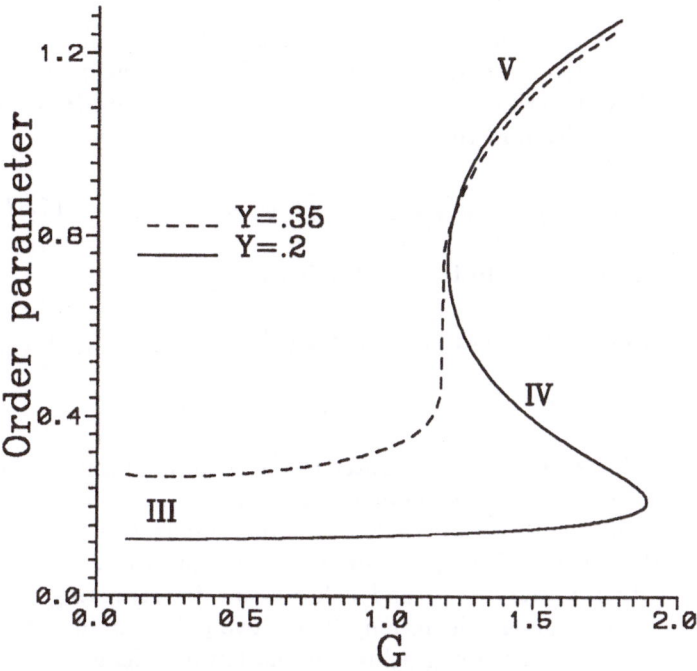

Fig. 7.10. Order parameter for the phases III, IV and V.

7.4 Comparison with Other Approaches

At the first glance above conclusion disagrees with the result of the lattice calculations [73] which claims that even very small Yukawa coupling generates a nonsero fermion mass. In this section we would like to clarify a relationship between results of the lattice approach and the OR method. The central point here is a basic difference between the renormalized and regularized (like the lattice QFT) formulations of the quantum field theory. In order to explain what we mean let us compare the main ideas of calculations in these two formulations of QFT for the simplest Yukawa model with the classical Lagrangian

$$L(x) = \bar{\psi}(x)i\hat{\partial}\psi(x) + \frac{1}{2}\varphi(x)\left(\Box - m^2\right)\varphi(x) + y\varphi(x)\bar{\psi}(x)\psi(x)$$

in the two-dimensional space-time. For this purpose it is more convenient to deal with the functional integral approach.

7.4.1 The Renormalized Formulation

First of all let us reformulate our method of description of the phase structure in application to the functional integral. We will do this quite schematically that is sufficient for above mentioned comparison. The vacuum amplitude for the model (7.37) can be written in the form

$$Z = \lim_{\Lambda \to \infty} \text{reg } N \int \delta\psi\delta\bar{\psi}\delta\varphi \exp\left\{iA_R[\psi, \bar{\psi}, \varphi]\right\}, \qquad (7.37)$$

where the action A_R corresponds to the renormalized Lagrangian

$$L_R(x) = \bar{\psi}(x)i\hat{\partial}\psi(x) + \frac{1}{2}\varphi(x)\left(\Box - m^2\right)\varphi(x) + y\varphi(x)\bar{\psi}(x)\psi(x)$$
$$-\frac{1}{2}\delta m^2 \varphi^2(x) - \delta E,$$

in which the mass and vacuum energy counter-terms corresponding to the divergent diagrams in Fig. 7.1 are incorporated. The quantity m is the renormalized mass of the scalar field φ within the on-shell renormalization scheme. Definition (7.37) of the functional integral implies some appropriate ultraviolet regularization with a parameter Λ and definite rule for removing this regularization at the final stage of calculations. These two points are denoted in (7.37) by the sign $\lim_{\Lambda \to \infty}$ reg. Integrating out the fermion fields we can represent the vacuum amplitude Z as

$$Z = \lim_{\Lambda \to \infty} \text{reg } N \int \delta\varphi \exp\left\{iA_R^{\text{eff}}[\varphi]\right\},$$

$$A_R^{\text{eff}}[\varphi] = \int d^2x \left[\frac{1}{2}\varphi(x)\left(\Box - m^2\right)\varphi(x) - \frac{1}{2}\delta m^2 \varphi^2(x)\right] \qquad (7.38)$$

$$+ \text{Tr}\ln\left(i\hat{\partial} + y\varphi\right). \qquad (7.39)$$

Now we look for a constant field configuration $\varphi(x) = \phi_0$ =const which minimizes the action (7.38). We have to solve the equation

$$\frac{dA_R^{\text{eff}}[\phi_0]}{d\phi_0} = m^2\phi_0 + \delta m^2\phi_0 - y\text{Tr}\frac{1}{i\hat{\partial} + y\phi_0} = 0. \qquad (7.40)$$

This equation is divergence-free since the ultraviolet divergences in the last two terms eliminate each other. Formally equation (7.40) coincides with the last equation (7.12) for $g = 0$. Equation (7.40) has two solutions for all y:

$$\phi_0 = 0 \quad \text{and} \quad \phi_0^2 = \frac{m^2}{y^2} \exp\left\{-\frac{2\pi m^2}{y^2}\right\}. \tag{7.41}$$

Now we have to study which of the solutions (7.41) provides a minimum of the free energy density. In order to do this self-consistently we should change the integration variable in the functional integral (7.38) $\varphi \to \Phi + \phi_0$ and, at the same time, take into account the quantum corrections to the boson mass. In other words we have to find how the boson mass depends on the coupling constant y in the representations with $\phi_0 = 0$ and $\phi_0 \neq 0$. This dependence can be calculated approximately (taking into account leading corrections) by the use of demand of the correct form of the total Lagrangian. In the simple case under consideration this demand results in the equation

$$m^2 - M^2 + \delta m^2(m, y) - \delta M^2(M^2, y, \phi_0) = 0, \tag{7.42}$$

where M is the renormalized boson mass and δM^2 is the counter-term corresponding to the diagram (a) in Fig. 7.1 in which the fermion propagators contain the mass term $y\phi_0$ and the external momentum is subjected to the condition $p^2 = M^2$. Equation (7.42) is divergence-free and is equivalent to the third equation (7.12) for $g = 0$. The general form of the vacuum amplitude Z can be written as

$$Z = \exp\left\{-iTV \cdot E(y)\right\} \lim_{\Lambda \to \infty} \text{reg } N' \int \delta\Psi \delta\bar{\Psi} \delta\Phi \exp\left\{iA_R^{\text{new}}[\Psi, \bar{\Psi}, \Phi]\right\},$$

$$L_R^{\text{new}}(x) = \bar{\Psi}(x)\left(i\hat{\partial} - M_F\right)\Psi(x) + \frac{1}{2}\Phi(x)\left(\Box - M^2\right)\Phi(x)$$

$$+ y\Phi(x)\bar{\Psi}(x)\Psi(x) - \frac{1}{2}\delta M^2 \Phi^2(x) - y\Phi\text{Tr}\frac{1}{i\hat{\partial} - M_F} - \delta E'. \tag{7.43}$$

In this representation the fermion mass M_F is

$$M_F^2 = y^2\phi_0^2 = \begin{cases} 0 & \text{for } \phi_0 = 0, \\ m^2 \exp\left\{-\frac{2\pi m^2}{y^2}\right\} & \text{for } \phi_0 \neq 0. \end{cases} \tag{7.44}$$

The free energy density $E(y)$ in (7.43) is ultraviolet finite, can be computed and looks like the energy density in the subsection 7.3.1. Different solutions of the coupled system of equations (7.40) and (7.42) give physically different representations for Z and describe possible phases of the system. Comparing the energy densities $E(y)$ corresponding to the solutions of (7.40) and (7.42) we choose the phase which has minimal energy and, hence, is realized for given value of y. Even for $\phi_0 = 0$ equation (7.42) can have several solutions. In the case under consideration two such solutions exist for all y and they have the lower energy than the phase with $\phi_0 \neq 0$ (like solutions I and II in section 7.3).

Thus, solving equation (7.42) for $\phi_0 = 0$ and $\phi_0 \neq 0$ and comparing the free energy densities for different solutions we see that the phase with massive fermion has larger free energy than the symmetric phases and, hence, is not realized for all y. The phase with massive fermion is not realized in the system.

The following should be stressed here.

– We have two coupled equations (7.40) and (7.42) describing different phases of the system. This equations take into account leading quantum contributions both to the fermion and boson masses.
– The phase structure of the system is described in terms of the renormalized (physical) parameters. In particular, the fermion mass (7.44) is expressed through renormalized mass of the scalar field.
– The fermion is massles for any y.

Now let us consider the regularized formulation.

7.4.2 The Regularized Formulation

The vacuum amplitude for the model (7.37) in the regularized formalizm can be represented in the form

$$Z = \text{reg } N \int \delta\psi \delta\bar\psi \delta\varphi \exp\left\{ iA[\psi, \bar\psi, \varphi] \right\}, \tag{7.45}$$

where the action A corresponds to the Lagrangian

$$L(x) = \bar\psi(x) i\hat\partial \psi(x) + \frac{1}{2}\varphi(x)\left(\Box - m_0^2\right)\varphi(x) + y\varphi(x)\bar\psi(x)\psi(x)$$

with the bare boson mass m_0. Some regularization is implied in (7.45) but the rule for its removing is not defined. For example, the lattice approximation of the integral can be used, that is equivalent, roughly speaking, to cutoff of the integrals in the momentum space. An integration over the fermion fields leads to the expression

$$Z = \text{reg } N \int \delta\varphi \exp\left\{ iA^{\text{eff}}[\varphi] \right\},$$

$$A^{\text{eff}}[\varphi] = \int d^2x \left[\frac{1}{2}\varphi(x)\left(\Box - m_0^2\right)\varphi(x)\right] + \text{Tr}\ln\left(i\hat\partial + y\varphi\right). \tag{7.46}$$

Looking for a constant field $\varphi(x) = \phi_0 =$const which minimizes the action (7.46) one has to solve the equation

$$\frac{dA^{\text{eff}}[\phi_0]}{d\phi_0} = m_0^2\phi_0 - y\text{Tr}\frac{1}{i\hat\partial + y\phi_0} = 0. \tag{7.47}$$

It is easy to check that equation (7.47) has two solutions for all y: $\phi_0 = 0$ and

$$\phi_0^2 = \frac{\Lambda^2}{y^2} \exp\left\{-\frac{2\pi m_0^2}{y^2}\right\}.$$

Here we used the regularization by cutoff of the momentum integrals at the scale Λ. As the next step one shifts the field $\varphi \rightarrow \Phi + \phi_0$ and gets the fermion mass in the form

$$M_{0F}^2 = \Lambda^2 \exp\left\{-\frac{2\pi m_0^2}{y^2}\right\}, \tag{7.48}$$

which is analogous to corresponding expression written in the paper [73]. In the new representation the vacuum functional Z takes the form

$$Z = \text{reg } N' exp\left\{-iTV \cdot E_0(y)\right\} \int \delta\Psi \delta\bar{\Psi} \delta\Phi \exp\left\{iA^{\text{new}}[\Psi, \bar{\Psi}, \Phi]\right\},$$

$$L^{\text{new}}(x) = \bar{\Psi}(x)\left(i\hat{\partial} - M_{0F}\right)\Psi(x) + \frac{1}{2}\Phi(x)\left(\Box - m_0^2\right)\Phi(x)$$

$$+ y\Phi(x)\bar{\Psi}(x) Psi(x). \tag{7.49}$$

The vacuum energy density $E_0(y)$ in (7.49) is a bare quantity depending on the regularization parameter Λ. We see that

– only one equation (7.47) describes different phases of the system. This equation takes into account quantum contributions to the fermion mass but the boson mass m_0 is fixed;

– the phase structure of the system is described in terms of the bare mass m_0 of the scalar field and ultraviolet cutoff parameter Λ; in partiqular, the fermion mass (7.48) is expressed through the bare mass m_0 and parameter Λ;

– the fermion is massive for all y.

Comparing the content of the current and previous subsections (especially the concluding remarks) one could get a quite definite impression that comparison of the results obtained within the renormalized and regularized formulations of QFT is a subtle thing. In these two cases we have qualitatively different sets of possible phases. After all this is explained by the basic difference in definitions (7.37) and (7.45) of the vacuum amplitude Z. Thus we can finish this chapter by a conclusion that our results and the results of the lattice approach [73] neither agree nor contradict to each other.

.

References

1. S.Coleman, E.Weinberg, Phys.Rev. D, 1973, vol.7, p.1888
2. D.A. Kirghnits JETP Lett., 1972, vol.15, p.745 (in Russian)
3. A.D.Linde Rep.Prog.Phys., 1979, vol.42, p.389;
   A.D. Linde *Elementary Particle Physics and Inflationary Cosmology*, Nauka, Moscow, 1990 (in Russian)
4. L. Dolan, R. Jackiv Phys. Rev.D, 1974, vol.9, p.3320
5. R. Su, P. Bi, G. Ni J.Phys.A, 1983, vol.16, p.2445
6. V.A. Osipov, V.K. Fedyanin Theor. Math. Phys., 1987, vol.73, p.393 (in Russian)
7. I.Roditi Phys.Lett.B, 1986, vol.169, p.264
8. B. Simon *The $P(\phi)_2$ Euclidean (quantum) field theory*, Princeton University Press, Princeton, 1974
9. B. Simon, R. Griffitths Comm. Math. Phys., 1973, vol.33, p.145
10. J. Glimm, A. Jaffe *Quantum Physics. A functional integral point of view*, Berlin, Springer-Verlag, 1981
11. J. Glimm, A. Jaffe Phys.Rev.D, 1975, vol.10, p.536
12. O. Mc Bryan, J. Rosen Comm. Math. Phys., 1976, vol.51, p.97
13. K.G. Wilson Phys.Rev.D, 1972, vol.6, p.419;
    I.A. Fox, I.G. Halliday Phys.Lett.B, 1985, vol.159, p.148
14. M. Aizenman Phys.Rev.Lett., 1981, vol.47, No.1, p.1;
    J. Frohlich Nucl.Phys.B, 1982, vol.200, No.2, p.281;
    C. Arago de Carvalho, S. Caracciolo, J. Frohlich Nucl.Phys.B, 1983, vol.215, No.2, p. 209
15. T. Barnes, G.I. Chandour Phys.Rev.D, 1980, vol.22, No.4, p.924
16. W.A. Bardeen, M. Moshe Phys.Rev.D, 1983, vol.28, No.6, p.1372
17. P.M.Stevenson Phys.Rev.D, 1984, vol.30, No.8, p.1712;
    P.M.Stevenson Z.Phys.C, 1984, vol.24, No.1, p.87
18. M. Consoli, A. Giansito Nucl.Phys.B, 1985, vol.254, No.3&4, p.653;
    M. Consoli, A. Passarino Phys.Lett.B, 1985, vol.165, No.1,2&3, p.113;
    Y. Brihaye, M. Consoli Phys.Lett.B, 1985, vol.157, No.1, p.48
19. S.-J.Chang Phys.Rev.D, 1975, vol.12, No.4, p.1071
20. S.F. Magruder Phys.Rev.D, 1976, vol.14, No.6, p.1602
21. G. Baym, G. Grinstein Phys.Rev.D, 1977, vol.15, No.10, p.2897
22. F.Grassi, R.Hakim, H.D.Sivak Int.J.Mod.Phys.A, 1991, vol.6, No.26, p.4579
23. L. Polli, U. Ritchel Phys.Lett.B, 1989, vol.221, No.1, p.44
24. R.P. Feynman, *Difficulties in applying the variational principle to quantum field theories*, in Proseedings of the International Workshop: *Variational Calculations in Quantum Field Theory*, eds L.Polley and D.Pottinger, World Scientific, Singapore, 1988
25. J.Wudka Phys.Rev.D, 1988, vol.37, No.6, p.1464
26. R. Tarrach Class. Quantum Grav., 1986, vol.3, No.6, p. 1207

122   References

27. U. Ritschel Z.Phys.C, 1991, vol.51, No.3, p. 469
28. R.Munoz-Tapia, J.Taron, R.Tarrach Int.J.Mod.Phys.A, 1988, vol.3, No.9, p.2143
29. S.-J.Chang Phys.Rev.D, 1976, vol.13, No.10, p.2778
30. M.B. Einhorn, D.R.T. Jones Nucl. Phys. B, 1993, vol.398, No.3, p.611
31. P.M.Stevenson, B.Alles, R.Tarrach Phys.Rev.D, 1987, vol.35, No.8, p.2407
32. G.V.Efimov Int.J.Mod.Phys.A, 1989, vol.4, No.18, p.4977
33. G.V.Efimov, S.N.Nedelko Int.J.Mod.Phys.A, 1992, vol.7, p.987
34. G.V.Efimov, S.N.Nedelko 1991, Heidelberg university preprint No.693
35. G.V.Efimov, S.N.Nedelko J.Phys.A, 1992, vol.25, No.6, p.2721
36. G.V.Efimov, S.N.Nedelko Int.J.Mod.Phys.A, 1992, vol.7, No.19, p.4539
37. G.V.Efimov, S.N.Nedelko 1992, JINR preprint E2-92-287
38. P.M.Stevenson Phys.Rev.D, 1985, vol.32, No.6, p.1389
39. P.M.Stevenson, I.Roditi Phys.Rev.D, 1986, vol.33, No.8, p.2305
40. R.Tarrach, B.Alles Phys.Rev.D, 1986, vol.33, No.6, p.1718
41. A.Kovner, B.Rosenstein Phys.Rev.D, 1989, vol.40, No.2, p.504
42. S.Coleman, R.Jackiw, H.D.Politzer Phys.Rev.D, 1974, vol.10, No.8, p.2491; R.J.Root Phys.Rev.D, 1974, vol.10, No.10, p.3322
43. N.N. Bogoliubov, D.V. Shirkov *Introduction to the Theory of Quantized Fields*, Moskow, Nauka, 1984 (in Russian)
44. N.N. Bogoliubov, D.V. Shirkov *Quantized Fields*, Moscow, Nauka, 1980 (in Russian)
45. G.V. Efimov *Problems of Quantum Theory of Nonlocal Interactions*, Moscow, Nauka, 1985 (in Russian)
46. D. Collins *Renormalization*, Cambridge University Press, Cambridge, 1984
47. G.V. Efimov, S.N. Nedelko JINR preprint, E2-89-478, 1989
48. G. Barton *Introduction to Advanced Field Theory*, W.A.Benjamin.Inc, New York, Amsterdam, 1963
49. R. Coquereaux Ann.Phys. 1980, vol.125, No.2, p.401
50. G. t'Hooft Nucl.Phys.B, 1973, vol.61, p.455
51. O.I. Zavialov *Renormalized Feynman Diagrams*, Moscow, Nauka, 1979 (in Russian)
52. L. van Hove Physica, 1952, vol.18, No.3, p.145
53. K.O. Friedrichs *Mathematical aspects of the quantum theory of fields*, Interscience, New York, 1953
54. A.S. Wightman, S.S. Schweber Phys.Rev., 1955, vol.98, p.812
55. R. Haag Dan.Mat.Fys.Medd., 1955, vol.29, No.12, p.1
56. R. Haag, in *Lectures in Theoretical Physics, vol.III*, Interscience Publishers, New York, 1961, p.353
57. G.G. Emch, *Algebraic Methods in Statistical Mechanics and Quantum Field Theory*, Wiley-Interscience, New York, 1972
58. D. Hall, A.S. Wightman Dan.Mat.Fys.Medd., 1957, vol.31, No.5, p.1
59. O.W. Greenberg Phys.Rev., 1959, vol.115, No.3, p.706
60. A.S. Wightman *Introduction to Some Aspects of the Relativistic Dynamics of Quantized Fields*, in: Lecture notes of the French Summer School of Theoretical Physics, Cargese, Corsica, July 1964
61. H. Umezawa, H. Matsumoto, M. Tachiki *Thermo Field Dynamics and Condensed States*, North-Holland Publishing Company, Amsterdam, New York, Oxford, 1982
62. O.I. Zavialov, V.N. Sushko *Nonequivalent Representations of the Commutation Relations in Physics of Infinite Systems*, in: *Statistical Physics and Quantum Field Theory*, Ed. N.N. Bogoliubov, Moscow, Nauka, 1973 (in Russian)
63. N.N.Bogoliubov Nuovo Cim., 1958, vol.7, No.6, p.794

64. N.N. Bogoliubov JETPh, 1958, vol.34, No.1, p.73
65. J.Valatin Nuovo Cim., 1958, vol.7, No.6, p.843
66. G. Ganbold, G.V. Efimov JINR preprint, E2-92-176, 1992
67. N.P.Landsman, Ch.G. van Weert Phys.Rep., 1987, vol.145, No.3&4, p.141
68. H.Matsumoto, I.Ojima, H.Umezawa Ann.Phys., 1984, vol.152, No.2, p.348
69. D.I. Kazakov, D.I. Shirkov 1980, JINR preprint, P2-80-462
70. R. Tarrach Nucl.Phys.B, 1981, vol.183, No.3, p.384
71. J. Gasser, H. Leutwyler Phys.Rep., 1982, vol.87, No.3, p. 77
72. I.Montvay, "Higgs and Yukawa Theories on the Lattice", in the Proceedings of the International Symposium "Lattice 91", Nucl. Phys. (Proc. Suppl.) **B26** (1992) 57
73. A.K. De, *et al* Phys.Lett. **B308** (1993) 327
74. T.D. Lee, Phys. Rep. **C9** (1974) 144
75. L.D. Landau, E.M. Lifshitz, "Statistical Physics", vol.5 of Course of Theoretical Physics, Pergamon Press, London-Paris, 1958

Part II

The Gaussian Equivalent Representation of Functional Integrals in Quantum Physics

8. Path Integrals in Quantum Physics

In the previous part of this book we proposed the oscillator representation method based on the ideas and methods of quantum field theory was to investigate the strong coupling regime of quantum field models. Now we want to apply the main idea of this method to one of the basic problems of quantum physics – calculating functional integrals or path integrals. The implementation of this approach to path integrals requires some specific techniques which we call the method of **Gaussian-equivalent representation** path integrals. The development of this method and its application to various actual problems of modern quantum physics is the content of this part.

8.1 Gaussian Path Integrals

A great number of problems of modern physics can be formulated in terms of the path integral (PI) approach. These problems have a common feature: their solution can be obtained in the form of a functional integral, which is defined on the Gaussian measure. The most general form of a typical functional integral can be written as follows:

$$Z(g) = C_o \int \delta\varphi \, \exp\left\{ -\frac{1}{2} \left(\varphi D_o^{-1} \varphi\right) + g W[\varphi] \right\} . \qquad (8.1)$$

where

$$\left(\varphi D_o^{-1} \varphi\right) = \int_\Gamma dx \int_\Gamma dy \left(\varphi(x) \, D_o^{-1}(x,y) \, \varphi(y)\right)$$

and $\Gamma \in \mathbf{R}^d$ $(d = 1, 2, ...)$. The Gaussian functional measure,

$$d\sigma_o = C_o \, \delta\varphi \, \exp\left\{ -\frac{1}{2} \left(\varphi D_o^{-1} \varphi\right) \right\} ,$$

is defined by a Green function $D_o(x, y)$ corresponding to a differential operator $D_o^{-1}(x, y)$ with appropriate boundary conditions. The normalization constant C_o is chosen in such way that

$$\int d\sigma_o = 1 \qquad \text{or} \qquad Z(0) = 1 .$$

In the standard nonrelativistic quantum mechanics the interaction functional W is usually defined by a potential $gU(\varphi)$:

$$W[\varphi] = g \int_\Gamma dx U(\varphi(x)),$$

where the coupling constant g is real. In other and more interesting cases (for example, polaron, bound states in QFT, stochastic processes etc.) the interaction functional $W[\varphi]$ usually represents a more complicated dependence on $\varphi(x)$.

Until now exact calculations of functional integrals of this type are known [1] only for a quite limitid class of interaction functionals: for quadratic forms of interaction leading to the pure Gaussian integral and very limited numbers of potentials (Coulomb potential and some others), for which the path integral can be reduced to the Gaussian integral after a definite change of variables. For others various approximate methods should be applied.

Contemporary progress in computer hardware and effective software for the numerical simulation technique enable one to obtain numerical calculations of (8.1) with sufficient accuracy, although the practical implementation of this approach is very laborious. Besides, direct numerical (lattice) simulation is bound up with the difficulties of the continuous limit in lattice discretization and limited computer resources.

The development of analytical methods is very important because only analytical methods permit us to investigate qualitative features of quantum physical systems and indicate effective ways for improving the numerical algorithms. Much efforts has been devoted to constructing analytic methods for calculating characteristics of quantum system within the PI formalism.

Among numerous approximate analytical PI methods we can list the following popular approaches: the standard perturbation expansion over g, the quasi-classic WKB approximation [2], the $1/N$-expansion [3], the instanton approximation [4] and the variational methods (e.g. the Gaussian effective potential [5], space-time transformation [6]).

The standard perturbative method usually provides the perturbation series

$$Z(g) = \sum_{n=0}^{\infty} g^n Z_n$$

having the practical sense for a weak interaction $g \ll 1$ when only a few of the lowest terms Z_n is enough to obtain $Z(g)$ with an acceptable accuracy. In addition, the calculation of Z_n for large n is really no more simple then the calculation of the total $Z(g)$.

Our goal is to develop an universal method to calculate this path integral for any, and especially large, g. Sometimes it is possible to hear an opinion that in the strong coupling regime $g \to \infty$ the integral of the type (8.1) loses its Gaussian character and another non-Gaussian measure should be introduced. For example, it can be like this

$$d\sigma = C \, \delta\varphi \, \exp\left\{ -\int dx \varphi^4(x) \right\}.$$

We want to claim that it is not true in the case of integrals of the type (8.1) where

- the highest derivative of the differential operator $D_o^{-1}(x,y)$ is 2ν, i.e. $D_o^{-1}(x,y) \sim \partial^{2\nu}$ for $\nu \geq 1$,
- the interaction functional $W[\varphi]$ depends only on $\varphi(x)$ and does not contain derivatives like $\partial\varphi(x)$.

Let us introduce semiqualitative arguments. Let $f_n(x)$ is an orthonormal system of eigenfunctions of the operator D_o^{-1}:

$$D_o^{-1} f_n(x) = \frac{1}{D_n} f_n(x),$$

$$(f_n, f_m) = \int_\Gamma dx \, f_n(x) f_m(x) = \delta_{nm}.$$

The eigennumbers D_n satisfy the following asymptotic

$$D_n = O\left(\frac{1}{n^{2\nu}}\right) \qquad \text{for} \qquad n \to \infty.$$

Let us introduce the representation

$$\varphi(x) = \sum_n f_n(x)\sqrt{D_n}\, u_n,$$

where $\{u_n, \ (n=0,1,...)\}$ is new denumerate set of variables. We have

$$(\varphi D_o^{-1} \varphi) = \sum_n |u_n|^2,$$

$$Z(g) = C_o \int \prod_n du_n \exp\left\{ -\frac{1}{2}\sum_n |u_n|^2 + gW[\varphi] \right\}.$$

We expand $\varphi(x)$ as follows

$$\varphi(x) = \phi_N(x) + \phi_{>N}(x),$$

$$\phi_N(x) = \sum_{n<N} f_n(x)\sqrt{D_n}\, u_n,$$

$$\phi_{>N}(x) = \sum_{n>N} f_n(x)\sqrt{D_n}\, u_n \sim \sum_{n>N} \frac{f_n(x)}{n^\nu} u_n = O\left(\frac{1}{N^\nu}\right),$$

where N is a large number. Then we have

$$d\sigma = C_o \prod_n du_n \exp\left\{ -\frac{1}{2}\sum_n |u_n|^2 \right\} = d\sigma_N d\sigma_{>N},$$

$$d\sigma_N = C_N \prod_{n<N} du_n \exp \left\{ -\frac{1}{2} \sum_{n<N} |u_n|^2 \right\},$$

$$d\sigma_{>N} = C_{>N} \prod_{n>N} du_n \exp \left\{ -\frac{1}{2} \sum_{n>N} |u_n|^2 \right\},$$

$$\int d\sigma_N = \int d\sigma_{>N} = 1.$$

The interaction functional can be represented as follows

$$W[\varphi] = g \int_\Gamma dx U(\phi_N(x)) + O\left(\frac{1}{N^\nu}\right).$$

Thus the functional integral under consideration can be approximated,

$$Z(g) \sim Z_N(g) = \int d\sigma_N \exp\left\{gW[\phi_N]\right\},$$

and

$$Z(g) = \lim_{N \to \infty} Z_N(g).$$

One can see that the existence of this limit does not depend on the value of the coupling constant g and for any large g there exists a number $N(g)$ so that

$$Z(g) = Z_N(g) + O\left(\frac{1}{g}\right),$$

i.e. the functional measure can be considered as a Gaussian measure.

As a result we can conclude that the path interals of the type (8.1) for any g can be considered as functional integrals over a Gaussian measure. Thus we can expect that there exists a representation of the initial functional integral (8.1)

$$Z(g) = C_g \int \delta\varphi \exp\left\{ -\frac{1}{2}\left(\varphi D_g^{-1}\varphi\right) + W_g[\varphi] \right\} \tag{8.2}$$

with another C_g, D_g^{-1} and $W_g[\varphi]$, for which the main contributions from the interaction functional $gW[\varphi]$ should be accumulated in the operator D_g^{-1} and the perturbation corrections over the new interaction $W_g[\varphi]$ should be small. Our problem is to find this representation.

For this aim we shall use the idea formulated in the preface: *the normal ordering of the Hamiltonian means essentially that the main quantum contributions to the ground state or vacuum of the system are taken into account.*

In the language of functional integrals it means [7] that the conception of normal ordering with respect to a given Gaussian measure should be formulated and next problem is to represent the functional integral (8.1) in the form (8.2) where

– the Gaussian measure is defined by the operator D_g^{-1},

– the interaction functional $W_g[\varphi]$ is written in the normal form with respect to the Gaussian measure with D_g^{-1} and it does not contain quadratic terms over φ, i.e.

$$W_g[\varphi] = O(\varphi^4) \quad \text{for} \quad \varphi \to 0.$$

This representation we shall call the **Gaussian equivalent representation** (GER) of functional integrals. In the chapter 9 all the definitions will be formulated.

This method will be aplied to the following problems:

– investigation of the behaviour of the polaron in ionic crystals in quantum statistics,
– phase transitions and phase restructure in quantum field models,
– propagation of waves in a stochastic medium with stochastically distributed centres in radiophysics,
– solution of differential equations for Green functions for bound states both in quantum field theory and nonrelativistic quantum mechanics.

Before considering in detail the idea and technique of the GER method we would like to give a short historical review on the PI method. To briefly introduce the readers to the PI technique we also introduce Feynman's PI formalism in QM and the way to get the solution of the differential equation in the form of PI.

8.2 A Short Historical Review on the Path Integration Method

In modern theoretical physics the formulation of quantum theory that relies on the original classical system is often distinguished in two mutually complementary ways. One of them is the method of canonical quantization (CQ), where the field dynamic variables are considered as operators satisfying certain commutation relations and defined on a Hilbert space of states. Many papers have been devoted to CQ and for the history and details we refer readers to [8, 9, 10, 11, 12]. This approach was also considered in part I of the present book.

The second approach of quantization is Feynman's method of path integrals (PI) [13, 14]. The basic idea of the Feynman formulation of PI is that the quantum motion of a particle is considered as the sum of all quantum transitions along all possible classical trajectories with an amplitude proportional to

$$A[\mathbf{r}] \propto \exp\left(i\frac{S[\mathbf{r}]}{\hbar}\right),$$

where $S[\mathbf{r}]$ is the classical action taken on a given trajectory $\mathbf{r}(\tau)$. The total transition amplitude is supposed to be proportional to the integral

$$A \propto \int_{\Gamma} \delta \mathbf{r} \exp \left\{ \frac{i}{\hbar} S(t, \mathbf{r}) \right\} , \qquad (8.3)$$

where the classical action

$$S(t, \mathbf{r}) = \int_0^t d\tau \left\{ \frac{m^2}{2} \mid \dot{\mathbf{r}}(\tau) \mid^2 - V[\mathbf{r}(\tau)] \right\} \qquad (8.4)$$

is taken along the given path $\mathbf{r}(\tau)$. The integration in (8.3) is performed over the space Γ of all possible "path-trajectories" $\mathbf{r}(\tau)$ with $0 < \tau < t_0$ for which $\mathbf{r}(0) = \mathbf{x}_0$ and $\mathbf{r}(t) = \mathbf{x}$. This representation attracts much attention because it is close to the classical theory, having both the physical clarity and the fine compact mathematical formulation. These advantages stimulated applications of PI to various problems in quantum physics.

From a technical point of view, the PI formalism of quantization represents an essential attempt to go beyond the perturbation expansion and becomes effective for describing systems with infinite numbers of degrees of freedom.

Below we give a short historical review of the development of PI.

In mathematics Wiener [15] was the first to introduce, in 1920, the conception of PI to describe Brownian motion. Dirac first suggested a representation of a particle propagator in terms very closed to PI techniques [16]. The systematic development of QM within the PI approach belongs to Feynman [13].

In quantum physics, Feynman [13] formulated nonrelativistic QM in the language of PI (in other words, functional or continual integrals) and showed that this approach is completely equivalent to the solution of the Schrödinger equation. One of the main reasons for the popularity of "path integrals" is the understanding that classical mechanics becomes an approximation to QM in the Feynman formulation, if one uses the method of "stationary phases" to the latter. At the classical limit $\hbar \to 0$ the leading contribution to PI is given by the stationary points of the phase function $S(t, \mathbf{r})$ in (8.4), which is the solution of Newton's classical equation of motion.

In 1949 Feynman used the PI for the construction of covariant QED (Feynman diagrams). After then, acknowledging the dignity of this new approach, Kac [17] suggested a PI of Wiener's type for representation of the evolution operator in Euclidean space.

Path integration has come a long way since the 1950s. Probably the most famous early application of PI in statistical physics was to the **polaron** (a nonrelativistic electron, "dressed" by surrounding quanta of lattice vibrations in ionic crystals). In the polaron theory, PI not only helps to formulate the answer qualitatively, but also remains the best way to calculate the answer more exactly than other methods. It is a tractable field theory; the benefits obtained from using PI are entirely analogous to those obtained in QFT.

But in contrary to the polaron problem, all steps for QFT are more difficult because of the divergences, the vector character of the fields and also gauge problems.

A number of investigators [18, 19] independently came to the formulation of QFT in terms of the PI by considering variational estimates for Green functions.

A relatively simple way to represent the Green function of a quantized field within the PI was suggested in [20], where the equivalence of the PI over bosonic fields to averaging over vacuum states of these fields is proved.

A new understanding of the PI occurred in [21, 22], where the evolution operator of the model $P(\varphi)_2$ in Euclidean metrics was represented in PI form as follows:

$$\exp\{-\beta H\} = \int d\sigma_o \exp\left\{-\int dx : P(\varphi):\right\}, \qquad (8.5)$$

$$d\sigma_o = C\delta\varphi \exp\left\{\frac{1}{2}\int dx[(\nabla\varphi)^2 + m^2\varphi^2]\right\},$$

where $d\sigma_o$ is a Gaussian measure of integration, generated by the action $S_o(\varphi) = \frac{1}{2}\int dx[(\nabla\varphi)^2 + m^2\varphi^2]$ of the free bosonic field and $\int dx : P(\varphi) :$ introduced for a certain renormalization of the classical interaction $\int dx P(\varphi)$. This definition of the PI in (8.5) which allows the removal of interaction divergences coming from low-order "tadpole-type" diagrams is the essentially new and important feature of the construction by Glimm and Jaffe.

The next important step in the application of PI in QFT was made in the quantization of Yang–Mills fields. A consequent scheme of quantization for a massless Yang–Mills field was constructed in 1967 by Faddeev, Popov [23] and De Witt [24] within the PI approach. The PI turned out to be the shortest and most convenient method for constructing Feynman rules for the perturbation expansion in gauge field theories. This method played an important role in the investigations of Slavnov [25], Taylor [26], Lee and Zinn-Justin [27], and t'Hooft and Weltman [28]. In these papers a generalized Ward–Takakhashi identity was obtained, various methods of invariant regularization were developed and a procedure renormalizing the perturbation series was built. Within the PI there has also been an attempt made to construct a quantum theory of gravitation [29].

In the 1970s, techniques based on the original ideas of Peak and Inomata [30], and Duru and Kleinert [31] for solving certain non-Gaussian PIs occurring in QM attracted much attention. Standard examples of QM considered in this approach [32] are defined by using Bessel-type and Legendre-type diffusion processes, other than the Wiener process often used on these subjects. These results do not require the machinery of stochastic analysis and can be treated in a quick, transparent way. A development of this method is assumed in [6], where certain non-Gaussian integrals with potentials like $\sim 1/r^2$ or of the Morse type have been derived rigorously by using the techniques of changing dimension and time in PIs.

Excellent monographs and review papers have been devoted to the PI in quantum theory [33, 34, 35, 36, 37, 38].

Although many points concerning the correct mathematical definition and practical calculation of the PI still remain open, it becomes clear that the description of quantum system within the PI is as convenient as using linear operators acting on vectors of Hilbert space in CQ.

We summarize the above, stressing in particular that:

− the PI is a *convenient conception for the qualitative consideration* of quantum theories owing to the simplicity of using the WKB approximation, the evident relativistic covariance of the formulation and the ease with which some specific constraints can be taken into account (e.g. introduction of "ghosts" in Yang–Mills theory).
− the PI can serve as a *practical tool for the quantitative estimation* of characteristics of quantum systems because of the possibilities of reducing some dynamical variables by exact integration (e.g. in the polaron problem), changing space/time (for the inverse-square potential in QM) and the convenience of computer calculations for imaginary-time sum over paths [39], etc.

In this book we consider mainly the second aspect of the application of PIs in quantum physics.

8.3 The Feynman Path Integral Formalism in Quantum Mechanics

Feynman's method of path integrals [40] is one of the approaches in the formulation of a quantum theory. This method is based on the idea that the quantum motion of a particle may be considered as the interference of all quantum transitions along the possible "paths", each of which has a weight factor proportional to the exponential of the classical action taken over it. This formulation is equivalent to the more familiar method of canonical quantization, but much more convenient to use for theories with gauge fields, nonlocalities or special constraint conditions. Although this method is excellently described in many textbooks [41, 42], below we bring shortly main ideas and techniques of this formalism.

Let us consider nonrelativistic quantum mechanics with one degree of freedom. In the method of canonical quantization, we work in Hilbert space with operators $\hat{q}$ and $\hat{p}$ representing respectively the coordinate and conjugate momentum, as defined by the commutation relation

$$[\hat{p}, \hat{q}] = -i\hbar .$$

We denote the eigenstates of these operators by $|q'\rangle$ and $|p'\rangle$:

$$\hat{q}|q'\rangle = q'|q'\rangle, \qquad \hat{p}|p'\rangle = p'|p'\rangle,$$

and normalize them according to

$$\langle q''|q'\rangle = \delta(q'' - q'), \qquad \int\limits_{-\infty}^{\infty} dq'|q'\rangle\langle q'| = 1,$$

$$\langle p''|p'\rangle = 2\pi\hbar\delta(p'' - p'), \qquad \int\limits_{-\infty}^{\infty} \frac{dp'}{2\pi\hbar}|p'\rangle\langle p'| = 1,$$

$$\langle p|q\rangle = \exp\left(-\frac{i}{\hbar}pq\right).$$

The time evolution is given by the wave function

$$|q',t'\rangle = \exp(-\frac{i}{\hbar}t'\hat{H})|q'\rangle,$$

which corresponds to formal solution of the Schrödinger equation

$$i\hbar\frac{\partial}{\partial t}|q\rangle = \hat{H}|q\rangle,$$

where $\hat{H}$ is the time-independent Hamiltonian operator.

The dynamics of the system is completely specified by the transition amplitude governing the time evolution of the system:

$$\langle q'',t''|q',t'\rangle = \langle q''|\exp[-\frac{i}{\hbar}(t'' - t')\hat{H}]|q'\rangle.$$

We may regard $|q',t'\rangle$ as the eigenstate of the Heisenberg operator $\hat{q}(t)$ with the eigenvalue q':

$$\hat{q}(t) = \exp(\frac{i}{\hbar}t\hat{H})\hat{q}\exp(-\frac{i}{\hbar}t\hat{H}),$$

$$\hat{q}(t)|q',t'\rangle = q'|q',t'\rangle.$$

The object of the method of path integrals is to express the transition amplitude entirely in terms of a classical Hamiltonian $H(p,q)$, without reference to operators and states in Hilbert space. To proceed, Feynman divided the time evolution interval $(t'' - t')$ into N equal steps, and took the limit $N \to \infty$ later. Consider

$$\Delta t = (t'' - t')/N, \qquad \epsilon = \Delta t/\hbar.$$

We can write

$$\langle q'',t''|q',t'\rangle = \langle q''|e^{-i\hat{H}N\epsilon}|q'\rangle = \langle q''|(1 - i\epsilon\hat{H})^N|q'\rangle$$

$$= \int dq_1 \ldots dq_{N-1}\langle q''|(1 - i\epsilon\hat{H})|q_{N-1}\rangle \ldots \langle q_1|(1 - i\epsilon\hat{H})|q'\rangle.$$

The typical factor in the integrand can be rewritten as

$$\langle q_2|(1 - i\epsilon\hat{H})|q_1\rangle = \int\limits_{-\infty}^{\infty} \frac{dp_1}{2\pi\hbar} \langle q_2|p_1\rangle\langle p_1|(1 - i\epsilon\hat{H})|q_1\rangle\,,$$

and we define the classical Hamiltonian $H(p, q)$ by

$$\langle p|\hat{H}|q\rangle = \langle p|q\rangle H(p, q)\,. \tag{8.6}$$

This definition gives the usual connection between the classical and quantum mechanical Hamiltonians, provided $H(p, q)$ does not contain cross-products of p and q. Otherwise, $\hat{H}$ must be normally ordered. Using (8.6), we obtain

$$\langle q_2|(1 - i\epsilon\hat{H})|q_1\rangle = \int\limits_{-\infty}^{\infty} \frac{dp_1}{2\pi\hbar} \langle q_2|p_1\rangle\langle p_1|q_1\rangle[1 - i\epsilon H(p_1, q_1)]$$

$$= \int\limits_{-\infty}^{\infty} \frac{dp_1}{2\pi\hbar} e^{ip_1(q_2-q_1)/\hbar}[1 - i\epsilon H(p_1, q_1)]\,.$$

Hence

$$\langle q'', t''|q', t'\rangle = \int \frac{dq_1 dp_1}{2\pi\hbar} \cdots \frac{dq_{N-1} dp_{N-1}}{2\pi\hbar}$$

$$\cdot \exp\left[\frac{i}{\hbar}\sum_{n=0}^{N-1} p_n(q_{n+1} - q_n)\right] \prod_{n=1}^{N-1}[1 - i\epsilon H(p_n, q_n)]\,, \tag{8.7}$$

with the conditions

$$q_0 = q'\,, \qquad q_N = q''\,.$$

Now the key step comes in the development: we note that in (8.7) the factor $(1 - i\epsilon H)$ may be effectively replaced by $\exp(-i\epsilon H)$. The reason is that we assume that

$$\lim_{N\to\infty} \prod_{n=1}^{N}(1 + z_n/N) = \lim_{N\to\infty} \prod_{n=1}^{N} e^{z_n/N} = e^X\,,$$

$$X = \lim_{N\to\infty} \frac{1}{N}\sum_n z_n\,.$$

The replacement of $(1 - i\epsilon H)$ by $\exp(-i\epsilon H)$ has an uncertainty proportional to $\sim O(\epsilon^2)$. We can now express the amplitude in (8.7) as integrals over unitary amplitudes:

$$\langle q'', t''|q', t'\rangle = \int \frac{dq_1 dp_1}{2\pi\hbar} \cdots \frac{dq_{N-1} dp_{N-1}}{2\pi\hbar}$$

$$\times \exp\left\{\frac{i}{\hbar}\Delta t \sum_{n=1}^{N-1}\left[\frac{p_n(q_{n+1}-q_n)}{\Delta t}-H(p_n,q_n)\right]\right\}.$$

To approach the limit $N \to \infty$ (or $\Delta t \to 0$), we take the set of values $q_1, p_1, \ldots, q_{N-1}, p_{N-1}$ of certain functions $q(t), p(t)$, which may be discontinuous functions. Accordingly, we use the notation

$$t_n = t' + n\Delta t,$$

$$q_n = q(t_n),$$

$$p_n = p(t_n),$$

and write

$$\frac{q_{n+1}-q_n}{\Delta t} \to \dot{q}(t_n),$$

$$\sum_{n=1}^{N-1} f(t_n)\Delta t \to \int_{t'}^{t''} dt\, f(t)$$

as $\Delta t \to 0$. Thus, we can rewrite the transition amplitude in the form

$$\langle q'', t'' | q', t' \rangle = \int\!\!\int \delta q \, \delta p \exp\left\{\frac{i}{\hbar}\int_{t'}^{t''} dt[p\dot{q}-H(p,q)]\right\},$$

$$q(t') = q', \qquad q(t'') = q''. \tag{8.8}$$

This represents an integral over all paths $p(t)$ in momentum space, and all paths $q(t)$ in coordinate space, between the times t' and t'' with fixed values of the coordinates at the endpoints. The volume elements in path space are denoted by

$$\delta q = \prod_{n=1}^{N-1} dq(t_n), \qquad \delta p = \prod_{n=1}^{N-1} \frac{dp_n}{2\pi\hbar}.$$

The generalization of (8.8) to more than one degree of freedom is

$$\langle q_1'', \ldots, q_n''; t'' | q_1', \ldots, q_n'; t' \rangle$$

$$= \int \prod_\alpha \delta q_\alpha \delta p_\alpha \exp\left\{\frac{i}{\hbar}\int_{t'}^{t''} dt \left[\sum_\alpha p_\alpha \dot{q}_\alpha - H(p,q)\right]\right\}.$$

Feynman's formula for the transition amplitude is derived from (8.8) by restricting the classical Hamiltonian to the standard form

$$H(p,q) = \frac{p^2}{2m} + V(q). \tag{8.9}$$

One can perform the momentum integrations explicitly to obtain

$$\int \delta p \exp\left\{\frac{i}{\hbar}\int\limits_{t'}^{t''} dt[p\dot{q} - H(p,q)]\right\}$$

$$= \left[\frac{m}{2\pi\hbar}\right]^{d/2} \exp\left\{\frac{i}{\hbar}\int\limits_{t'}^{t''} dt L(q,\dot{q})\right\}, \tag{8.10}$$

where $L(q,\dot{q})$ is the classical Lagrangian:

$$L(q,\dot{q}) = \frac{1}{2}m\dot{q}^2 - V(q).$$

Substituting (8.10) into (8.8), we obtain

$$\langle q'', t''|q', t'\rangle = C \int \delta q \exp\left\{\frac{i}{\hbar}\int\limits_{t'}^{t''} dt L(q,\dot{q})\right\},$$

which is the Feynman formula. Here, C is a normalization constant which is usually infinite in the limit $N \to \infty$, but irrelevant to physical results because it is cancelled in matrix elements of the form

$$\langle q'', t''|O|q', t'\rangle / \langle q'', t''|q', t'\rangle.$$

For this reason, one need only define δq up to a multiplicative constant, which might even possibly be infinite.

Under the assumption in (8.9) one can show that

$$\langle q'', t''|T[\hat{q}(t_1)\dots\hat{q}(t_n)]|q', t'\rangle$$

$$= C \int \delta q\,[q(t_1)\dots q(t_n)] \exp\left\{\frac{i}{\hbar}\int\limits_{t'}^{t''} dt L(q,\dot{q})\right\}.$$

8.4 PIs as Solutions of Differential Equations

Feynman's starting-point was to give another interpretation of quantum mechanics. The equivalence of Feynman's PI approach with the standard CQ method was proved, in that it was shown that the Feynman PI satisfies the Schrödinger equation. In other words the solution of the Schrödinger equation can be represented in the form of a PI. Then a question arises: is it convenient to obtain the solution of the Schrödinger equation in PI form, considering this problem as the derivation of a differential equation? We would like to claim that in general the PI method can be considered a powerful method of solution of practically any linear differential equation. Now we want to demonstrate some functional technique to obtain the solution of the Schrödinger equation in the PI representation. The following computations are quite general and useful so that we would like to give full details here.

The wave function of a nonrelativistic spinless particle satisfies the following Schrödinger equation:

$$i\hbar \frac{\partial}{\partial t} \Psi(\mathbf{x}, t) = \hat{H} \Psi(\mathbf{x}, t) ,$$

$$\hat{H} = \frac{\mathbf{p}^2}{2m} + V(\mathbf{x}) = -\frac{\hbar^2}{2m} \frac{\partial^2}{\partial x^2} + V(\mathbf{x}). \qquad (8.11)$$

The noncommuting operators of coordinate $\mathbf{x}$ and momentum $\mathbf{p}$ satisfy the relation

$$[x_i, p_j] = \delta_{ij} .$$

The formal solution of the Cauchy problem, obeying the initial condition $\Psi(\mathbf{x}, t_o) = \Psi_o(\mathbf{x})$, is

$$\begin{aligned}
\Psi(\mathbf{x}, t) &= \exp[-\frac{i}{\hbar} \hat{H}(t - t_o)]\Psi(\mathbf{x}, t_o) \\
&= \int dx_o G(\mathbf{x}, t; \mathbf{x}_o, t_o)\Psi(\mathbf{x}_o, t_o) ,
\end{aligned}$$

where the Green function is defined as

$$G(\mathbf{x}, t; \mathbf{x}_o, t_o) = \Theta(t - t_o) \exp\left[-\frac{i}{\hbar} H(t - t_o)\right] \delta(\mathbf{x} - \mathbf{x}_o) \qquad (8.12)$$

and satisfies the following differential equation

$$\left(H - i\hbar \frac{\partial}{\partial t}\right) G(\mathbf{x}, t; \mathbf{x}_o, t_o) = -i\hbar\delta(\mathbf{x} - \mathbf{y})\delta(t - t_o) .$$

Our aim is to get the representation for the Green function in the form of a PI. For $t > t_o$ we can rewrite (8.12) in the following form:

$$G(\mathbf{x}, t; \mathbf{x}_o, t_o) = \exp[-\frac{i}{\hbar} H(t - t_o)]\delta(\mathbf{x} - \mathbf{x}_o)$$

$$= T_\tau \exp\left\{-i \int\limits_{t_o}^{t} d\tau \left[-\frac{1}{2m} \left(\frac{\partial}{\partial \mathbf{x}(\tau)}\right)^2 + V(\mathbf{x}(\tau))\right]\right\} \delta(\mathbf{x} - \mathbf{y}) . \qquad (8.13)$$

Here the "time-ordering" symbol T_τ is used for the time-ordering of noncommuting operators $\partial/\partial \mathbf{x}(\tau)$ and $\mathbf{x}(\tau)$. These operators commute under this symbol. Let us introduce the following integral representation:

$$\exp\left\{-\frac{i}{2} \int\limits_{0}^{\alpha} d\tau \left(\frac{\partial}{\partial \mathbf{x}(\tau)}\right)^2\right\}$$

$$= C \int \delta \nu \exp \left\{ -\frac{i}{2} \int_0^\alpha d\tau \nu^2(\tau) + \int_0^\alpha d\tau \nu(\tau) \frac{\partial}{\partial x(\tau)} \right\} .$$

Then, (8.13) may be rewritten as follows:

$$G(\mathbf{x}, t; \mathbf{x}_o, t_o)$$

$$= T_\tau \cdot C \int \delta \nu \exp \left\{ i \int_{t_o}^t d\tau \left[\frac{m\nu^2(\tau)}{2} + \nu(\tau) \frac{\partial}{\partial x_\tau} + V(\mathbf{x}_\tau) \right] \right\} \delta(\mathbf{x} - \mathbf{x}_o)$$

$$= C \int \delta \nu \exp \left\{ i \int_{t_o}^t d\tau \left[\frac{m\nu^2(\tau)}{2} + V \left(\mathbf{x} + \int_\tau^t d\tau' \nu(\tau') \right) \right] \right\} \qquad (8.14)$$

$$\times \delta \left(\mathbf{x} - \mathbf{x}_o + \int_{t_o}^t d\tau' \nu(\tau') \right) ,$$

where we have performed the action of the operators of displacement

$$\exp \left\{ \int_0^\alpha d\tau \nu(\tau) \frac{\partial}{\partial \mathbf{x}(\tau)} \right\}$$

and the time-ordering T_τ.

The normalization condition is

$$1 = C \int \delta \nu \exp \left\{ i \int_{t_o}^t d\tau \frac{m\nu^2(\tau)}{2} \right\} . \qquad (8.15)$$

It is convenient to introduce the notation

$$\mathbf{b}(\tau) = \mathbf{x} + \int_\tau^t d\tau' \nu(\tau'), \qquad \dot{\mathbf{b}}(\tau) = -\nu(\tau),$$

so that

$$\mathbf{b}(t_o) = \mathbf{x} + \int_{t_o}^t d\tau' \nu(\tau') = \mathbf{x}_o$$

according to the δ-function in (8.14). Then (8.14) takes the following compact form:

$$G(\mathbf{x}, t; \mathbf{x}_o, t_o) = C \int \delta \mathbf{b} \exp \left\{ i \int_{t_o}^t d\tau \left[\frac{m\dot{\mathbf{b}}^2(\tau)}{2} + V(\mathbf{b}(\tau)) \right] \right\} \qquad (8.16)$$

with $\mathbf{b}(t_o) = \mathbf{x}_o$ and $\mathbf{b}(t) = \mathbf{x}$.

Sometimes it is convenient to use another representation for the Green function (8.16). To this end in the integral (8.14) we can always represent any function as

$$\nu(\tau) = \nu_o + \dot{\rho}(\tau)$$

where ν_o does not depend on τ and

$$\int_{t_o}^{t} d\tau \dot{\rho}(\tau) = 0, \quad \text{or} \quad \rho(t) - \rho(t_o) = 0. \tag{8.17}$$

In particular, we can always choose $\rho(t_o) = \rho(t) = 0$. Then we have

$$\int_{t_o}^{t} d\tau \nu^2(\tau) = \int_{t_o}^{t} d\tau (\nu_o + \dot{\rho}(\tau))^2 = \int_{t_o}^{t} d\tau \dot{\rho}^2(\tau) + \nu_o^2(t - t_o).$$

Consider the measure

$$C\delta\nu \exp\left\{ i \int_{t_o}^{t} d\tau \frac{m\dot{\nu}^2(\tau)}{2} \right\}$$

$$= C_o C_\rho d\nu_o \delta\rho \exp\left\{ i \int_{t_o}^{t} d\tau \left[\frac{m\dot{\rho}^2(\tau)}{2} + \frac{m\nu_o^2(t - t_o)}{2} \right] \right\}.$$

The normalization condition (8.15) for the integration over $d\nu_o$ transforms to

$$1 = C_o \int d\nu_o \exp\left\{ i \frac{m\nu_o^2(t - t_o)}{2} \right\} = C_o \left(\frac{2\pi i}{m(t - t_o)} \right)^{3/2},$$

which gives

$$C_o = \left(\frac{m(t - t_o)}{2\pi i} \right)^{3/2}.$$

The Green function $G(\mathbf{x}, t; \mathbf{x}_o, t_o)$ in (8.14), defining the solution of the Schrödinger equation (8.11), is obtained in PI representation as follows:

$$G(\mathbf{x}, t; \mathbf{x}_o, t_o)$$

$$= C_o \int d\nu_o \exp\left\{ i \frac{m\nu_o^2(t - t_o)}{2} \right\} \delta\left(\mathbf{x} - \mathbf{x}_o + \nu_o(t - t_o)\right)$$

$$C_\rho \int \delta\rho \exp\left\{ i \int_{t_o}^{t} d\tau \left[\frac{m\dot{\rho}^2(\tau)}{2} + V\left(\mathbf{x} + \nu_o(t - \tau) - \rho(\tau)\right) \right] \right\}$$

$$= G_o(\mathbf{x}, t; \mathbf{x}_o, t_o) \cdot J(\mathbf{x}, t; \mathbf{x}_o, t_o), \tag{8.18}$$

where the first factor in (8.18) represents the "free" Green function, obtained as

$$G_o(\mathbf{x}, t; \mathbf{x}_o, t_o)$$

$$= C_o \int d\boldsymbol{\nu}_o \exp\left\{i\frac{m\nu_o^2(t - t_o)}{2}\right\} \delta\left(\mathbf{x} - \mathbf{x}_o + \boldsymbol{\nu}_o(t - t_o)\right)$$

$$= \left(\frac{m}{2\pi i(t - t_o)}\right)^{3/2} \cdot \exp\left\{i\frac{m(\mathbf{x} - \mathbf{x}_o)^2}{2(t - t_o)}\right\}. \tag{8.19}$$

Corrections to (8.19) due to interaction under a potential $V(\mathbf{x})$ are described by the function $J(\mathbf{x}, t; \mathbf{x}_o, t_o)$, written in the following PI form:

$$J(\mathbf{x}, t; \mathbf{x}_o, t_o) = C_\rho \int\limits_{\rho(t_o)=\rho(t)=0} \delta\rho \exp\left\{i\int\limits_{t_o}^{t} d\tau \left[\frac{m\dot{\rho}^2(\tau)}{2}\right.\right.$$

$$\left.\left. + V\left(\mathbf{x}_o \cdot \frac{t - \tau}{t - t_o} + \mathbf{x} \cdot \frac{\tau - t_o}{t - t_o} + \rho(\tau)\right)\right]\right\},$$

where

$$1 = C_\rho \int \delta\rho \exp\left\{i\int\limits_{t_o}^{t} d\tau \left[\frac{m\dot{\rho}^2(\tau)}{2}\right]\right\}.$$

Thus the Green function can be represented in two equivalent forms (8.16) and (8.18). These results will be useful for investigation of the wave dumping process in stochastically distributed media (see chapter 12).

8.5 Path Integrals in Quantum Field Theory

The above formulation of the PI approach obtained as a solution of the Schrödinger equation in QM may be extended to QFT straightforwardly; one merely allows the number of coordinates to become nondenumerably infinite. We shall illustrate this extension for the case of one boson field $\varphi(x)$. Generalization to more than one boson field is straightforward. The case of fermion fields [43] will not be considered here.

Let the classical Lagrangian density be

$$\mathcal{L}(x) = \frac{1}{2}\left[\left(\frac{\partial\varphi(x)}{\partial x_\mu}\right)^2 - m^2\varphi^2(x)\right] - g\varphi^4(x).$$

The S-matrix can be written as

$$S[\varphi] = T\exp\left\{-g\int dx\, \varphi^4(x)\right\}, \tag{8.20}$$

where the "time-regulator" T corresponds to Wick ordering. In particular, it can be written in the functional derivative form

$$T = \exp\left\{\frac{1}{2} \iint dx_1 dx_2\, D(x_1 - x_2)\frac{\delta^2}{\delta\varphi(x_1)\delta\varphi(x_2)}\right\} . \qquad (8.21)$$

In (8.21) the Green function in Minkowski space has the standard form

$$D(x) = \frac{1}{i}\int \left(\frac{dk}{2\pi}\right)^4 \frac{e^{ikx}}{m^2 - k^2 - i0} . \qquad (8.22)$$

Let us introduce the functional delta-function

$$\delta[\varphi - \phi] = \prod_x (\varphi(x) - \phi(x))$$

$$= C_1 \int \delta a \exp\left\{i\int dx\, a(x)(\varphi(x) - \phi(x))\right\} ,$$

where C_1 is an appropriate normalization constant. Then (8.20) reads

$$S[\varphi] = T\int \delta\phi\, \delta[\varphi - \phi]\exp\left\{-ig\int dx\, \phi^4(x)\right\}$$

$$= C\int \delta\phi \int \delta a \exp\left\{-\frac{1}{2}\iint dx_1 dx_2\, a(x_1)D(x_1 - x_2)a(x_2)\right\} \qquad (8.23)$$

$$\times \exp\left\{i\int dx\, a(x)(\varphi(x) - \phi(x)) - ig\int dx\, \phi^4(x)\right\} .$$

Notice that the inner PI in (8.23) over δa is of a Gaussian-type. Integrating explicitly over δa one gets

$$S[\varphi] = C\int \delta\phi\, \exp\left\{-ig\int dx\, \phi^4(x)\right\}\exp\left\{-\frac{1}{2}\iint dx_1 dx_2\right.$$

$$(\varphi(x_1) - \phi(x_1))\, D^{-1}(x_1 - x_2)\,(\varphi(x_2) - \phi(x_2))\} ,$$

where the normalization constant C is defined by the condition

$$S[\varphi]|_{g=0} = 1 .$$

The operator $D^{-1}(x_1 - x_2)$ corresponding to the Green function in (8.22) is

$$D^{-1}(x_1 - x_2) = -i(\Box - m^2)\delta(x_1 - x_2) .$$

Transforming the variable of integration

$$\phi(x) \rightarrow \phi(x) + \varphi(x)$$

we obtain the PI representation for the normal symbol for the S-matrix as follows:

$$S[\varphi] =: C \int \delta\phi \exp\left\{i \int dx\right.$$

$$\left.\left[\frac{1}{2}\phi(x)(\Box - m^2)\phi(x) - g\left(\phi(x) + \varphi(x)\right)^4\right]\right\} : . \tag{8.24}$$

In particular, the vacuum-energy density in the theory under consideration is produced as

$$E_{\text{vac}} = i \lim_{V \to \infty} \frac{1}{V} \ln S_V[0],$$

$$S_V[0] = \int \delta\phi \exp\left\{i \int_V dx \left[\frac{1}{2}\phi(x)(\Box - m^2)\phi(x) - g\phi^4(x)\right]\right\}. \tag{8.25}$$

In representations (8.24) and (8.25) one can go to the Euclidean metrics

$$x_o \to -ix_4,$$

and then these formulae look like

$$S[\varphi] =: C \int \delta\Phi \exp\left\{\int dx\right.$$

$$\left.\left[\frac{1}{2}\Phi(x)(\Box - m^2)\Phi(x) - g\left(\Phi(x) + \varphi(x)\right)^4\right]\right\} : ,$$

$$E_{\text{vac}} = - \lim_{\Omega \to \infty} \frac{1}{\Omega} \ln S_\Omega[0], \tag{8.26}$$

$$S_\Omega[0] = \int \delta\Phi \exp\left\{\int_\Omega dx \left[\frac{1}{2}\Phi(x)(\Box - m^2)\Phi(x) - g\Phi^4(x)\right]\right\},$$

where $\Phi(x) = \phi(-ix_4, \mathbf{x})$ and Ω is a volume in Euclidean space.

In chapter 11 this representation will be used for investigation of the models $g\varphi_2^4$ and $g\varphi_3^4$.

9. The Gaussian Equivalent Representation of Functional Integrals

The main content of this chapter is the development of the method of the **Gaussian equivalent representation (GER)** of PIs and its application to the investigation of the ground state (vacuum) of various QFT and QM models in order to study the nonperturbative phenomena such as the strong coupling regime, bound state formation, phase structure and phase transitions. In this chapter we give the general description of the GER method in QFT and its special implementation into nonrelativistic QM.

The GER method generalizes both the variational and perturbation techniques, but in contrast the latters, it is free from the defects of these methods and can be considered as the next step in the development of approximate calculation methods of PIs. This method is characterized by the high accuracy of the lowest approximation, which can be obtained by simple and rapid calculations. It also gives a regular prescription for calculation of the highest order corrections to the lowest approximation.

This method is efficient for considering strong coupling regimes of QFT models with ultraviolet (UV) divergences and for theories with non-Hermitean and nonlocal action (stochastic and dissipative processes), where the traditional variational and perturbation techniques are not applicable.

9.1 A General Description of the Method

Considering many theoretical problems in statistical physics, quantum field theory and mathematical physics one deals with a class of functional integrals defined on a Gaussian measure like (8.1). Let us consider the generating functional $Z_\Gamma(g)$ in the one-component scalar field theory as follows:

$$
\begin{aligned}
Z_\Gamma(g) &= C_o \int \delta\varphi \exp\left\{-\frac{1}{2}(\varphi\, D_o^{-1}\, \varphi) + g\, W_o[\varphi]\right\} \\
&= \int d\sigma_o \exp\left\{g\, W_o[\varphi]\right\} .
\end{aligned}
\tag{9.1}
$$

Here we have introduced the following notation for the Gaussian measure:

$$
d\sigma_o = C_o \delta\varphi \exp\left\{-\frac{1}{2}(\varphi D_o^{-1}\varphi)\right\}
\tag{9.2}
$$

$$= \frac{1}{\sqrt{\det D_o}} \prod_x d\varphi(x) \exp\left\{-\frac{1}{2}\iint_\Gamma dx dy\, \varphi(x) D_o^{-1}(x,y)\varphi(y)\right\}.$$

The Gaussian measure is normalized in such way that $\int d\sigma_o = 1$. The integration in (9.1) is performed over functions $\varphi(x)$ defined on a region $\Gamma \subseteq \mathbf{R}^d (d = 1, 2, \ldots)$. Usually the region Γ is chosen as a multidimensional box:

$$\Gamma = \{x : a_j \le x_j \le b_j, \quad j = 1, \ldots, d\}.$$

A differential operator $D_o^{-1}(x,y)$ is defined on functions $\varphi(x)$ with appropriate boundary conditions. For example, the operator

$$D_o^{-1}(x,y) = \left(-\frac{\partial^2}{\partial x^2} + m_o^2\right)\delta(x-y)$$

acts on functions satisfying some periodic boundary conditions. The corresponding Green function $D_o(x,y)$ satisfies the equation

$$\int_\Gamma dy\, D_o^{-1}(x,y)\, D_o(y,z) = \delta(x-z)$$

and ensures definite boundary conditions.

The parameter g is a coupling constant. The interaction functional $W_o[\varphi]$ can be written in a general form:

$$W_o[\varphi] = \int d\mu_a\, e^{i(a\varphi)},$$

where we have introduced the notation

$$(a\varphi) = \int_\Gamma dy\, a(y)\, \varphi(y),$$

and $d\mu_a$ is a functional measure. For example, for the local theory φ^4 we have

$$W_o[\varphi] = \int_\Gamma dx\, \varphi^4(x) = \int_\Gamma dx \left(\frac{d}{d\xi}\right)^4 e^{i\xi\varphi(x)}|_{\xi=0}$$

$$= \int_\Gamma dx \left(\frac{d}{d\xi}\right)^4 \exp\left\{i\xi\int_\Gamma dy\, \varphi(y)\,\delta^d(x-y)\right\}|_{\xi=0},$$

and for a potential $U[\varphi(x)]$ having the Fourier transform one can write

$$W_o[\varphi] = \int_\Gamma dx\, U[\varphi(x)]$$

$$= \int_{\Gamma} dx \int \frac{dk}{2\pi}\, \tilde{U}(k)\, \exp \left\{ i \int_{\Gamma} dy\, k\, \varphi(y)\, \delta(x-y) \right\}.$$

The PI in representation (9.1) is well defined as a perturbation expansion over the coupling constant g. Thus physically acceptable results can be obtained only in the weak coupling regime $g \ll 1$. In this case the Gaussian measure $d\sigma_o$ (9.2) gives the main contribution in the PI and corrections can be calculated by using a perturbation expansion.

The problem is to give a representation of this integral in the strong coupling regime [44]. Our idea is that the PI beyond the perturbation regime remains of the Gaussian type but with another Green function in the Gaussian measure. In other words, we want to obtain a representation in which all main contributions of the strong interaction are concentrated in the Gaussian measure.

Let us perform the following transformations of the integral (9.1):

$$\begin{aligned}
\varphi(x) &\longrightarrow \varphi(x) + b(x), \\
D_o^{-1}(x,y) &\longrightarrow D^{-1}(x,y),
\end{aligned} \qquad (9.3)$$

where $b(x)$ is an arbitrary function and $D(x,y)$ is an appropriate Green function of the differential operator D^{-1}:

$$\int_{\Gamma} dy\, D^{-1}(x,y)\, D(y,z) = \delta(x-z),$$

providing the same boundary conditions.

Transformations (9.3) represent in a certain sense a functional analogue of standard canonical transformations made in the Hamiltonian formalism. The functional integral (9.1) takes the form

$$Z_\Gamma(g) = \sqrt{\det \frac{D}{D_o}}\, \exp \left\{ -\frac{1}{2}\, (b\, D_o^{-1} b) \right\} \cdot \int d\sigma\, \exp\left\{ g W_1[\varphi, b, D] \right\}, \qquad (9.4)$$

where

$$d\sigma = C\, \delta\varphi \exp \left\{ -\frac{1}{2}\, (\varphi\, D^{-1}\, \varphi) \right\},$$

$$g W_1[\varphi, b, D] = g\, W[\varphi + b] - (b\, D_o^{-1} \varphi) - \frac{1}{2}\, (\varphi\, [\, D_o^{-1} - D^{-1}]\, \varphi),$$

with the normalization condition $\int d\sigma = 1$.

The tadpole Feynman diagrams give the main quantum contributions to the background energy of the system under consideration or, in other words, into the formation of the background state or vacuum. The mathematical problem is to take them into account correctly. In the quantum theory (see part I) the main divergences given by tadpole vacuum diagrams are efficiently eliminated from consideration if the normally ordered product of operators is introduced into the interaction Hamiltonian. Following this the interaction functional in (9.4) should be written in the normally ordered form. Thus we should introduce in W_1 the concept of the normal product according to the given Gaussian measure $d\sigma$. This can be done in the following way

$$: e^{i(a\varphi)} := e^{i(a\varphi)} e^{\frac{1}{2}(aDa)}.$$

This definition leads to the following relations:

$$\int d\sigma : e^{i(a\varphi)} := 1 \qquad \int d\sigma : \varphi(x_1)\ldots\varphi(x_n) := 0.$$

After these transformations the functional in the integrand can be rewritten as

$$gW_1 = g\int d\mu_a e^{i(ab)-\frac{1}{2}(aDa)} : e_2^i(a\varphi) : \tag{9.5}$$

$$+ \left[g\int d\mu_a e^{i(ab)-\frac{1}{2}(aDa)} - \frac{1}{2}\left([D_o^{-1}-D^{-1}]D\right)\right]$$

$$+ \left[g\int d\mu_a e^{i(ab)-\frac{1}{2}(aDa)} \cdot i(a\varphi) - (bD_o^{-1}\varphi)\right]$$

$$- \frac{1}{2} : \left[g\int d\mu_a e^{i(ab)-\frac{1}{2}(aDa)} \cdot (a\varphi)^2 + (\varphi[D_o^{-1}-D^{-1}]\varphi)\right] :,$$

where $e_2^z = e^z - 1 - z - \frac{z^2}{2}$.

Now we introduce the concept of the "correct form" of the action in the PI as it was done for the total Hamiltonian in part I. We demand that the linear and quadratic terms on the integration variables $\varphi(x)$ should be absent in the interaction functional W_1 in (9.5). This requirement is argued in the same way. The system under consideration should be near its equilibrium point so that any linear terms on the variable $\varphi(x)$ must be absent. The quadratic configurations $\sim \varphi^2$ determine the Gaussian oscillator character of the equilibrium point and all of them are concentrated in the Gaussian measure $d\sigma$ only. Therefore they should not appear in the interaction functional and

$$W_I \sim O(\varphi^3) \qquad \text{for} \qquad \varphi \to 0.$$

Thus the "correct form" requirement is satisfied if the following equations are held:

$$g\int d\mu_a\, i a(x)\, e^{i(ab)-\frac{1}{2}(aDa)} - \int_\Gamma dy\, D_o^{-1}(x,y)\, b(y) = 0, \quad nonum \tag{9.6}$$

$$g\int d\mu_a\, a(x)\, a(y)\, e^{i(ab)-\frac{1}{2}(aDa)} \tag{9.7}$$

$$+ D_o^{-1}(x,y) - D^{-1}(x,y) = 0.$$

These equations provide the removal of the linear and quadratic terms from the interaction functional. Let us introduce the following functional and its correlation functions:

$$\hat{W}[b] = \int d\mu_a \exp\left\{i(ab) - \frac{1}{2}(aDa)\right\},\qquad(9.8)$$

$$w_n(x_1, ..., x_n) = \frac{\delta^n}{\delta b(x_1) \cdot ... \cdot \delta b(x_n)}\hat{W}[b].$$

Equations (9.7) can be written in the form

$$b(x) = g\int_\Gamma dy D_o(x, y)w_1(y),\qquad(9.9)$$

$$D(x_1, x_2) = D_o(x_1, x_2) + g\iint_\Gamma dy_1 dy_2\, D_o(x_1, y_1)w_2(y_1, y_2)D(y_2, x_2).$$

These equations determine the new Green function $D(x_1, x_2)$ and the function $b(x)$ in (9.5).

Finally the new representation for the PI in (9.1) can be rewritten in the form

$$Z_\Gamma(g) = \exp\{E_o\}\int d\sigma \exp\{g\, W_I[\varphi]\},\qquad(9.10)$$

where

$$E_o = \frac{1}{2}\ln\det\left(\frac{D}{D_o}\right) - \frac{1}{2}(bD_o^{-1}b) - \frac{1}{2}\left([D_o^{-1} - D^{-1}]D\right) + g\hat{W}[b],$$

$$gW_I[\varphi] = g\int d\mu_a\, e^{i(ab)-\frac{1}{2}(aDa)} : e_2^{i(a\varphi)} : .\qquad(9.11)$$

The representation of the interaction functional in the normal product form means that

$$\int d\sigma\, W_I[\varphi] = 0.$$

The function E_o defines the "energy" of the zero approximation. Next corrections to the leading term in (9.10) can be calculated by using a perturbation expansion over the new interaction functional W_I.

It should be stressed that representations (9.1) and (9.10) are equivalent. Therefore the mathematical object $Z_r(g)$ has at least two different representations (9.1) and (9.10). In principle other representations may exist if (9.9) has a more distinct solution. In this case we give preference to the representation in which the perturbation corrections connected with gW or gW_I are minimal for given parameters.

All our transformations are valid for real and complex functions and functionals in the PI.

In the case of a real PI representation (9.10) leads to the following conclusion. Using the Jensen's inequality one can get

$$Z_\Gamma(g) \geq \exp\{E_o\},\qquad(9.12)$$

so that E_o defines the lowest estimation for our PI.

On the other hand, one can easily check that (9.9) defines the minimum of the functional E_o. Thus inequality (9.12) is the variational estimate of the initial PI. Moreover, representation (9.10) makes it possible to calculate the perturbation corrections to E_o by developing the functional integral in (9.10) over W_I.

9.2 The GER Method for Quantum Statistics

In this section we develop the main techniques of the GER method especially for the case of quantum statistics (QS). In other words, we deal with integrals where the field variable is the coordinate of a particle $r(t)$ which is parameterized by the one dimensional parameter t. For simplicity one can choose the symmetrical interval $-T < t < T$. The parameter T is connected with the "time" in QM or the inverse "temperature" $2T = \beta$ in QS.

The partition function plays an important role in QS. For a wide class of quantum mechanical and quantum statistical problems describing the interaction of a quantum particle with a field or the propagation of waves and quantum particles through a medium with random or stochastic admixtures the partition function can be represented in the form of a PI of the following general type:

$$Z_T(g) = C_o \int_{\mathbf{r}(-T)=\mathbf{r}(T)} \delta r \exp \left\{ -\frac{1}{2} \int_{-T}^{T} dt\ \dot{\mathbf{r}}^2(t) \right.\qquad(9.13)$$

$$\left. +\frac{g}{2} \iint_{-T}^{T} dt ds\ V(\mathbf{r}(t) - \mathbf{r}(s); t - s) \right\}.$$

The standard normalization is $Z_T(0) = 1$.

The integration in (9.13) is performed over all "paths" in a d-dimensional space satisfying periodic boundary conditions.

The kinetic term in the Gaussian measure can be written in the form

$$\int_{-T}^{T} dt\ \dot{\mathbf{r}}^2(t) = \iint_{-T}^{T} dt ds\ \mathbf{r}(t) D_o^{-1}(t, s)\mathbf{r}(s),$$

$$D_o^{-1}(t, s) = -\frac{\partial^2}{\partial t^2}\delta(t - s).\qquad(9.14)$$

The Green function $D_o(t, s)$ corresponding to the differential operator $D_o^{-1}(t, s)$ and satisfying the periodic boundary conditions is

$$D_o(t, s) = -\frac{1}{2}|t - s| - \frac{ts}{2T} . \tag{9.15}$$

In the following we shall be interested in considering of the limit $T \to \infty$. Hence the parameter T is supposed to be asymptotically large. In this limit we get

$$D_o(t, s) \xrightarrow{T \to \infty} D_o(t - s) = -\frac{1}{2}|t - s| . \tag{9.16}$$

The Fourier transform of this Green function is

$$\tilde{D}_o(p^2) = \int_{-\infty}^{\infty} dt e^{ipt} D_o(t) = \frac{1}{2}\left[\frac{1}{(p + i0)^2} + \frac{1}{(p - i0)^2}\right] \to \frac{1}{p^2} .$$

The normalization condition $Z_T(0) = 1$ leads to

$$C_0 = \frac{1}{\sqrt{\det D_o}} = \prod_p \sqrt{p^2} .$$

In QM and QS, the potentials describing the influence of a field interaction or medium on a quantum particle usually have the general form $V(\mathbf{r}-\mathbf{r}';t-t')$. So, we will consider this class of potentials further. The potential $V(\mathbf{r}(t) - \mathbf{r}(s); t - s)$ in (9.13) is supposed to have the Fourier representation

$$V(\mathbf{R}(t, s); t - s) = \int \left(\frac{d\mathbf{k}}{2\pi}\right)^d \tilde{V}(\mathbf{k}; t - s) \, e^{i\mathbf{k}\mathbf{R}(t,s)}$$

$$= \int d\mathcal{K}(\mathbf{k}; t - s) \, e^{i\mathbf{k}\mathbf{R}(t,s)} ,$$

$$d\mathcal{K}(\mathbf{k}; t - s) = \left(\frac{d\mathbf{k}}{2\pi}\right)^d \tilde{V}(\mathbf{k}; t - s) , \quad \mathbf{R}(t, s) = \mathbf{r}(t) - \mathbf{r}(s) . \tag{9.17}$$

Thus the initial PI in (9.13) can be rewritten as

$$Z_T(g) = \int d\sigma_o \exp\{gW_o[\mathbf{r}]\} , \tag{9.18}$$

where

$$d\sigma_o = C_o \delta \mathbf{r} \exp\left\{-\frac{1}{2}\iint_{-T}^{T} dt ds \, \mathbf{r}(t) D_o^{-1}(t, s)\mathbf{r}(s)\right\} ,$$

$$gW_o[\mathbf{r}] = \frac{g}{2}\iint_{-T}^{T} dt ds \int d\mathcal{K}(\mathbf{k}; t - s) \, e^{i\mathbf{k}\mathbf{R}(t,s)}$$

and the normalization condition is $\int d\sigma_o = 1$.

Now we are ready to apply the GER method to this PI. Note that for spherically symmetric potentials $V(\mathbf{r}; t)$ of type (9.17) we do not need to introduce the function $\mathbf{b}(t)$, i.e., we choose $\mathbf{b}(t) = 0$. According to the GER method a new Gaussian measure should be introduced into the integral (9.18) as follows:

$$d\sigma = C\delta\mathbf{r}\exp\left\{-\frac{1}{2}\iint_{-T}^{T} dt\,ds\ \mathbf{r}(t)D^{-1}(t-s)\mathbf{r}(s)\right\} . \qquad (9.19)$$

The normalization constant C is chosen in such a way that

$$\int d\sigma = 1 \quad \text{or} \quad C = \frac{1}{\sqrt{\det D}} .$$

Then, we introduce the "normally-ordered" form of potential (9.17) in the following way:

$$e^{i\mathbf{k}\mathbf{R}(t,s)} =: e^{i\mathbf{k}\mathbf{R}(t,s)} : \exp[-\mathbf{k}^2 F(t-s)] , \qquad (9.20)$$

where

$$\int d\sigma\ R_i(t,s)R_j(t,s) = 2\delta_{ij}\,F(t-s) , \qquad F(t-s) = D(0) - D(t-s) .$$

In particular the following relations are valid:

$$\int d\sigma : e^{i\mathbf{k}\mathbf{R}(t,s)} := 1 ,$$

$$r_i(t)r_j(s) =: r_i(t)r_j(s) : +\delta_{ij}\,D(t-s) , \qquad i,j = 1\ldots N ,$$

The functional $\hat{W}[b]$ in (9.8) becomes

$$\hat{W}[b] = \frac{1}{2}\iint_{-T}^{T} dt\,ds$$

$$\int d\mathcal{K}(\mathbf{k}; t-s)\ \exp[-\mathbf{k}^2 F(t-s)]\ e^{i\mathbf{k}(\mathbf{b}(t)-\mathbf{b}(s))} . \qquad (9.21)$$

Its second correlation function is

$$\frac{\delta^2}{\delta b_i(t)\delta b_j(s)}\hat{W}[b]|_{b=0} = \delta_{ij}\ w_2(t-s)$$

$$= -\delta_{ij}\left[\delta(t-s)\int_{-\infty}^{\infty} d\tau\Phi(\tau) - \Phi(t-s)\right] ,$$

where

$$\Phi(\tau) \;=\; \frac{1}{d} \int d\mathcal{K}(\mathbf{k};\tau)\mathbf{k}^2 \, \exp[-\mathbf{k}^2 F(\tau)] \, . \tag{9.22}$$

Let us define the function

$$\tilde{\Sigma}(p^2) = -\tilde{w}_2(p^2) = \int\limits_{-\infty}^{infty} d\tau \, [1 - \cos(p\tau)] \, \Phi(\tau) \, . \tag{9.23}$$

Equation (9.7) defining the correct form of the interaction functional by excluding from it terms proportional to $\sim \mathbf{r}^2$ now looks like

$$\int\limits_{-T}^{T}\!\!\int dt \, ds \left\{ : \mathbf{r}(t)[D_o^{-1}(t,s) - D^{-1}(t,s)]\mathbf{r}(s) : \right.$$

$$\left. +g \int d\mathcal{K}(\mathbf{k};t-s)\exp[-\mathbf{k}^2 F(t,s)] : \frac{(\mathbf{k}\mathbf{R}(t,s))^2}{2} : \right\} = 0 \, ,$$

which leads to

$$D_o^{-1}(t-s) - D^{-1}(t-s)$$

$$+g \left\{ \delta(t-s) \int\limits_{-\infty}^{\infty} d\tau \, \Phi(\tau) - \Phi(t-s) \right\} = 0 \, . \tag{9.24}$$

This equation defines the unknown Green function $D(\tau)$. It can be rewritten in the form

$$\tilde{D}(p^2) = \tilde{D}_0(p^2) - \tilde{D}_0(p^2)g\tilde{\Sigma}(p^2)\tilde{D}(p^2) \, ,$$

or

$$\tilde{D}(p^2) = \frac{1}{p^2 + g\tilde{\Sigma}(p^2)} \, ,$$

which gives

$$F(\tau) \;=\; D(0) - D(\tau) = \int\limits_{0}^{\infty} \frac{dp}{\pi} \frac{1 - \cos(p\tau)}{p^2 + g\tilde{\Sigma}(p^2)} \, . \tag{9.25}$$

Equations (9.22)–(9.25) define the functions $D(\tau)$, $\Phi(\tau)$, $F(\tau)$ and $\Sigma(p^2)$. For the asymptotic cases of weak ($g \to 0$) and strong ($g \to \infty$) interaction regimes, these equations may admit analytic solutions because we are interested only in their behaviour within an accuracy of the first few leading order terms such as $\sim g, g^2$ or, $\sim 1/g, 1/g^2$. In general, these are not solvable analytically, being nonlinear integral equations over functionals, and their solutions may be obtained by developing some numerical techniques. For example, the fixed-point method of consequent iterations can be used. Starting from a guessed function $\tilde{\Sigma}_o(p^2)$ we can calculate iterations:

$$\tilde{\Sigma}_{n+1}(p^2) = \frac{1}{d} \int\limits_{-\infty}^{\infty} d\tau \, [1 - \cos(p\tau)] \int d\mathcal{K}(\mathbf{k}; \tau) \, \mathbf{k}^2 \exp[-\mathbf{k}^2 F_n(\tau)] \, ,$$

+1
$$F_n \ (\tau) = \int\limits_{0}^{\infty} \frac{dp}{\pi} \frac{1 - \cos(p\tau)}{p^2 + g\tilde{\Sigma}_n(p^2)} \, . \tag{9.26}$$

This procedure can be developed for the numerical solution of (9.23) and (9.25). In this case, however, the initial guessed functions $F_o(\tau)$ and $\tilde{\Sigma}_o(p^2)$ should be chosen reasonably, i.e. the iteration process (9.26) has to converge to solutions

$$\tilde{\Sigma}(p^2) = \tilde{\Sigma}_\infty(p^2) = \lim_{n\to\infty} \tilde{\Sigma}_n(p^2),$$
$$F(t) = F_\infty(t) = \lim_{n\to\infty} F_n(t).$$

For a reasonable choice of guessed functions, it is useful to investigate the asymptotics of solutions of (9.23) and (9.25). An example of the analytic and numerical solution of (9.23) and (9.25) is given in chapter 10 within the polaron problem.

Substitution of (9.19)–(9.25) into (9.18) and the requirement that the new interaction functional to be written in "the correct form" (see section 9.1) leads to the following new representation of the initial PI:

$$Z_T(g) = \exp(-2TE(g)) \cdot J_T(g) \, , \tag{9.27}$$

$$J_T(g) = \int d\sigma \exp\{gW_I[\mathbf{r}]\} \, ,$$

where the interaction functional is

$$gW_I[\mathbf{r}] = \frac{g}{2} \int\limits_{-T}^{T}\!\!\int dt\,ds$$

$$\int d\mathcal{K}(\mathbf{k}; t-s) \, \exp[-\mathbf{k}^2 F(t-s)] : e_2^{i\mathbf{k}\mathbf{R}(t,s)} : . \tag{9.28}$$

The function $E(g)$ is the "leading-order energy" or the energy in the zero approximation, given by

$$2TE(g) = \frac{1}{2} \ln \det \left(\frac{D_o}{D}\right) + \frac{1}{2} \int\limits_{-T}^{T}\!\!\int dt\,ds \, [D_o^{-1}(t-s) - D^{-1}(t-s)]D(t-s)$$

$$+ \frac{g}{2} \int\limits_{-T}^{T}\!\!\int dt\,ds \int d\mathcal{K}(\mathbf{k}; t-s) \, \exp[-\mathbf{k}^2 F(t-s)],$$

and in the limit $T \to \infty$ it becomes

$$E(g) = d \int\limits_0^\infty \frac{dp}{2\pi} \left[\ln \left(\frac{\tilde{D}_o(p^2)}{\tilde{D}(p^2)} \right) + p^2 \tilde{D}(p^2) - 1 \right] \qquad (9.29)$$

$$+ \frac{g}{2} \int\limits_{-\infty}^\infty d\tau \int d\mathcal{K}(\mathbf{k}; \tau) \exp[-\mathbf{k}^2 F(\tau)].$$

Thus the GER of the initial PI in (9.18) is defined by (9.27)–(9.29). For a given potential $V(\mathbf{r})$ we have the pure mathematical problem of solving the (9.23) and (9.25) and finding the Green function $D(t, s)$. Then we can compute the leading-order energy $E(g)$ (9.29) and the highest corrections to it by perturbation calculations over the new interaction functional W_I (9.28).

Below, in the following chapters of this part of the book, we apply the GER method to different problems of theoretical physics:

– the problem of the polaron in QS,
– the phase transition phenomenon in the QFT model,
– the solution of the wave differential equation.

Each of these subjects reflects a feature of the GER method. High accuracy is reached in calculation of the ground state energy of the d-dimensional Fröhlich polaron. One effective scheme of mass renormalization in the $g\varphi_{2,3}^4$ theory, suggested within the GER method, leads to the correct prediction of the nature of phase transitions in this theory. Finally an estimate of non-Hermitean path integral arising in the theory of wave propagation in media with Gaussian noise is obtained. The reduction of the initial PI to the new representation generates a certain constraint equation determining this state and one should give preference to the representation, that is efficient for solving a given task.

10. The Polaron Problem

The study of the physical properties of a particle interacting with a quantum medium is common to many branches of physics. A classic example of this kind is the Fröhlich model of the **polaron**, – an electron moving with the polarization distortion of ions in an crystal. Polaron's popularity as a model is due to its similarity to many field-theoretical constructions where bosons couple linearly to fermions (the meson-nucleon interactions inside nuclei, the "dressing" of quarks in the nonperturbative vacuum of QCD et cet.). The polaron problem is treated most straightforwardly in the PI formalism which allows one to reduce this problem to an effective one-particle task and, leads to new results not given by other conventional techniques. However, despite its long history and importance, the exact solution of the Fröhlich Hamiltonian is still lacking due to a high nonlocality (in time) and a Coulomb-like singularity in the polaron action. The application of the GER method to the d-dimensional polaron in this chapter results in highly accurate estimations of the main quasi-particle characteristic of the polaron – its ground-state energy.

10.1 Introduction

The polaron problem embraces a wide range of questions concerning the conductivity electrons in polar crystals [45, 46]. The first field-theoretical formulation of polaron theory was proposed by Fröhlich [47, 48] to describe the interaction of a single band electron with phonons – quanta associated with the longitudinal optical branch of lattice vibration. Since that time, the Fröhlich polaron model has attracted interest as a testing ground of various nonperturbative methods [49, 50, 51, 52], and it has influenced development in other areas of physics [53, 54] in quantum physics. We refer readers to conference proceedings [55, 56, 57, 58] and review articles in [59, 60, 61], which cover a wide field of investigations on this subject.

The main quasi-particle characteristics of the polaron are its ground-state energy (GSE) $E(\alpha)$, effective mass m_{eff} and life-time. These depend on the electron–phonon coupling constant α. The problem of finding the GSE of the Fröhlich Hamiltonian assumes considerable significance because one can suppose that in comparing two approximate methods the one giving the better

$E(\alpha)$ will likely give the better m_{eff}, which can be measured directly [62]. Second, experiments on the ionization energy of bound polarons [63] require theoretical estimation of the free-polaron GSE. Therefore, it is important to have good values for the polaron GSE.

Historically, the GSE of the polaron was investigated in the strong coupling regime [64, 65], weak [66, 67] and intermediate [68, 69] coupling regimes by using different methods. The first attempt to construct polaron theory, valid for arbitrary values of α, was made by Feynman [40] within the PI formalism. The Feynman approach to the polaron has an advantage because of the adequate elimination of the phonon coordinates, and as a consequence, the polaron problem is reduced to an effective one-particle problem with retarded interaction in real three-dimensional ($d = 3$) space. Besides, the path integral formalism allows one to build a class of exactly solvable models, corresponding to quadratic functionals. Then one can use these functionals as approximants for variational estimations. As a result, Feynman's PI approach has given good upper bound estimations of $E_o(\alpha)$ over the entire range of α in a unified way.

First, there arises the question, whether Feynman's estimation of the polaron GSE can be improved by introducing some trial actions, more general than the quadratic action with two variational parameters, used in [40]? This question, in particular, has been studied within the variational ansatz [70, 71, 72]. But giving variational answers, it could not estimate the next corrections to the obtained values and besides are given only for $d = 3$.

Second, the polaron problem has been traditionally investigated for $d = 3$. In recent years, however, polaron effects have been observed in low-dimensional systems [73, 58], and certain physical problems have been mapped into a two-dimensional ($d = 2$) polaron theory [74]. The possibility that an electron may be trapped at the surface of a dielectric material has been attracting much interest [57].

In the present chapter, we want to investigate the GSE of the polaron in arbitrary space dimensions $d > 1$ by using the GER method. We try not only to *improve Feynman's upper bound* in our lowest approximation, but also to estimate the next corrections that allow one to test the accuracy and reliability of the obtained values of $E_o(\alpha)$. Moreover, there is an indication [75] that taking into account the second-order correction could result in the *new lower bound* of the polaron GSE for $d > 1$. To this end, we transform Feynman's PI of the polaron to the representation construct so that all the quadratic part of the polaron's action is concentrated in the Gaussian measure of PI, which is defined by certain equations.

The organization of this chapter is as follows. In section 10.2 we formulate a generalization of Feynman's PI to the case of arbitrary space dimension $d > 1$. In section 10.3 we have obtained a convenient representation for this path integral. The necessary equations defining the explicit polaron correlation function and the measure of PI in this representation are obtained here. In section 10.4 the leading-order term of the polaron GSE, which represents upper bounds is obtained in d–dimensions for arbitrary $\alpha > 0$. In section 10.5, we show that a simple approximation of this leading term leads to the reproduction of Feynman's variational upper bound to the polaron GSE in d–dimensions. The exact values of our leading term differ from that obtained by Feynman and improve it for $d = 2$ and $d = 3$. In section 10.6, we have obtained next corrections to the leading term of polaron self-energy. In section 10.7, we have obtained scaling relations connecting our key equations between two- and three-dimensional spaces. The numerical results, obtained within our method in the whole range of $\alpha = 0 \div \infty$ for $d = 2$ and $d = 3$, are given in section 10.8 compared with known data in each case.

10.2 The Polaron Path Integral Formulated in d-dimensions

The Fröhlich longitudinal–optical (LO) polaron model for $d = 3$ is described by the Hamiltonian

$$H_F = \frac{1}{2m}\mathbf{p}^2 + \hbar\omega \sum_{\mathbf{k}} a_{\mathbf{k}}^\dagger a_{\mathbf{k}} + \frac{1}{\sqrt{\Omega}} \sum_{\mathbf{k}} g_{\mathbf{k}} \left(a_{\mathbf{k}}^\dagger e^{-i\mathbf{k}\mathbf{x}} - a_{\mathbf{k}} e^{i\mathbf{k}\mathbf{x}} \right) , \quad (10.1)$$

$$[a_{\mathbf{k}}, a_{\mathbf{k}'}^\dagger] = \delta_{\mathbf{k}\mathbf{k}'} ,$$

which models the interaction of an electron (position and momentum vectors $\mathbf{x}$ and $\mathbf{p}$, band mass m) with the phonon field (creation and annihilation operators $a_{\mathbf{k}}^\dagger$, $a_{\mathbf{k}}$, quantization volume Ω, Plank constant $\hbar$) associated with an LO branch of lattice vibrations (wave vector $\mathbf{k}$ and frequency ω) in a polar crystal. The electron–phonon interaction coefficient for coupling with a wave vector $\mathbf{k}$ in (10.1) is defined as follows:

$$g_{\mathbf{k}} = \frac{i\hbar\omega(\hbar/2m\omega)^{1/4}(4\pi\alpha)^{1/2}}{|\mathbf{k}|} .$$

The dimensionless Fröhlich coupling constant reads

$$\alpha = \frac{e^2}{2} \left(\frac{1}{\epsilon_\infty} - \frac{1}{\epsilon_o} \right) \frac{1}{\hbar\omega} \left(\frac{2m\omega}{\hbar} \right)^{1/2} , \quad (10.2)$$

where e is the electron charge, ϵ_∞ and ϵ_o are the high frequency and static dielectric constants of the crystal. In most of the real ionic crystals it takes the value $\alpha \sim 1 \div 20$ (e.g. $\alpha \simeq 5$ for sodium chloride). In the following, units will be chosen such that $\hbar = m = \omega = 1$.

Until now, no nontrivial solution of $H_F \Psi_n = E_n \Psi_n$ is known. It has been shown [61] for generalized Fröhlich models that the function $E_o(\alpha)$ has no points of nonanalyticity for an arbitrary $\alpha \geq 0$. Various methods [76, 77,

78, 79, 80] have been used to calculate approximately the spectrum of H_F, especially to obtain its GSE E_o for selected (weak, intermediate or, strong) regions of α.

To extend the Fröhlich Hamiltonian (10.1) written for $d = 3$ to arbitrary dimension d, we follow a physical approach [81, 82] inspired by the formulation of a lower-dimensional polaron problem as obtained from the Fröhlich Hamiltonian of a higher-dimensional system by integrating out one or more dimensions. In this approach the basic interaction characterizing the electron motion in d-dimensions remains Coulomb-like ($\sim 1/r$), i.e. the same as for $d = 3$ space. In particular cases of $d = 2$ and $d = 3$, this definition of d-dimensional polaron Hamiltonian reduces to the standard expressions for the Fröhlich Hamiltonian. Following [82] we assume that the form of the Fröhlich Hamiltonian in d–dimensional space is the same as in (10.1) except that now all vectors and operators are d–dimensional and the electron–phonon interaction coefficient $g_\mathbf{k}$ is redefined as follows:

$$|g_\mathbf{k}|^2 = \frac{\lambda^2}{|\mathbf{k}|^{d-1}}, \qquad \lambda^2 = \Gamma\left(\frac{d-1}{2}\right) 2^{(d-3/2)} \pi^{(d-1)/2} \alpha.$$

In particular,

$$|g_\mathbf{k}|^2 = \frac{\sqrt{2}\pi\alpha}{|\mathbf{k}|} \quad \text{for} \quad d = 2 \quad \text{and} \quad |g_\mathbf{k}|^2 = \frac{\sqrt{8}\pi\alpha}{|\mathbf{k}|^2} \quad \text{for} \quad d = 3.$$

Accordingly, we write the PI representation of the free-energy $F(\beta)$ of a polaron with given temperature $\Theta = 1/\beta$ as follows:

$$\exp(-\beta F) = \text{Tr}[\exp(-\beta H_F)], \tag{10.3}$$

where the Hamiltonian H_F in (10.1) should be written in terms of the coordinates and momenta. The procedure "Trace" $\text{Tr} = \text{Tr}_{el}\text{Tr}_{ph}$ here is assumed to be taken over the whole space of states of the "electron + phonon" system.

It is well known from the famous paper by Feynman [40] that the path integral approach to the polaron has an advantage because the phonon trace Tr_{ph} in (10.3) can be adequately eliminated and as a consequence, the polaron problem is reduced to an effective one-particle problem with retarded interaction. The result reads

$$Z_\beta(\alpha) = \exp(-\beta F) = \int\limits_{\mathbf{x}(0)=\mathbf{x}(\beta)} \delta\mathbf{x}\exp(S[\mathbf{x}]), \tag{10.4}$$

where the action $S[\mathbf{x}]$ is

$$S[\mathbf{x}] = -\frac{1}{2}\int\limits_0^\beta dt\dot{\mathbf{x}}^2(t) + \frac{\lambda^2}{8\pi}\int\limits_0^\beta\int\limits_0^\beta dtds\frac{G(t-s)}{|\mathbf{x}(t)-\mathbf{x}(s)|}. \tag{10.5}$$

In (10.5), $G(t)$ is the temperature-dependent Green function of a harmonic oscillator

$$G(t) = \frac{e^{|t|} + e^{\beta - |t|}}{e^{\beta} - 1} .$$ (10.6)

The free energy $F(\beta)$ tends to the GSE as $\beta \to \infty$ (zero temperature case):

$$E_o = - \lim_{\beta \to \infty} \frac{1}{\beta} \ln[Z_\beta(\alpha)] .$$

The path integral in (10.4) is not explicitly solvable. For the variational estimation for $d = 3$ Feynman suggested [40] a quadratic two-body trial action S_F instead of S as follows:

$$S[\mathbf{x}] \longrightarrow S_F[\mathbf{x}] = -\frac{1}{2} \int\limits_0^\beta dt \dot{\mathbf{x}}^2(t)$$

$$+ \frac{C}{2} \int\limits_0^\beta\!\!\int\limits_0^\beta dt ds [\mathbf{x}(t) - \mathbf{x}(s)]^2 \exp\{-w|t - s|\} ,$$ (10.7)

where constants C and w are variational parameters. With the trial action S_F one gets an exact solution for path integral in (10.4). A variation for finding the absolute minimum of $E_o^F(\alpha) = F_F(\alpha)$ for $\beta \to \infty$ over parameters C and w leads to a rigorous upper bound of the polaron GSE at arbitrary α, that is Feynman's well known result [40].

10.3 Application of the GER Method to the d–dimensional Polaron

Here we will show that the application of the GER method improves Feynman's estimation. We consider the polaron GSE in the case of an arbitrary space dimension $d > 1$ and start again from the PI in (10.4)–(10.5).

For further convenience, to get a symmetrical region over t, we change the variable of the PI in (10.4)

$$\mathbf{x}(t) \longrightarrow \mathbf{r}(t - T), \qquad T = \beta/2$$

with electron motion $\mathbf{r}(t)$ embedded in d–dimensional space. Accordingly, the GSE of the Fröhlich polaron $E_o(\alpha)$ (it will hereafter be denoted $E(\alpha)$) can be defined as follows:

$$E(\alpha) = - \lim_{T \to \infty} \frac{1}{2T} \ln Z_T(\alpha),$$ (10.8)

where a PI is introduced:

$$Z_T(\alpha) = C_o \int\limits_{\mathbf{r}(-T)=\mathbf{r}(T)} \delta\mathbf{r}$$

$$\exp\left\{ -\frac{1}{2}(\mathbf{r}D_o^{-1}\mathbf{r}) + \frac{\alpha}{2} \int\limits_{-T}^{T}\!\!\int dtds V[\mathbf{r}(t) - \mathbf{r}(s); t-s] \right\}, \qquad (10.9)$$

$$C_o = \sqrt{\det D_o^{-1}}, \qquad (\mathbf{r}\,D_o^{-1}\mathbf{r}) = \int\limits_{-T}^{T}\!\!\int dt\,ds\,\mathbf{r}(t)\,D_o^{-1}(t,s)\,\mathbf{r}(s).$$

The standard normalization $E(0) = 0$ in (10.8) is satisfied under the condition $Z_T(0) = 1$.

The free-electron system is described by the kinetic term $(\mathbf{r}D_o^{-1}\mathbf{r})$, where the differential operator D_o^{-1} and its Green function D_o are given by (9.14)–(9.16) in the previous section as $T \to \infty$.

The Coulomb-like interaction part, reflecting the electron self-interaction, is given by the retarded potential

$$V[\mathbf{R}(t,s); t-s] = \frac{\Gamma(\frac{d-1}{2})}{4\sqrt{2}\pi^{\frac{d+1}{2}}} \int\limits_{-T}^{T}\!\!\int dt\,ds e^{-|t-s|}\int \frac{d\mathbf{k}}{|\mathbf{k}|^{d-1}} \exp(i\mathbf{k}\mathbf{R}(t,s)), \quad (10.10)$$

$$\mathbf{R}(t,s) = \mathbf{r}(t) - \mathbf{r}(s)$$

with the electron position vector $\mathbf{r}(t)$ belonging to d-dimensions.

10.4 Bounds for the Polaron Ground-State Energy in d Dimensions

The polaron partition function $Z_T(g)$ in (10.9) with its kinetic and interaction parts given by (10.9), (10.10) coincides exactly with the general form of the PI in (9.13), where the constant g should be replaced by α and $d\mathcal{K}(\mathbf{k}; t-s)$ becomes

$$d\mathcal{K}(\mathbf{k}; t-s) = \frac{\Gamma(\frac{d-1}{2})}{4\sqrt{2}\pi^{\frac{d+1}{2}}} e^{-|t-s|}\frac{d\mathbf{k}}{|\mathbf{k}|^{d-1}}.$$

If α is not too large, the PI in the initial representation (10.9) can be estimated by using a perturbation expansion over α. The problem is to calculate $Z_T(\alpha)$ beyond the weak coupling regime.

Now we want to apply the GER method to the polaron PI defined in (10.9).

Our key steps will be the same, that as those in chapter 9. We remember that these are:

- *the introduction of new Gaussian measure* $d\sigma$ *(9.19), which forms the kinetic part of the PI in a new representation, and*
- *the requirements of the "normally ordered" and "correct" form of the interaction part of the PI in this representation.*

This program results in a new representation of the initial PI: an exponential with the leading term of the energy is factorized out as a free multiplicand (9.29) and all the corrections to it are defined by another PI in (9.27).

Performing this scheme and using formulae (9.23), (9.25) and (9.27)–(9.29), we obtain the GSE of the optical polaron within the GER method as follows:

$$E(\alpha) = E_o(\alpha) + \Delta E(\alpha), \qquad (10.11)$$

where the function $E_o(\alpha)$ being the "leading-order energy", or the GSE in the zeroth approximation, is (see Eq. (9.29))

$$E_o(\alpha) = -d \left\{ \frac{1}{2\pi} \int\limits_0^\infty dk \left[\ln \left(k^2 \tilde{D}(k) \right) - k^2 \tilde{D}(k) + 1 \right] \right. \qquad (10.12)$$

$$\left. + \frac{\alpha_d}{3\sqrt{2\pi}} \int\limits_0^\infty dt \, \frac{\exp(-t)}{F^{1/2}(t)} \right\}$$

Here we have introduced "the effective coupling constant"

$$\alpha_d = \alpha \cdot R_d, \qquad R_d = \frac{3\sqrt{\pi}\,\Gamma\left(\frac{d-1}{2}\right)}{2d\,\Gamma\left(\frac{d}{2}\right)}.$$

The high-order corrections $\Delta E(\alpha)$ in (10.11) can be obtained by evaluating the PI

$$\exp\left\{-2T \cdot \Delta E(\alpha)\right\} = C \int\limits_{\mathbf{r}(-T)=\mathbf{r}(T)} \delta\mathbf{r}$$

$$\exp\left\{ -\frac{1}{2} \iint\limits_{-T}^{T} dt\, ds\, \mathbf{r}(t)\, D^{-1}(t,s)\, \mathbf{r}(s) + W[\mathbf{r}] \right\}$$

$$= \int d\sigma \exp\{W[\mathbf{r}]\}. \qquad (10.13)$$

Here, the interaction functional written in the new representation is

$$W[\mathbf{r}] = \alpha_d \cdot \frac{\Gamma(d/2)\,d}{6\sqrt{2}\pi^{d/2+1}} \iint\limits_{-T}^{T} dt\, ds\, e^{-|t-s|} \qquad (10.14)$$

$$\cdot \int \frac{d\mathbf{k}}{|\mathbf{k}|^{d-1}} \exp\left\{-\mathbf{k}^2 F(t-s)\right\} : e_2^{i\mathbf{k}[\mathbf{r}(t)-\mathbf{r}(s)]} : .$$

where $e_2^x = e^x - 1 - x - x^2/2$. The function $F(t)$ in (10.12) and (10.14) is defined by the equations (see Eqs. (9.23) and (9.25))

$$\tilde{\Sigma}(p) = -\tilde{\omega}_2(p) = \frac{1}{3\sqrt{2\pi}} \int_0^\infty dt\, e^{-t}\, \frac{1-\cos(pt)}{F^{3/2}(t)}, \tag{10.15}$$

$$F(t) = D(0) - D(t) = \int_{-\infty}^\infty \frac{dp}{2\pi} \tilde{D}(p)\left(1 - e^{ipt}\right) \tag{10.16}$$

$$= \frac{1}{\pi} \int_0^\infty dp\, \frac{1-\cos(pt)}{p^2 + \alpha_d \tilde{\Sigma}(p)}.$$

Equations (10.15) and (10.16) provide that in the representation (10.13) all the quadratic terms in the polaron action functional are concentrated only in the new Gaussian measure $d\sigma$ and so the interaction functional $W[\mathbf{r}]$ behaves like $\sim \mathbf{r}^3$ as $|\mathbf{r}| \to 0$.

It should be stressed that representation (10.11) is completely equivalent to the initial representation (10.8) for asymptotically large $T \to \infty$. The Gaussian equivalent representation (10.11) gives the origin of various approximations differing from each other in the accuracy of deriving equations (10.15)–(10.16).

The problem is to solve the nonlinear integral equations (10.15) and (10.16), defining the desired two-point correlation function $D(x)$. Unfortunately, we are not able to solve them exactly in analytic form except for the asymptotic cases $\alpha \to 0$ and $\alpha \to \infty$. In the intermediate region of $\alpha = 1 \div 40$ we solve them by developing a numerical iteration scheme by developing a fixed-point method of consequent approximation like that suggested in (9.26). We can easily find the asymptotical behaviour of $F(t)$ and $\tilde{\Sigma}(p)$:

$$F(t) \sim t \quad \text{as} \quad t \to 0, \qquad F(t) \sim F_o \quad \text{as} \quad t \to \infty,$$

$$\tilde{\Sigma}(p) \sim p^2 \quad \text{as} \quad p \to 0, \qquad \tilde{\Sigma}(p) \sim \tilde{\Sigma}_o \quad \text{as} \quad p \to \infty,$$

where F_o and $\tilde{\Sigma}_o$ are constants. This behaviour will be useful for performing the correct numerical iterations for solutions of (10.15) and (10.16).

Note that in the particular case of $d = 3$, equations similar to (10.15) and (10.16) were obtained in other approaches [83, 71] from stationary conditions.

The leading term (or the zero-order approximation) $E_o(\alpha)$ gives an *upper bound* to the exact GSE of a polaron $E(\alpha)$. Actually, applying the Jensen's inequality to (10.13) one gets

$$\exp\left\{-2T \cdot \Delta E(\alpha)\right\} \leq \exp\left\{\int d\sigma W[\mathbf{r}]\right\} = 1.$$

Consequently,
$$E(\alpha) \le E_o(\alpha).$$

The numerical calculations brought in section 10.8 show that the obtained upper bound (10.12) improves Feynman's often quoted result [40]. Meanwhile one can note that a natural way to check to error of any upper bound is to compare the latter with a suitable lower bound. We have an indication that the second-order correction $\Delta E_2(\alpha)$ to $E_o(\alpha)$ may result in a new *lower bound* to the polaron GSE. This deduction is argued below in section 10.6.

The numerical results $E_o(\alpha)$ obtained for $d = 2$ and $d = 3$ compared with known estimates are displayed in tables II.1–II.6.

10.5 Particular Case of Feynman's Estimation

As a simple approximation of $\Sigma(p)$ obeying the asymptotics in (10.15), one can choose the function

$$\tilde{\Sigma}^F(p) = \frac{\mu^2}{\alpha_d} \cdot \frac{p^2}{\xi^2 + p^2}, \tag{10.17}$$

where μ and ξ are parameters. Then, (10.16) gets the form:

$$F^F(t) = \frac{1}{\pi} \int_0^\infty dp \frac{1 - \cos(pt)}{p^2} \left[1 - \frac{\mu^2}{\mu^2 + \xi^2 + p^2} \right]$$

$$= \frac{\mu^2}{2\lambda^3} \left(1 - e^{-\lambda t} + \lambda t \frac{\xi^2}{\mu^2} \right), \tag{10.18}$$

$$\lambda = \sqrt{\mu^2 + \xi^2}.$$

These approximate solutions of (10.15), (10.16) provide the leading term of the polaron self-energy in the following form

$$E_o^F(\alpha, \mu, \xi) = -\frac{d}{2} \cdot \left[\xi - \lambda + \frac{\mu^2}{2\lambda} \right] - \frac{\alpha_d d \lambda^{\frac{3}{2}}}{3\mu\sqrt{\pi}} \int_0^\infty \frac{dt e^{-t}}{\sqrt{1 - \exp(-\lambda t) + \lambda t \frac{\xi^2}{\mu^2}}}. \tag{10.19}$$

Minimizing the obtained function over the parameters μ and ξ, one easily finds Feynman's variational upper bound [82] in d-dimensions. For $d = 3$ ($\alpha_3 = \alpha$) it explicitly reproduces Feynman's well-known variational upper bound to the polaron GSE [42]:

$$E^F(\alpha) = \min_\mu \min_\xi E_o^F(\alpha, \mu, \xi)|_{d=3}. \tag{10.20}$$

We stress that the extremal conditions on parameters μ, ξ are equivalent to a particular choice of the functions $\tilde{\Sigma}^F(p)$ and $F^F(t)$ in (10.17)–(10.18). However, the functions in (10.17)–(10.18) are not the exact solutions of (10.15) and (10.16). It means, that Feynman's trial quadratic action does not represent entirely the Gaussian part of the polaron action for $d = 3$.

Exact numerical solution of equations (10.15) and (10.16) by iteration, as in 9.2, allows us to obtain $E_o(\alpha)$ more exactly and improves Feynman's result $E^F(\alpha)$ in the whole region of α. A comparison of numerical results for $E_o(\alpha)$ and $E^F(\alpha)$ is given in tables II.1–II.6.

10.6 Corrections to the Leading Term of the Energy

In paper [85] it has been shown that the main contribution to $E(\alpha)$ comes from $E_o(\alpha)$ for $\alpha \to \infty$. In the weak coupling limit $\alpha \to 0$ the contribution of $E_o(\alpha)$ proportional to α is also dominant because the corrections generated by the functional integral in (10.13) vanish because they are proportional to α^2. Then we can suppose that in the intermediate range of α, $E_o(\alpha)$ also gives the main contribution to the polaron self-energy. The correction $\Delta E(\alpha)$ should be evaluated from the functional integral in (10.13) by expanding e^W in a series as follows:

$$\Delta E(\alpha) = \sum_{n=1}^{\infty} \Delta E_n(\alpha) = - \lim_{T \to \infty} \frac{1}{2T} \sum_{n=1}^{\infty} \frac{1}{n!} \int d\sigma \, \{W[\mathbf{r}]\}^n_{\text{connected}} \cdot \quad (10.21)$$

We stress that (10.21) is not the standard perturbation series in the coupling constant α as α enters W not only directly as a multiplicand, but also indirectly through the function $F(t)$. The first term in (10.21) with $n = 1$ is equal to zero $\Delta E_1(\alpha) \equiv 0$ owing to normally–ordering in (10.14). Nontrivial corrections are given by terms with $n \geq 2$. The N-th approximation of the GER method for the polaron GSE is given by

$$E^{(N)}(\alpha) = E_o(\alpha) + \sum_{n=2}^{N} \Delta E_n(\alpha) \,.$$

For the second-order correction to $E_o(\alpha)$ we get

$$\Delta E_2(\alpha) = - \lim_{T \to \infty} \frac{1}{2T} \left\{ \frac{\alpha^2_d}{2!} \left[\frac{d\Gamma(\frac{d}{2})}{6\sqrt{2}\pi^{\frac{d}{2}+1}} \right]^2 \cdot \iint_{-T}^{T} dt \, ds \iint_{-T}^{T} dx \, dy e^{-|t-s|-|x-y|} \right.$$

$$\times \int \frac{d\mathbf{k}}{|\mathbf{k}|^{d-1}} \int \frac{d\mathbf{p}}{|\mathbf{p}|^{d-1}} \exp(-\mathbf{k}^2 F(t-s)) \exp(-\mathbf{p}^2 F(x-y))$$

$$\times : e_2^{(i\mathbf{k}[\mathbf{r}(t)-\mathbf{r}(s)])} : \cdots : e_2^{(i\mathbf{p}[\mathbf{r}(x)-\mathbf{r}(y)])} : \Big\} \,.$$

We calculate

$$\int d\sigma : \exp(i\mathbf{k}[\mathbf{r}(t) - \mathbf{r}(s)]) : \cdots : \exp(i\mathbf{p}[\mathbf{r}(x) - \mathbf{r}(y)]) := \exp(-\mathbf{kp} \cdot \varXi),$$

where we introduce the four-point correlation function

$$\Xi(t, s, x, y) = D(t - x) + D(s - y) - D(s - x) - D(t - y)$$
$$= -[F(t - x) + F(s - y) - F(s - x) - F(t - y)].$$

Expanding $\exp(-\mathbf{kp} \cdot \Xi)$ into a series and taking into account the relation

$$\iint d^d k\, d^d \mathbf{p}\; f(\mathbf{k}^2, \mathbf{p}^2)\,(\mathbf{kp})^{2n}$$

$$= \frac{4\pi^{d-\frac{1}{2}}\Gamma(n+\frac{1}{2})}{\Gamma(\frac{d}{2})\Gamma(n+\frac{d}{2})} \int_0^\infty du \int_0^\infty dv\,(uv)^{d-1+2n}\;f(u^2, v^2),$$

one finally gets

$$\Delta E_2(\alpha) = -\lim_{T\to\infty}\frac{1}{2T}\frac{\alpha_d^2\,d^2\Gamma(\frac{d}{2})}{144\pi^{\frac{3}{2}}}\cdot\int_{-T}^{T}\!\!\int\! dt\,ds \int_{-T}^{T}\!\! dx\,dy\,e^{-|t-s|-|x-y|}$$

$$\cdot\sum_{n=2}^{\infty}\frac{[\Xi(t, s, x, y)]^{2n}\,\Gamma(n+\frac{1}{2})\,[(2n-1)!!]^2}{(2n)!\,4^n\,\Gamma(n+\frac{d}{2})\,[F(t-s)\times F(x-y)]^{n+\frac{1}{2}}}\,. \tag{10.22}$$

The function $F(t)$ is the solution of (10.15)–(10.16). Using the symmetry behaviour of the functions $F(t - s) \cdot F(x - y)$ and $\Xi(t, s, x, y)$ in (10.22) we can reduce the 4-dimensional integration volume to the 3-dimensional domain, and (10.22) can be simplified to

$$\Delta E_2(\alpha) = -\alpha_d^2 \cdot \frac{d^2\Gamma(\frac{d}{2})}{18\pi^{\frac{3}{2}}}\sum_{n=2}^{\infty} Q_n\,A_n\,, \tag{10.23}$$

$$Q_n = \frac{(2n)!\,\Gamma(n+\frac{1}{2})}{16^n(n!)^2\Gamma(n+\frac{d}{2})}\,,$$

$$A_n = \iiint_0^\infty da\,db\,dc \cdot \left\{ e^{-a-c} \cdot \frac{[F(a+b) + F(b+c) - F(a+b+c) - F(b)]^{2n}}{[F(a)\cdot F(c)]^{n+1/2}}\right.$$

$$+ e^{-a-2b-c} \cdot \frac{[F(a) + F(c) - F(a+b+c) - F(b)]^{2n}}{[F(a+b)\cdot F(b+c)]^{n+1/2}}$$

$$\left.+ e^{-a-2b-c} \cdot \frac{[F(a+b) + F(b+c) - F(a) - F(c)]^{2n}}{[F(a+b+c)\cdot F(b)]^{n+1/2}}\right\}.$$

We stress that the expression for A_n in (10.23) can be simplified, but we retain this form for clarity.

Finally, we get the following expression for the self-energy of the polaron:

$$E^{(2)}(\alpha) = E_o(\alpha) + \Delta E_2(\alpha)\,, \tag{10.24}$$

which can be evaluated numerically for arbitrary α and different space dimensions d.

For general values of α there various lower bounds – without proof – are available [69, 86] on the polaron GSE. However, the only available proven lower bound obtained in [68] is not very accurate beyond the weak-coupling regime and recent rigorous result [87] is inferior to Lieb and Yamazaki lower bound. We have made preliminary estimations which indicate that the decreasing series in (10.21) is alternating. Then one can expect that of the third-order correction $\Delta E_3(\alpha)$ may slightly decrease the value of $E^{(2)}(\alpha)$ and the conclusion of higher order corrections $\Delta E_{n>2}(\alpha)$ might result in an insignificant oscillation of $E^{(n>2)}(\alpha)$ between $E_o(\alpha)$ and $E^{(2)}(\alpha)$. In other words, the $E^{(2)}(\alpha)$ obtained may be accepted as a *lower bound* of the ground state energy of the polaron. In table 10.6 and figure 10.2 we displayed a comparison of some numerical results on both upper and lower bounds to the polaron GSE for $d = 3$.

10.7 Scaling Relations

The theory under consideration has two independent parameters α and d. In general, all our expressions should depend on both of them. Notice, however, that our key expressions in (10.12)–(10.16) completely defining the d–dimensional polaron GSE at arbitrary α depend only on the effective coupling constant α_d. This means that the following relations

$$F^{[d_1]}(\alpha_{d_2}, t) = F^{[d_2]}(\alpha_{d_1}, t) , \qquad \tilde{\Sigma}^{[d_1]}(\alpha_{d_2}, p) = \tilde{\Sigma}^{[d_2]}(\alpha_{d_1}, p) ,$$

take a place, where the numbers of space dimensions d_1 and d_2 are given in square brackets [...]. In the particular case of $d = 2$ and $d = 3$, we found

$$F^{[2]}(\alpha, t) = F^{[3]}\left(\frac{3\pi\alpha}{4}, t\right) , \qquad \tilde{\Sigma}^{[2]}(\alpha, k) = \tilde{\Sigma}^{[3]}\left(\frac{3\pi\alpha}{4}, k\right) .$$

Then, considering (10.12) one finds that this scaling relation is also valid for $\frac{1}{d} E_o(\alpha_d)$. Consequently, these relations lead to

$$E_o^{[2]}(\alpha) = \frac{2}{3} E_o^{[3]}\left(\frac{3\pi\alpha}{4}\right) . \tag{10.25}$$

Note that this relation in (10.25) was obtained earlier in [82, 83]. But this scaling is not valid beyond E_o because the interaction functional $W[\mathbf{r}]$ depends not only on α_d, but also on d in a complicated way. We can rewrite (10.23) for $d = 2$ and $d = 3$:

$$\Delta E_2^{[2]}(\alpha) = -\frac{\pi\alpha^2}{8} \sum_{n=2}^{\infty} Q_n^{[2]} A_n(\alpha),$$

$$\Delta E_2^{[3]}(\alpha) = -\frac{9\pi\alpha^2}{32} \sum_{n=2}^{\infty} Q_n^{[3]} A_n(\alpha) \,.$$

One easily finds that

$$\frac{15}{8} \leq \frac{Q_n^{[2]}}{Q_n^{[3]}} = \frac{(2n+1)!}{4^n(n!)^2} \leq 2\sqrt{\frac{n}{\pi}} \,.$$

This leads to the inequality

$$\Delta E_2^{[2]}(\alpha) \leq \frac{5}{6} \Delta E_2^{[3]}\left(\frac{3\pi\alpha}{4}\right) , \tag{10.26}$$

which shows that the scaling relation (10.25) is no longer satisfied beyond $E_o(\alpha)$.

Let us consider the asymptotic limits of spatial dimensions d at fixed finite α. We get

$$\lim_{d\to 1} \alpha_d = \frac{3\alpha}{d-1} \to \infty \,, \qquad \lim_{d\to\infty} \alpha_d = \frac{3\alpha\sqrt{\pi e}}{\sqrt{2}\, d^{3/2}} \to 0 \,. \tag{10.27}$$

Taking into account (10.27) we can conclude that as d becomes larger, α_d decreases rapidly and in fact we deal with the effective weak–coupling regime $\alpha_d \ll 1$ even for α not too small. For example, the second–order corrections $\Delta E_2(\alpha)$ behaves as follows:

$$\Delta E_2(\alpha)\xrightarrow[d\to\infty]{} -\frac{1}{8\pi} \alpha_d^2 \to 0 \,.$$

In other words, our leading–order energy term $E_o(\alpha)$ tends to the exact GSE $E(\alpha)$ as d grows because the role of $\Delta E(\alpha)$ becomes insignificant.

10.8 Numerical Results

In this section, we demonstrate numerical values of $E_o(\alpha)$ and $E^{(2)}(\alpha)$ calculated within the GER method and compare them with known results obtained at various (weak, strong and intermediate) ranges of α. The results obtained are given in tables 10.1–10.6 and shown in figures 10.1,10.2.

10.8.1 The Weak Coupling Limit

Among known numerical results concerning the GSE of the polaron, the more accurate are those obtained for $\alpha \to 0$. In this limit the problem was investigated in a pioneering work [66]. Lee, Low and Pines [76] applied to this problem a variational principle and the Tomonaga method. As $\alpha \to 0$ one can apply either the standard perturbation approach [48, 55] or canonical transformations of the Hamiltonian with consequent variational estimations [88, 89, 67, 69].

Below, we calculate the exact GSE of the d–dimensional polaron for the order α^2 in the weak coupling limit and compare the accuracy of the obtained results with exact perturbation estimations presented in [69, 42, 90, 91, 92] for $d = 2$ and $d = 3$.

If α is not too large, the polaron self-energy $E(\alpha)$ has the form

$$E(\alpha) = \alpha \cdot C_{w1} + \alpha^2 \cdot C_{w2} + O(\alpha^3). \tag{10.28}$$

The coefficients C_{w1} and C_{w2} are known with good accuracy for $d = 2$ [82] and $d = 3$ [92, 93]. In our approach, the coefficient C_{w1} arises only from $E_o(\alpha)$ in (10.12) whereas the C_{w1}, from both $E_o(\alpha)$ and $\Delta E_2(\alpha)$ in (10.23).

Since we are interested in calculating only the coefficients C_{w1} and C_{w2}, it is enough to solve the functions $F(t)$ and $\tilde{D}(p)$ with an accuracy up to α. From (10.16) and (10.15) we get

$$\tilde{\Sigma}(p) = \frac{2}{3\sqrt{\pi}} \int_0^\infty dt\, e^{-t}\, \frac{1 - \cos(pt)}{t^{\frac{3}{2}}} + O(\alpha),$$

$$F(t) = \frac{1}{\pi} \int_0^\infty dp\, [1 - \cos(pt)]\, \tilde{D}(p) = \frac{t}{2} - \alpha_d\, f(t) + O(\alpha^2), \tag{10.29}$$

where

$$\tilde{D}(p) = \frac{1}{p^2} \left[1 - \alpha_d \frac{\tilde{\Sigma}(k)}{p^2} \right] + O(\alpha^2),$$

$$f(t) = \frac{1}{\pi} \int_0^\infty dp\, \frac{1 - \cos(pt)}{p^2}\, \frac{\tilde{\Sigma}(p)}{p^2} + O(\alpha^2).$$

Substituting these expressions into (10.12) we get

$$E_o(\alpha) = -\alpha \cdot \frac{dR_d}{3} - \alpha^2 \cdot \frac{dR_d^2}{36} \left[1 - \frac{8}{3\pi} \right] + O(\alpha^3), \tag{10.30}$$

where the following relation has been used:

$$\frac{2\sqrt{\pi}}{3} \int_0^\infty dt\, \exp(-t)\, \frac{f(t)}{t^{\frac{3}{2}}} = \int_0^\infty dp\, \left[\frac{\tilde{\Sigma}_2(p)}{p^2} \right]^2 = \frac{\pi}{9} \left[1 - \frac{8}{3\pi} \right] + O(\alpha).$$

Equation (10.30) defines the coefficient C_{w1} exactly and contributes also to C_{w2}. Concerning the coefficient C_{w2}, we should also take into account corrections coming from $\Delta E_2(\alpha)$. Inserting (10.29) into (10.23) and going to variables $x = 1 + a/b$ and $y = 1 + c/b$, one gets

$$\Delta E_2(\alpha) = -\alpha^2 \cdot \frac{d^2 R_d^2 \Gamma(\frac{d}{2})}{9\pi^{\frac{3}{2}}} \sum_{n=2}^{\infty} \frac{(2n)! \, \Gamma(n+\frac{1}{2})}{4^n \, (n!)^2 \, \Gamma(n+\frac{d}{2})} \cdot B_n + O(\alpha^3), \quad (10.31)$$

where

$$B_n = \iint\limits_{1}^{\infty} dx dy \, \frac{1}{(x+y)^2} \left[\frac{1}{(x \cdot y)^{n+\frac{1}{2}}} + \frac{1}{(x+y-1)^{n+\frac{1}{2}}} \right].$$

Higher order corrections $\Delta E_{n>2}(\alpha)$ are proportional $\sim \alpha^3$ so they do not contribute to C_{w1} and C_{w2}. Then, we get the coefficients C_{w1} and C_{w2} of the d–dimensional polaron GSE exactly as follows

$$C_{w1} = -\frac{dR_d}{3}$$

and

$$C_{w2} = -\frac{dR_d^2}{36} \left[1 - \frac{8}{3\pi} \right] - \frac{d^2 R_d^2 \Gamma(\frac{d}{2})}{9\pi^{\frac{3}{2}}} \sum_{n=2}^{\infty} \frac{(2n)! \, \Gamma(n+\frac{1}{2})}{4^n \, (n!)^2 \, \Gamma(n+\frac{d}{2})} \cdot B_n \,.$$

The behaviour of these coefficients with respect to the space-dimension number d is shown in figure 10.1. Note that $C_{w2} = -O(1/d)$ as $d \to \infty$, which reflects the fact that polaron effects decrease in large space dimensions. On the other hand, this effect great for $d = 2$. We stress that our results for C_{w1} and C_{w2} coincide with those obtained in [82].

Table 10.1. Comparison of known weak coupling results for the polaron ground-state energy $E(\alpha) = \alpha \cdot C_{w1} + \alpha^2 \cdot C_{w2} + O(\alpha^3)$ in two-dimensions.

Authors	C_{w1}	C_{w2}
S.Das Sarma, B.Mason [?]	$-\pi/2$	-0.062
R.Feynman's theory [?]	$-\pi/2$	-0.04569
Wu Xiaoguang, ... [?]	$-\pi/2$	-0.06397
O.Hipolito [?]	$-\pi/2$	-0.0245
Present $E_o(\alpha)$	$-\pi/2$	-0.046626
Present $E_o(\alpha) + \Delta E_2$	$-\pi/2$	-0.063974

The Two-Dimensional Polaron. In two-dimensions $R_2 = 3\pi/4$, and hence (10.30) is reduced to

$$E_o(\alpha) = -\alpha \cdot \frac{\pi}{2} - \alpha^2 \cdot \frac{9\pi^2}{288} \left[1 - \frac{8}{3\pi} \right] + O(\alpha^3) = -\frac{\alpha}{2} - \alpha^2 \cdot 0.046626 + O(\alpha^3).$$

$$(10.32)$$

The second correction to $E_o(\alpha)$ in (10.31) becomes

$$\Delta E_2(\alpha) = -\alpha^2 \cdot \frac{\pi}{4} \sum_{n=2}^{\infty} \frac{[(2n)!]^2}{16^n \, (n!)^4} \cdot B_n + O(\alpha^3) = -\alpha^2 \cdot 0.017341 + O(\alpha^3).$$

(10.33)

Adding this to (10.32), one obtains

$$E^{(2)}(\alpha) = -\alpha \cdot \frac{\pi}{2} - \alpha^2 \cdot 0.063967 + O(\alpha^3)$$

(10.34)

as the exact GSE of the three-dimensional polaron up to the order of α^2. The estimated uncertainty in (10.34) is ± 1 unit in the last digit. For comparison, we give in table 10.1 the known results for $d = 2$ as $\alpha \to 0$. One can see that our C_{w2} obtained only from $E_o(\alpha)$ improves Feynman's estimate by about 2 per cent. Adding the next correction calculated from $\Delta E^{(2)}$ results in $C_{w2} = -0.063974$, which is in good agreement with the exact value in [82]. Note that, for $d = 2$, $\Delta E^{(2)}$ contributes about 27 per cent to the total value of C_{w2}.

Table 10.2. Comparison of known weak coupling results for the polaron ground-state energy $E(\alpha) = \alpha \cdot C_{w1} + \alpha^2 \cdot C_{w2} + O(\alpha^3)$ in three-dimensions.

Authors	C_{w1}	C_{w2}
S.Das Sarma, B.Mason [92]	−1	−0.016
R.Feynman,s theory [91]	−1	−0.012347
J.Röseler [67]	−1	−0.0159196
T.Lee, ... [76]	−1	−0.014
D.Larsen [69]	−1	−0.016
Present $E_o(\alpha)$	−1	−0.012598
Present $E_o(\alpha) + \Delta E_2$	−1	−0.015919

The Three-Dimensional Polaron. For $d = 3$ highly accurate results have been obtained in [67, 82]. Notice that $R_3 = 1$. Then, from (10.30) and (10.31) we get

$$E_o(\alpha) = -\alpha - \alpha^2 \cdot \frac{1}{12} \left[1 - \frac{8}{3\pi} \right] + O(\alpha^3) = -\alpha - \alpha^2 \cdot 0.012597803 + O(\alpha^3),$$

$$\Delta E_2(\alpha) = -\frac{\alpha^2}{\pi} \sum_{n=2}^{\infty} \frac{(2n)!}{4^n \, (n!)^2 \, (2n+1)} \cdot B_n + O(\alpha^3) = -\alpha^2 \cdot 0.003322 + O(\alpha^3).$$

(10.35)

Summing these, we have

$$E^{(2)}(\alpha) = -\alpha - \alpha^2 \cdot 0.015919 + O(\alpha^3).$$

(10.36)

Our results are displayed in table 10.2 together with known results of the polaron GSE for $d = 3$ in the weak coupling limit. Our leading term of

$$E^{(2)}(\alpha) = -\alpha - \alpha^2 \cdot 0.015919 + O(\alpha^3). \qquad (10.36)$$

Our results are displayed in table 10.2 together with known results of the polaron GSE for $d = 3$ in the weak coupling limit. Our leading term of the energy $E_o(\alpha)$ improves Feynman's variational estimation by 2 per cent in value of C_{w2}. The next correction coming from $\Delta E^{(2)}$ results in $C_{w2} = -0.015919$, which is in good agreement with the exact value in [82]. Note that, for $d = 3$, our $\Delta E^{(2)}$ contributes about 21 per cent (smaller than for $d = 2$) to the total value of C_{w2}. Comparing (10.26), (10.33) and (10.35) with the results for $d = 2$ and $d = 3$, we conclude that the higher-order corrections (the second-order in our case) coming from $J_T(\alpha)$ are substantially more important for $d = 2$ than for $d = 3$. In other words, the polaron effect is stronger in low space dimensions. This effect was noted earlier in [82, 83].

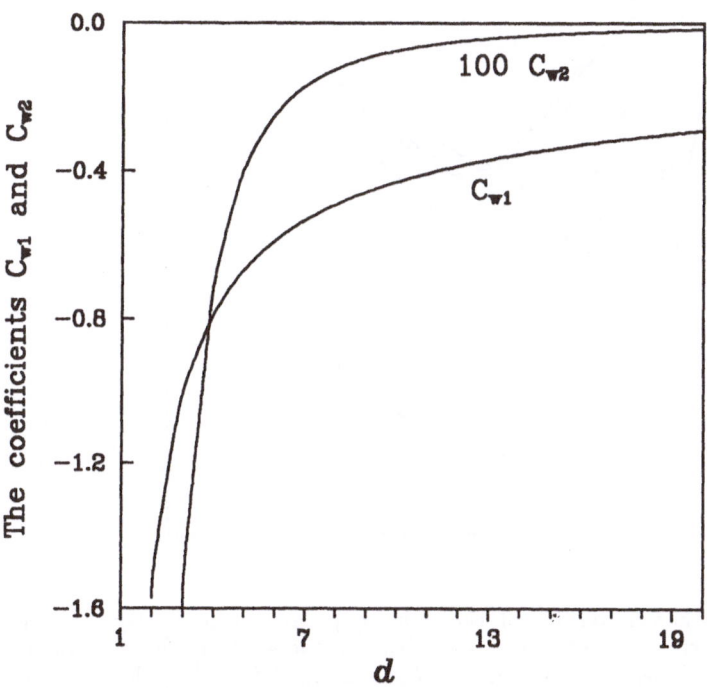

Fig. 10.1. The behavior of the coefficients C_{w1} and C_{w2} of the polaron ground state energy $E(\alpha) = \alpha \cdot C_{w1} + \alpha^2 \cdot C_{w2} + O(\alpha^3)$ at the weak coupling limit $\alpha \to 0$ in dependence on the number of space-dimensions d. For clarity, the plotted coefficient C_{w2} is multiplied by the factor 100.

10.8.2 The Strong Coupling Regime

The GSE of the polaron in the strong electron–phonon coupling regime was considered first by Landau and Pekar [94] in the Born–Oppenheimer approach and by Bogolubov [95] – in the adiabatic approximation. Special forms of perturbation analysis also can be performed as $\alpha \to \infty$ [96, 97, 98, 65, 84, 99]. It is well known that in this limit

$$E(\alpha) = \alpha^2 \cdot C_s + O(1) \,.$$

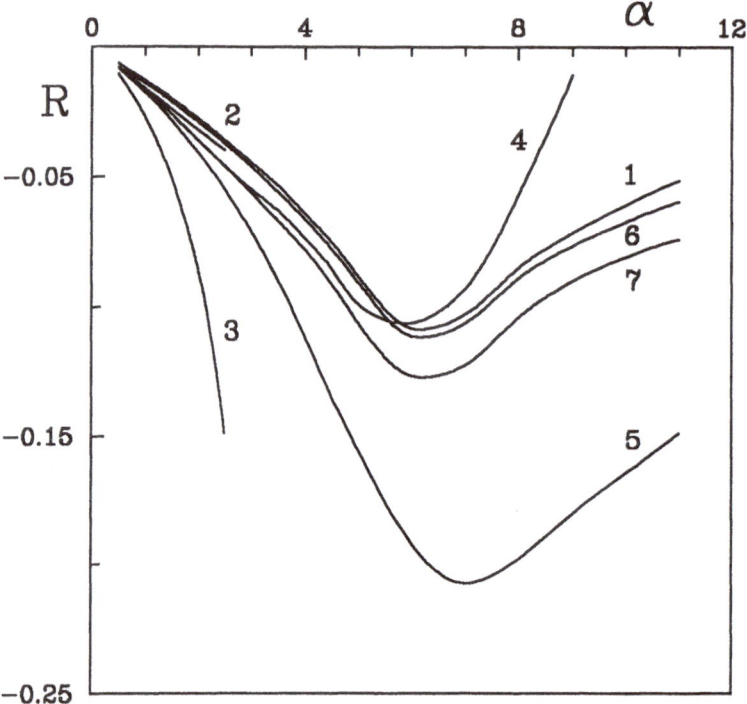

Fig. 10.2. Some known results of the polaron ground state energy $E(\alpha)$ (in three-dimensional space) displayed as a function of the electron–phonon coupling constant α. For clarity, the ratio $R = (E_* - E_{\text{harm}})/|E_{\text{harm}}|$ is shown, where E_* are estimations obtained in [71, 42, 86, 69] and E_{harm} is the 'harmonic-oscillator'approximation [71]. In these units the curve for E_{harm} coincides with the abscissa axis. Curves correspond to estimations: 1–Feynman's upper; 2/3–Larsen's upper/lower; 4/5–Smondyrev's upper/lower; 6–our $E_0(\alpha)$ and 7–our $E^{(2)}(\alpha)$.

At large α, the functions $F(t)$ and $\tilde{\Sigma}(p)$ behave as constants except for small regions near the points $t = 0, p = 0$. Starting with the guessed functions

$$F(t) = A, \qquad \tilde{\Sigma}(p) = B$$

and substituting these into (10.15)–(10.16), one gets the following equations for the constants A and B:

$$A = \frac{1}{\pi}\int\limits_0^\infty dp \frac{1}{p^2 + \alpha R_d B} = \frac{1}{2\mu},$$

$$B = \frac{1}{3\sqrt{2\pi}}\int\limits_0^\infty dt e^{-t} \frac{1}{A^{\frac{3}{2}}} = \frac{2\mu^{\frac{3}{2}}}{3\sqrt{\pi}},$$

where

$$\mu = \frac{4\alpha_d^2}{9\pi}.$$

Substituting these functions into (10.15) and (10.16) we have

$$F(t) = \frac{1}{\pi}\int\limits_0^\infty dp \frac{1 - \cos(pt)}{p^2 + \alpha_d B} = \frac{1 - \exp(-\mu|t|)}{2\mu}, \qquad (10.37)$$

so that

$$\tilde{D}(p) = \frac{1}{p^2 + \mu^2}, \qquad \tilde{\Sigma}(p) = B\frac{p^2}{1 + p^2}. \qquad (10.38)$$

We substitute these into (10.12) and obtain the leading term of the polaron GSE for the strong coupling regime as follows:

$$E_o(\alpha) = -\frac{d}{2\pi}\cdot\int\limits_0^\infty dp \left[\ln\left(\frac{1 + p^2}{1 + p^2 + \mu^2}\right) + \frac{\mu^2}{1 + p^2 + \mu^2}\right]$$

$$-\frac{\alpha_d\sqrt{\mu}}{3\sqrt{\pi}}\cdot\int\limits_0^\infty dt \frac{\exp(-t)}{\sqrt{1 - \exp(-\mu t)}} = -\alpha_d^2\cdot\frac{d}{9\pi} + O(1). \qquad (10.39)$$

Concerning $\Delta E_2(\alpha \to \infty)$ we have found that only the first term in curly brackets $\{..\}$ in (10.23) could give a nonvanishing contribution as $\mu \to \infty$, and there we can make the following substitution

$$\exp(-a - c)\frac{[F(a + b) + F(b + c) - F(a + b + c) - F(b)]^{2n}}{[F(a)\cdot F(c)]^{n+1/2}}$$

$$\longrightarrow 8\mu\exp(-a - c - 2n\mu b) + O(1)$$

in (10.23). Integrating over da, db, dc one gets

$$\Delta E_2(\alpha) = -\alpha_d^2 \cdot \frac{2d^2\Gamma(\frac{d}{2})}{9\pi^{\frac{3}{2}}}\sum_{n=2}^\infty \frac{(2n)!\,\Gamma(n + \frac{1}{2})}{16^n(n!)^2\,n\,\Gamma(n + \frac{d}{2})} + O(1). \qquad (10.40)$$

Adding it to (10.39) we finally obtain

$$E^{(2)}(\alpha) = -\alpha_d^2 \cdot \left\{\frac{d}{9\pi} + \frac{2d^2\Gamma(\frac{d}{2})}{9\pi^{\frac{3}{2}}}\cdot\sum_{n=2}^\infty \frac{(2n)!\,\Gamma(n + \frac{1}{2})}{16^n(n!)^2\,n\,\Gamma(n + \frac{d}{2})}\right\} + O(1).$$

$$(10.41)$$

Table 10.3. Comparison of estimations of the coefficient C_s of the polaron ground-state energy $E(\alpha) = \alpha^2 \cdot C_s + O(1)$ obtained for $d = 2$ as $\alpha \to \infty$.

Authors	C_s
S.Das Sarma, B.Mason [92]	−0.392699
R.Feynman,s theory	−0.392699
W.Xiaoguang, ... [91]	−0.4047
O.Hipolito [90]	−0.392699
Present $E_o(\alpha)$	−0.392699
Present $E_o(\alpha) + \Delta E_2$	−0.400538

The Two-Dimensional Polaron. For $d = 2$ (10.41) becomes

$$E^{(2)}(\alpha) = -\alpha^2 \cdot \left\{ \frac{\pi}{8} + \frac{\pi}{2} \sum_{n=2}^{\infty} \frac{[(2n)!]^2}{64^n (n!)^4 n} \right\} + O(1) = -\alpha^2 \cdot 0.400538 + O(1) .$$

For comparison, we give in table 10.3 our result with the known results of the polaron GSE for $d = 2$ in the strong coupling regime $\alpha \to \infty$.

Table 10.4. Comparison of estimations of the coefficient C_s of the polaron ground-state energy $E(\alpha) = \alpha^2 \cdot C_s + O(1)$ obtained for $d = 3$ as $\alpha \to \infty$.

Authors	C_s
Feynman, Schultz [100]	−0.1061
Pekar(by Miyake) [65]	−0.108504
Miyake [65]	−0.108513
Luttinger, Lu [77]	−0.1066
Marshall, Mills [49]	−0.1078
Sheng, Dow [101]	−0.1065
Adamowski,... [71]	−0.1085128
Feranchuk, Komarov [50]	−0.1078
Efimov, Ganbold [85]	−0.10843

The Three-Dimensional Polaron. In three-dimensions $R_3 = 1$ and from (10.41) we get

$$E^{(2)}(\alpha) = -\alpha^2 \cdot \left\{ \frac{1}{3\pi} + \frac{2}{\pi} \sum_{n=2}^{\infty} \frac{(2n)!}{16^n (n!)^2 n (2n + 1)} \right\} + O(1)$$

$$= -\alpha^2 \cdot 0.107766 + O(1), \quad C_s = -0.107766 .$$

This value is obtained by developing a specific perturbation expansion in (10.21) up to the order $\sim 1/2! \int d\sigma \, \{W^2[\mathbf{r}]\}_{\text{connected}}$.

The next correction for the coefficient C_s was obtained by us earlier in [44], where the details of estimation can be found. The result is [85]

$$C_s \leq -0.108431 .$$

The exact coefficient, obtained numerically by Miyake in [65] is

$$C_s^M = -0.108513.$$

A comparison of known results for the coefficient C_s for $d = 3$ is displayed in table 10.3.

Table 10.5. The estimations of the polaron ground-state energy $E_o(\alpha)$ and $E^{(2)}(\alpha)$ obtained for $d = 2$ in the intermediate range of α compared with known results [90, 102, 92].

					Present	
α	Feynman	[90]	[102]	[92]	E_o	$E_o + E_2$
0.6364	−1.0198	−1.0266	−1.0201	−1.0405	−1.020	−1.028
1.909	−3.2247	−3.2263	−3.2263	−3.5690	−3.231	−3.250
3.183	−5.9191	−6.0902	−5.9193	−6.9688	−5.928	−6.039
4.450	−9.6935	−9.8723	−9.7154	−11.388	−9.710	−9.871

10.8.3 The Intermediate Coupling Range

In the intermediate-coupling regime the main tool for obtaining an upper bound for the polaron energy is the variational approach [40, 59]. In the particular case of $d = 3$, the Feynman variational method based on a trial oscillator-type action gives the most successful upper bound of the polaron free energy, valid for arbitrary α. Generalizations of the Feynman action for $d = 3$ to the arbitrary density function [70] and arbitrary quadratic action [71] have improved this upper bound. In our opinion, the result [71] obtained for $d = 3$ is the best variational upper bound on the whole range of α. But this variational method does not give subsequent corrections to this bound. Other numerical methods dealing with this problem [77, 78] require specific complicated schemes of calculations which may introduce statistical errors. Lower bounds are harder to obtain due to the Coulomb-like singularity in the polaron action. The obtained lower bounds are either less stringent [69, 86] or not very accurate [68, 87]. Estimations of both the upper and lower bounds for the polaron self-energy should be improved.

Considering intermediate values of α, we have derived (10.15), (10.16) numerically, by the following iteration scheme:

$$F_{n+1}(t) = \Phi_t[\tilde{\Sigma}_n],$$
$$\tilde{\Sigma}_n(p) = \Omega_p[F_n], \quad n \geq 0, \tag{10.42}$$

starting from reasonable guessed functions $F_o(t)$ and $\tilde{\Sigma}_o(p)$ (see Eqs. (10.37) and (10.38)). Both the series $F_n(t)$ and $\tilde{\Sigma}_n(p)$ converge quickly to exact solutions $F(t)$, $\tilde{\Sigma}(p)$. We restrict ourselves by considering only up to the 5th iteration step in (10.42) because the value of the leading term $E_o(\alpha)$ does not change after $n \geq 6$. The results for $E_o(\alpha)$ and $E^{(2)}(\alpha)$ in two-dimensions are presented in table 10.5. The values of $E_o(\alpha)$ and $E^{(2)}(\alpha)$ for $d = 3$ are given in table 10.6 and displayed in figure 10.2 in comparison with the known data [71, 42, 86, 69]. For clarity, in figure 10.2 we showed only the deviation of quoted polaron energies from the standard "oscillator-potential" approximation result.

Our $E_o(\alpha)$ for $d = 3$ coincides with the upper bound obtained in [71] and improves the variational results calculated in [52]. Note that numerical results obtained in [80] at three points ($\alpha = 1, 3, 5$) by the method of "partial averaging" lie exactly between our curves for $E_o(\alpha)$ and $E^{(2)}(\alpha)$. Recent exact Monte–Carlo calculations [51] are in good agreement with our results for $d = 3$.

Table 10.6. The estimations of the polaron ground-state energy $E_o(\alpha)$ and $E^{(2)}(\alpha)$ obtained for $d = 3$ in the intermediate range of α compared with known results obtained in [42, 86, 69]. Our $E_o(\alpha)$ coincides exactly with the upper bound obtained in [71].

α	[42] upper	[86] upper	[86] lower	[69] upper	[69] lower	Present E_o	Present $E^{(2)}$
0.5	−0.5032	−0.5041	−0.5041	−0.5040	−0.5052	−0.504	−0.505
1.0	−1.0130	−1.0167	−1.0175	−1.0160	−1.0270	−1.014	−1.017
1.5	−1.5302			−1.5361	−1.576	−1.532	−1.539
2.0	−2.0554			−2.0640	−2.172	−2.058	−2.071
2.5	−2.5894			−2.5995	−2.872	−2.593	−2.614
3.0	−3.1333	−3.1645	−3.2122	−3.1421		−3.138	−3.167
4.0	−4.2565			−4.2771		−4.265	−4.305
5.0	−5.4401	−5.4945	−5.7767			−5.452	−5.528
7.0	−8.1127	−8.0406	−8.8832			−8.137	−8.255
9.0	−11.486	−10.834	−12.654			−11.54	−11.69
11.	−15.710	−13.905	−17.165			−15.83	−16.04
20.	−45.283					−45.33	−45.99
30.	−98.328					−98.52	−99.86
40.	−172.60					−173.4	−175.1

Our results [75] obtained with the proposed method provide a reasonable description of both two– and three–dimensional polarons at an arbitrary coupling α. The consideration could be extended to computing the other characteristics of the polaron, the effective mass and the average number of phonons, as well as to estimating the energy of the polaron in the presence of a magnetic field due to the validity of the proposed method for the complex functionals.

The successful completion of our calculations of the polaron GSE stimulates the fortcoming investigations of similar nonlocal actions arising in other areas of physics, namely in considering the solution of the Green function for wave propagation in randomly distributed media and the bound states in nonrelativistic quantum mechanics, considered in chapters 12 and 13.

11. The Character of the Phase Transition in Two- and Three-Dimensional φ^4 Theory

The phenomenon of spontaneous symmetry breaking, or in other words, vacuum structure rearrangement, is an important part of many quantum field constructions. Earlier investigations of the triviality problem have shown that for dimensions $d > 4$ the scalar theory of self-interaction behaved either unstable or trivial. Much interest is attracted to the cases $d = 2$ and $d = 3$ demonstrating nontrivial phase restructure. In this chapter, we will investigate an example of this phenomenon within the GER method. The problem, of course, can also be studied within the canonical quantization method, which is given in part I. However, the functional representation has the advantage of calculating the whole effective potential (EP) in this theory, which allows one to get more information about the phase structure and phase transitions in the system under consideration.

11.1 A Statement of the Problem

The scalar φ^4 theory in two and three dimensions has been intensively investigated [103, 104, 105] as a simple, but nontrivial example, with which the problem of spontaneous symmetry breaking or, in other words, the phase structure of quantum field models is studied. It has been found [106] that the highest order quantum corrections can give rise to the instability of the classical symmetrical vacuum. There are two phases in this system and phase transition phenomena occur at certain coupling strengths. The most difficult problem here is to determine the order of the phase transition (PT).

The simplest example, where the vacuum exhibits a nontrivial structure, is the φ_2^4 theory. Many papers [107, 108, 109] are devoted to the investigation of the nature of the PT in this model. We shall briefly treat some nonperturbative methods that seem to be basic to most investigations in this subject. An original approximation [110] using Hartree-type renormalization exhibits a first-order PT in this theory. A similar result was obtained [111] within the Gaussian EP approach. The dimensionless critical coupling constant, for which the first-order PT takes place, is $G = 1.62$ in both papers. These conclusions disagree with the mathematical theorems [103, 105] proving that the second-order PT should occur in the φ_2^4 model. There are papers [112, 113] where different variational methods have been used for solving this problem and the second-order PT has been observed in the region $G \sim 1$. In earlier studies [114, 115, 116], we have shown that the critical coupling constant leading to a second order PT cannot exceed the value $G_0 = 1.4392$ and may be found near $G_{\text{crit}} \sim 0.53$.

We study this problem using the method of EP. The absolute minimum of the EP $V(\varphi_0)$ at the point $\varphi_0 = \varphi_c$ determines the true ground state (vacuum) of the theory. If a PT takes place at a certain coupling $g = g_c$, then for $g < g_c$ the system is still in the original unbroken symmetry phase with $\varphi_c = 0$. On reaching $g = g_c$ the origin $\varphi_0 = 0$ is no longer the absolute minimum of $V(\varphi_0)$ and the system goes to a new state with $\varphi_c \neq 0$ corresponding to the lower energy. The first-order PT means that the point $\varphi = 0$ remains local, but not the absolute minimum of $V(\varphi_0)$. In other words, the first derivative of $V(\varphi_0)$ is zero and the second one is positive at the origin $\varphi_0 = 0$. In the case of the second-order transition, the point $\varphi_0 = 0$ is a local maximum of EP at $g > g_c$. The second derivative of $V(\varphi_0)$ at $\varphi_0 = 0$ becomes negative. Thus, the coefficient $\alpha(g)$ in the representation of $V(\varphi_0)$ for small φ_0

$$V(\varphi_0) = E(g) + \alpha(g) \cdot \varphi_0^2 + O(\varphi_0^4)$$

plays an important role in the determination of the character of the PT. If $\alpha(g)$ is zero at a certain $g = g_c$ and negative for $g > g_c$ up to $g \to \infty$, then one can say that the second-order PT appears here. On the other hand, the positiveness of $\alpha(g)$ for any g excludes the second-order transition. Rigorous calculation of $\alpha(g)$ at an arbitrary coupling constant is a complicated problem. However, we know that at large g, the coefficient $\alpha(g)$ remains negative in the case of the second-order PT and is positive if the transition is of the first order.

In this chapter, we shall study this problem qualitatively by using the GER method, described in chapter 9. We will show the possibility of the second-order PT in $g\varphi_2^4$ and give an estimation for the corresponding critical coupling constant g_c. For the model $g\varphi_3^4$ our result excludes the occurrence of the second-order PT.

11.2 The Renormalized Lagrangian of the $\varphi_{2,3}^4$ – Model

We consider the $g\varphi^4$ scalar field model in two- and three-dimensions. We will use throughout this chapter the Euclidean form of the model[1]. This theory contains ultraviolet divergences, but it is superrenormalizable, i.e., it has only a finite number of divergent Feynman diagrams. In order to remove these divergences we should introduce appropriate counter-terms into the

---

[1] In the case of the Euclidean metrics a separation of the coordinates into space and time coordinates is not important, so the accepted notation for the "space-time" is R^d, where d relates to the number of space coordinates plus Euclidean (imaginary) time as well.

Lagrangian. In this section we consider the superrenormalized scalar field theory with the Lagrangian density [114]

$$\mathcal{L}(x) = \frac{1}{2}\varphi(x)[\Box - m^2]\varphi(x) - \frac{g}{4}N_m\{\varphi^4(x)\} - R_m , \tag{11.1}$$

where we have introduced a "normally-ordered" form of interaction as follows:

$$N_m\{\varphi^4(x)\} = \varphi^4(x) - 6\varphi^2(x)D_m(0) + 3D_m^2(0),$$

$$D_m(x) = \int \frac{d^d k}{(2\pi)^d} \frac{\exp\{ikx\}}{m^2 + k^2} .$$

Here $x \in \Omega$, Ω is a large finite volume in R^d $(d = 2, 3)$ and m and g are the mass and the self-coupling constant, respectively. In two-dimensions $(d = 2)$ all divergences are only of the "tadpole-type" and are readily removed by introducing the normal product N_m of the fields $\varphi(x)$ into (11.1). In this case $R_m = 0$. In the three-dimensional theory there arise additional divergences which are cancelled by counter-terms

$$R_m = \frac{1}{2}A_m N_m\{\varphi^2(x)\} + \delta E_m ,$$

where

$$A_m = 6g^2 \int d^3x D_m^3(x) ,$$

$$\delta E_m = \frac{3}{4}g^2 \int d^3x D_m^4(x) - \frac{3}{2}g^3 \int \frac{d^3k}{(2\pi)^3} \left\{\int d^3x e^{ikx} D_m^2(x)\right\}^3 . \tag{11.2}$$

At small g the Lagrangian (11.1) describes a system invariant with respect to the transformation $\varphi \leftrightarrow -\varphi$. The question is whether this symmetry remains for increasing g.

11.3 The Effective Potential in $\varphi^4_{2,3}$ – Theory

The EP is defined as

$$V(\varphi_0) = - \lim_{\Omega \to \infty} \frac{1}{\Omega} \ln I_\Omega(\varphi_0),$$

$$I_\Omega(\varphi_0) = C_m \int \delta\varphi \delta\left(\varphi_0 - \frac{1}{\Omega}\int d^d x \varphi(x)\right) \exp\left\{\int_\Omega d^d x \mathcal{L}(x)\right\} , \tag{11.3}$$

$$C_m = \sqrt{\det\{-\Box + m^2\}}.$$

All integrations are performed in Euclidean metrics.

According to the GER method, we transform the field variable as

$$\varphi(x) = \phi_0 + b(x) + \phi(x), \tag{11.4}$$

where ϕ_0 is a constant, the new field variable $\phi(x)$ corresponding to the new mass μ and the function $b(x)$ satisfy the conditions

$$\int_\Omega d^d x\, \phi(x) = 0, \qquad \int_\Omega d^d x\, b(x) = 0, \qquad b^2(x) = b^2 = \text{const}.$$

Let us go over to normal ordering in the new field $\phi(x)$ using the well-known formula [106]

$$N_m\left\{\exp\{\beta\varphi(x)\}\right\} = N_\mu\left\{\exp\{\beta(\phi_0 + b(x) + \phi(x)) + \frac{\beta^2}{2}\Delta(m,\mu)\}\right\},$$

$$\Delta = \Delta(m,\mu) = D_m(0) - D_\mu(0), \tag{11.5}$$

$$D_\mu(x) = \int \frac{d^d k}{(2\pi)^d}\frac{\exp\{ikx\}}{\mu^2 + k^2} - \frac{1}{\mu^2\Omega}.$$

Let us substitute (11.4) and (11.5) into (11.3) and integrate over $d\phi_0$. Then, following the key steps of the GER method, we obtain the formula for the EP as follows:

$$V(\varphi_0) = V_0(\varphi_0) + V_{sc}(\varphi_0). \tag{11.6}$$

The leading order term $V_0(\varphi_0)$ of the EP in (11.6) is

$$V_0(\varphi_0) = -\frac{1}{2}\int \frac{d^d k}{(2\pi)^d}\left[\ln\left(1 + \frac{m^2 - \mu^2}{\mu^2 + k^2}\right) - \frac{m^2 - \mu^2}{\mu^2 + k^2}\right] + \frac{m^2}{2}(\varphi_0^2 + b^2)$$

$$+ \frac{g}{4}(\varphi_0^4 + 6\varphi_0^2 b^2 + b^4 - 6\Delta(\varphi_0^2 + b^2) + 3\Delta^2) \tag{11.7}$$

$$+ \frac{\varphi_0^2 + b^2}{2}(A_m - A_\mu) + (\delta E_m - \delta E_\mu - \frac{1}{2}A_m\Delta).$$

High-order contributions $V_{sc}(\varphi_0)$ for the leading-order potential (11.7) are defined by:

$$V_{sc}(\varphi_0) = -\lim_{\Omega\to\infty}\frac{1}{\Omega}\ln J_\Omega(\varphi_0), \tag{11.8}$$

where

$$J_\Omega(\varphi_0) = e^{-\Omega V_{sc}(\varphi_0)} = \int d\sigma_\mu \exp\left\{\int_\Omega d^d x\right.$$

$$\times N_\mu\left\{-\frac{g}{4}\left[\phi^4(x) + 4\phi^3(x)(\varphi_0 + b(x)) + 12\varphi_0 b(x)\phi^2(x)\right]\right. \tag{11.9}$$

$$\left.\left. - \left[\frac{1}{2}A_\mu\phi^2(x) + A_\mu b(x)\phi(x) + \delta E_\mu + \frac{1}{2}(b^2 + \varphi_0^2)A_\mu\right]\right\}\right\}.$$

The normalization is

$$\int d\sigma_\mu = C_\mu \int \delta\phi \exp\left\{-\frac{1}{2}\int_\Omega d^d x\,\phi(x)(-\Box + \mu^2)\phi(x)\right\} = 1,$$

where

$$\int_\Omega d^d x \phi(x) = 0 \,.$$

The requirement that the linear term $N_\mu\{\phi\}$ must not arise in the interaction and the quadratic field configurations be concentrated in the Gaussian measure $d\sigma_\mu$ leads to the following constraint equations for the parameters $b(x)$ and μ:

$$b(x) \cdot [-m^2 + 3g(\Delta - \varphi_0^2) - gb^2 - A_m + A_\mu] = 0 \,,$$

$$\mu^2 - m^2 + 3g(\Delta - \varphi_0^2 - b^2) - A_m + A_\mu = 0 \,. \qquad (11.10)$$

Equations (11.6)–(11.10) completely define the GER of the initial EP at an arbitrary coupling g.

11.4 The Leading-Order Term of the EP as the "Cactus-type" Potential in φ_2^4

Below we will investigate the EP in (11.7), whose parameters $b(x)$ and μ are limited by the constraints (11.10). For further consideration, it will be convenient to work in units of m dealing with numerical results. We define

$$G = g/2\pi m^{4-d}, \quad \xi = (\mu/m)^{4-d}, \quad \Phi_0^2 = 4\pi m^{2-d}\varphi_0^2 \quad \text{and} \quad B^2 = 4\pi m^{2-d}b^2 \,.$$

In two-dimensions, the leading-order part of the EP becomes

$$V_0(\Phi_0) = \frac{m^2}{8\pi} \left\{ \xi - 1 - \ln\xi + \Phi_0^2 + B^2 \right.$$

$$\left. + \frac{G}{4} \left[\Phi_0^4 + B^4 + 3\ln^2\xi + 6(B^2\Phi_0^2 - B^2\ln\xi - \Phi_0^2\ln\xi) \right] \right\} \,. \qquad (11.11)$$

We note that the potential (11.11) is invariant for $\Phi_0 \leftrightarrow B$.

The parameters ξ and B in (11.11) are determined by the following equations:

$$\begin{cases} B^2(\xi - GB^2) = 0 \,, \\ 2\xi - 2 + 3G(\ln\xi - \Phi_0^2 - B^2) = 0 \,. \end{cases} \qquad (11.12)$$

Let us consider the constraint (11.12). A pair of "trivial" solutions

$$B = 0 \quad \text{and} \quad \xi = 1 - \frac{3G}{2}(\ln\xi - \Phi_0^2) \,, \qquad (11.13)$$

can be found for an arbitrary coupling constant G. Substituting these solutions to (11.11) we see that our leading-order term $V_0(\Phi_0)$ becomes exactly the "cactus-type" (or, so-called "Gaussian") EP [107, 111].

However, for $G > G_0 = 1.4392$ an additional pair of "nontrivial" solutions

$$B = \frac{\xi}{G} \quad \text{and} \quad \xi = -2 + \frac{3G}{2}(\ln \xi - \Phi_0^2) \qquad (11.14)$$

appears here too. So for $G < G_0$ the only solution to be substituted into (11.11) is the "trivial" one, but since $G > G_0$ there is an alternative: one can choose either (11.13) or (11.14). We choose the pair obeying the lowest value of $V_0(\Phi_0)$ for certain fixed Φ_0.

All necessary calculations can be performed numerically. The potential $V_0(\Phi_0)$ obtained is plotted in figure 11.1 . Near the origin $\Phi_0 = 0$ the potential $V_0(\Phi_0)$ is represented by the "nontrivial" branch (if $G > G_0$) $B \neq 0$ as it is situated lower than the "trivial" one. But for larger values of Φ_0 the "trivial" solution $B = 0$ provides the lowest value of the potential. This picture leads to an interesting result. Let us consider the local minima of both branches. For $B = 0$ the minimum point Φ_0 in figure 11.1 is given by the equations

$$\begin{cases} B = 0, \\ 2 - 3G \ln \xi + G\Phi_0^2 = 0. \end{cases} \qquad (11.15)$$

On the other hand, the minimum of the "nontrivial" branch $B \neq 0$ is fixed at the origin $\Phi_0 = 0$ for any $G > G_0$ and (11.12) becomes

$$\begin{cases} \Phi_0 = 0, \\ 2 - 3G \ln \xi + GB^2 = 0. \end{cases} \qquad (11.16)$$

Owing to the invariance of the potential $V_0(\Phi_0, B)$ in (11.11) for $\Phi_0 \leftrightarrow B$ our (11.15) and (11.16) are identical. In other words, the minima of the potential (11.11) corresponding to different solutions of (11.12) are equal. The vacuum with $\langle \Phi(x) \rangle = \Phi_0 \neq 0$ is not lower than the initial one located at the origin point $\langle \Phi(x) \rangle = \Phi_0 = 0$. There is no reason for an occurrence of the first-order PT.

11.5 The Non-Gaussian Correction in φ_2^4

In the previous section, we derived the expression for the EP consisting of two parts. Considering only the "leading" term $V_0(\varphi_0)$ one can say nothing about the nature of the PT in the theory. To answer this question one should also consider the remaining part of the EP, $-V_{sc}(\varphi_0)$, defined in (11.9). In the weak coupling limit one can estimate it by expanding the exponential in (11.9) in a perturbative series. In general, explicit calculations of this non-Gaussian PI at arbitrary values of g and φ_0 is a complicated problem. However, our aim is to estimate it for infinitesimal values of φ_0 at arbitrary g. This task is solvable if we use the following technique.

We rewrite (11.9) in the form correct for infinitesimal φ_0:

$$J_\Omega(\varphi_0) = \int d\sigma_\mu \exp\left\{-\frac{g}{4}\int_\Omega d^dx N_\mu\left[\phi^4(x) + 4b(x)\phi^3(x)\right]\right.$$

$$\left. + \frac{g^2\varphi_0^2}{2}\left[\int_\Omega d^dx N_\mu(\phi^3(x) + 3b(x)\phi^2(x))\right]^2\right\}. \tag{11.17}$$

This representation can easily be obtained owing to the validity of the following transformation in the integrand of (11.9):

$$\exp(-\varphi_0 W) = \cosh(\varphi_0 W) \simeq \exp\left\{\frac{1}{2}\varphi_0^2 W^2 + O(\varphi_0^4)\right\}$$

for infinitesimal φ_0 and a finite functional W.

Applying to (11.17) Jensen's inequality we get an upper bound

$$V_{sc}(\varphi_0) \le V_{sc}^+(\varphi_0) = -\frac{g^2\varphi_0^2}{2\Omega}\int_\Omega d^dx \int_\Omega d^dy \int d\sigma_\mu$$

$$\times \left\{N_\mu\phi^3(x)N_\mu\phi^3(y) + 9b(x)b(y)N_\mu\phi^2(x)N_\mu\phi^2(y)\right\}. \tag{11.18}$$

It is easy to get that

$$\int d\sigma_\mu N_\mu\phi^3(x)N_\mu\phi^3(y) = 6D_\mu^3(x-y),$$

$$\int d\sigma_\mu N_\mu\phi^2(x)N_\mu\phi^2(y) = 2D_\mu^2(x-y).$$

Then, we rewrite (11.18) in the form

$$V_{sc}^+(\varPhi_0) = -\frac{m^2}{8\pi}\frac{3G^2\varPhi_0^2}{2\xi}(Q + 3B^2), \tag{11.19}$$

$$Q = \iiint\limits_0^1 d\alpha d\beta d\gamma \frac{\delta(1-\alpha-\beta-\gamma)}{\alpha\beta+\alpha\gamma+\beta\gamma}$$

$$= \frac{4\pi\ln 2}{3\sqrt{3}} - 4\int\limits_0^1 \frac{du}{u^2+3}\ln(1-u^2) = 2.3439.... \,.$$

Substituting the parameters ξ and B in either (11.13) or (11.14) into (11.19) one gets the behavior of $V_{sc}^+(\varPhi_0)$ for small values $\varPhi_0 \sim 0$. Omitting the details of the calculations we write the results

$$V_{sc}^+(\varPhi_0) = -\frac{m^2}{8\pi}\left\{-\frac{3Q}{2}G^2\varPhi_0^2 + O(\varPhi_0^4)\right\} \quad \text{for} \quad G < G_*$$

and

$$V_{sc}^+(\varPhi_0) = -\frac{m^2}{8\pi}\left\{-\left[\frac{3QG^2}{2\xi} + \frac{9G}{2}\right]\varPhi_0^2 + O(\varPhi_0^4)\right\} \quad \text{for} \quad G > G_*,$$

$$3G \ln \xi - \xi - 2 = 0 \,.$$

From (11.11) we get the following asymptotic behaviour:

$$V_0(\Phi_0) = \frac{m^2}{8\pi} \left\{ \Phi_0^2 + O(\Phi_0^4) \right\}$$

as $\Phi_0 \to 0$ at any G.

Finally, taking into account (11.8) we obtain the following behaviour for an upper bound of the EP in the region of small $\Phi_0 \sim 0$:

$$V^+[\Phi_0] = V_0[\Phi_0] + V_{sc}^+[\Phi_0] = \frac{m^2}{8\pi} \left[\alpha(G)\Phi_0^2 + O(\Phi_0^4) \right] \,,$$

where

$$\alpha(G) = \begin{cases} \alpha_1(G) = 1 - 3QG^2/2 \,, & G \le 1.6251 \,, \\ \alpha_2(G) = 1 - 3QG^2/(2\xi) - 9G/2 \,, & G > 1.6251 \,, \end{cases} \tag{11.20}$$

$$3G \ln \xi - \xi - 2 = 0 \,.$$

One can easily check that the coefficient $\alpha_1(G)$ in (11.20) becomes negative as $G > G_{\text{crit}} = 0.5333$ and remains negative for increasing G. The coefficient $\alpha_2(G)$ is negative for arbitrary $G > 1.4392$. In our opinion, this indicates the occurrence of the second-order PT in the model under consideration.

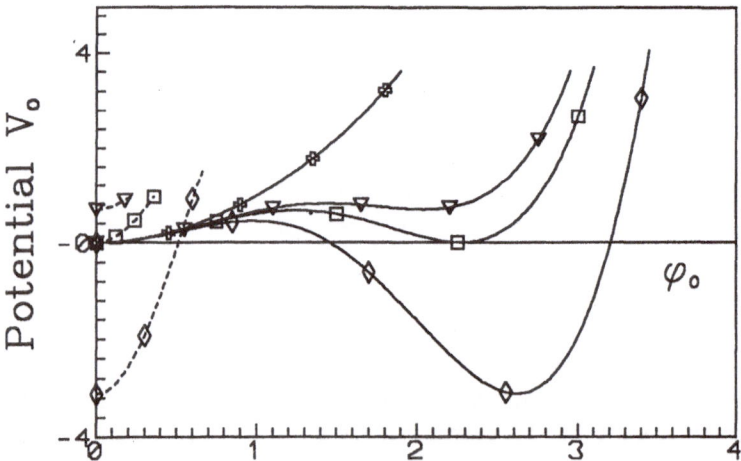

Fig. 11.1. The Gaussian part $V_0(\Phi_0)$ (in units of $m^2/8\pi$) of the effective potential in two-dimension as a function of Φ_0 for different values of the coupling constant: crosses, $G = 0.5$; triangles, $G = 1.5$; squares, $G = 1.6251$ and rhombs, $G = 2.0$. The dashed lines represent the "nontrivial" branches. The "trivial" branches are denoted by the solid lines.

11.6 The Strong Coupling Regime in φ_3^4

In the three-dimensional case the counter-terms defined by (11.2) play an important role in the behaviour of the EP in the strong coupling regime. We have

$$D_m(x) = \frac{\exp\{-\mu|x|\}}{4\pi|x|}, \quad \Delta = \frac{m}{4\pi}(\xi - 1). \tag{11.21}$$

Substituting (11.21) into (11.10) we get

$$B \cdot [2 + 3G(\Phi_0^2 - \xi + 1) + GB^2 + 3G^2 \ln\xi] = 0,$$

$$-2\xi^2 + 2 + 3G(\Phi_0^2 - \xi + 1) + 3GB^2 + 3G^2 \ln\xi = 0.$$

A nontrivial solution $B \neq 0$ exists only for $0 < \xi < 1$. Let us consider the solution $B = 0$. In the strong coupling regime we obtain

$$\xi = G\sqrt{\frac{3}{2}\ln G} + O(G\sqrt{\ln\ln G}).$$

In other words, the effective coupling constant

$$G_{eff} = \frac{g}{2\pi\mu} = \frac{G}{\xi} = \sqrt{\frac{2}{3\ln G}}\left\{1 + O\left(\frac{\ln\ln G}{\ln G}\right)\right\}$$

is small as $G \to \infty$ and actually we deal with the weak coupling regime. Then one can successfully develop perturbation expansion in G_{eff} series for the functional integral (11.9):

$$V_{sc}(\varphi_0) = \sum_{n=1}^{\infty} G_{eff}^n V_{sc}^{(n)}(\varphi_0).$$

Here $V_{sc}^{(1)} = 0$ owing to normal ordering in the exponential in (11.9). After some calculations we obtain

$$V_{sc}^{(1)}(\varphi_0) = V_{sc}^{(2)}(\varphi_0) = 0,$$

$$V_{sc}^{(3)}(\varphi_0) = \frac{m^3}{8\pi}\frac{18C_1}{\xi}G^3\Phi_0^2,$$

where the constant is

$$C_1 = \frac{1}{2\pi^6}\iiint\frac{d^3k\,d^3p\,d^3q}{(1+k^2)(1+p^2)(1+q^2)(1+(k+p)^2)(1+(k+q)^2)}$$

$$= \frac{16}{\pi}\int_0^{\infty}\frac{du}{1+4u^2}(\arctan u)^2 = 1.7593\ldots.$$

Taking into account the "cactus-type" potential

$$V_0(\Phi_0) = \frac{m^3}{8\pi} \left\{ E_c(G) + \frac{3G}{2}(G\ln\xi - \xi)\Phi_0^2 + O(\Phi_0^4) \right\},$$

we finally obtain the EP

$$V(\Phi_0) = V_0(\Phi_0) + V_{sc}(\Phi_0) = \frac{m^3}{8\pi} \left\{ E(G) + \alpha\Phi_0^2 + O(\Phi_0^4) \right\},$$

where the desired coefficient

$$\alpha(G) = \frac{3G^2}{2}\ln G \left\{ 1 + \frac{\sqrt{96}C_1}{(\ln G)^{3/2}} + O\left(\frac{1}{(\ln G)^{5/2}}\right) \right\} \qquad (11.22)$$

is positive. This result excludes the second-order PTs in the ϕ_3^4 model. It can be accepted as an argument in favor of either the existence of the only first-order transition or the absence of any PT with spontaneous breaking of symmetry in the three-dimensional case.

Comparing the results (11.20), (11.22) for $d = 2$ and $d = 3$ we find that the effective mass renormalization is crucial for this problem. In two-dimensions the mass renormalization includes the "tadpole" divergences only and the behaviour of $\alpha(G)$ in (11.20) is a favor of the second-order PT in ϕ_2^4. For $d = 3$ the mass renormalization contains an additional term of the second perturbative expansion's order which has the opposite sign comparable with a "tadpole" contribution. As a result, the function $\alpha(G)$ in (11.22) remains positive for all $G > 0$.

12. Wave Propagation in Randomly Distributed Media

Theoretical investigation of the propagation properties of waves in a stochastically distributed environment reflects a certain interest owing to its many practical applications, including calculation of electronic conductance in crystals [117], wave localization [118] and dumping of signals in the atmosphere or water [119]. A series of different methods has been applied to this problem, among which path integral techniques [120, 121, 122] are of considerable interest. In this chapter we investigate wave transmission in a randomly distributed medium using the GER method.

12.1 The Green Function of the Wave Equation

The propagation of a wave $u(\mathbf{x})$ (e.g., electromagnetic) in a time-independent environment can be described by the wave equation given in the real 3-dimensional space $\mathbf{x} \in R^3$:

$$[\Delta + \omega^2(1 + \varepsilon(\mathbf{x}))]\, u(\mathbf{x}|\varepsilon) = J(\mathbf{x}),\qquad (12.1)$$

$$\omega \neq 0.$$

The constant ω is the "dielectric constant" and defines the frequency of unperturbed waves. $J(\mathbf{x})$ is the source function.

The random noise is described by a random stationary field $\varepsilon(\mathbf{x})$, which is assumed to vary stochastically with a certain correlation function $\langle \varepsilon(\mathbf{x})\varepsilon(\mathbf{y})\rangle$. For simplicity we shall consider the Gaussian noise

$$\langle \varepsilon(\mathbf{x})\varepsilon(\mathbf{y})\rangle_\varepsilon = \lambda P(\mathbf{x} - \mathbf{y})$$

$$= \lambda \exp\left(-\frac{(\mathbf{x} - \mathbf{y})^2}{4l^2}\right) = \lambda \int \frac{d\mathbf{k}}{\pi^{3/2}} \cdot \exp\left\{-\mathbf{k}^2 + i\mathbf{k}\frac{\mathbf{x} - \mathbf{y}}{l}\right\},\qquad (12.2)$$

where the interaction coefficient λ shows the intensity of noise described by the distribution function $P(\mathbf{x} - \mathbf{y})$ with a correlation length l. These two constants define the influence of the Gaussian noise on the propagation of waves in media.

The solution of (12.1) can be represented in the form

$$u(\mathbf{x}|\varepsilon) = \int d\mathbf{y}\, G(\mathbf{x}, \mathbf{y}|\varepsilon) J(\mathbf{y}) ,$$

where $G(\mathbf{x}, \mathbf{y}|\varepsilon)$ is the Green function of wave equation (12.1):

$$[\Delta + \omega^2(1 + \varepsilon(\mathbf{x}))]\, G(\mathbf{x}, \mathbf{y}|\varepsilon) = \delta(\mathbf{x} - \mathbf{y}) . \tag{12.3}$$

The problem is to find the solution of (12.1) and then average it over random fields $\varepsilon(\mathbf{x})$ to find the wave amplitude:

$$u(\mathbf{x}) = \langle u(\mathbf{x}|\varepsilon)\rangle_\varepsilon .$$

For this the Green function should be averaged over random fields $\varepsilon(\mathbf{x})$:

$$G(\mathbf{x} - \mathbf{y}) = \langle G(\mathbf{x}, \mathbf{y}|\varepsilon)\rangle_\varepsilon .$$

Thus we consider this problem solved if the averaged Green function $G(\mathbf{x})$ is found and its asymptotic behaviour for large distances $|\mathbf{x}| \to \infty$ can be calculated.

Let us proceed to solve the equation (12.3) for the Green function. It is essential that the operator

$$K = \Delta + \omega^2(1 + \varepsilon(\mathbf{x}))$$

is not definitely positive. We shall consider the solution

$$G(\mathbf{x}, \mathbf{y}|\varepsilon) = \frac{1}{K + i0}\delta(\mathbf{x} - \mathbf{y}),$$

corresponding to the so-called causal Green function. This solution can be written in integral representation like (8.13) as follows:

$$G(\mathbf{x}, \mathbf{y}|\varepsilon) = -\frac{i}{2}\int\limits_0^\infty du\, e^{\frac{i}{2}(K+i0)u} \cdot \delta(\mathbf{x} - \mathbf{y})$$

$$= -\frac{i}{2}\int\limits_0^\infty du\, \mathrm{T}_\tau \exp\left\{\frac{i}{2}\int\limits_0^u d\tau\left[\left(\frac{\partial}{\partial \mathbf{x}_\tau}\right)^2 + \omega^2(1 + \varepsilon(\mathbf{x}_\tau))\right]\right\}\delta(\mathbf{x} - \mathbf{y}).$$

Here we have used the "time-ordering" operator T_τ introduced in section 8.4. Repeating all the calculations of chapter 9, one gets

$$|\varepsilon) = -G(\mathbf{x}, \mathbf{y} \qquad \frac{1}{2}\int\limits_0^\infty \frac{du}{(2\pi i u)^{3/2}}\exp\left[-\frac{i}{2}\left(\omega^2 u + \frac{(\mathbf{x} - \mathbf{y})^2}{u}\right)\right]\cdot I_u(\mathbf{x}, \mathbf{y}|\varepsilon) ,$$

where a PI is introduced:

$$I_u(\mathbf{x}, \mathbf{y}|\varepsilon) = \int d\sigma_0 \exp\left\{\frac{i}{2}\omega^2\int\limits_0^u d\tau\, \varepsilon\left(\mathbf{x}\frac{\tau}{u} + \mathbf{y}\left(1 - \frac{\tau}{u}\right) + \boldsymbol{\nu}(\tau)\right)\right\} , \tag{12.4}$$

with the measure defined as

$$d\sigma_o = C_0 \delta\nu \exp\left\{ \frac{i}{2} \int\limits_0^u d\tau \dot{\nu}^2(\tau) \right\} .$$

The integration in (12.4) is taken over "paths" ν obeying the condition

$$\nu(0) = \nu(u) = 0 .$$

Here the normalization is chosen as

$$\int d\sigma_o = 1 \qquad \text{or,} \qquad I_u(\mathbf{x}, \mathbf{y}|\varepsilon)|_{\varepsilon=0} = 1 .$$

Now we can average the functional $I_u(\mathbf{x}, \mathbf{y}|\varepsilon)$ over the random fields $\varepsilon(\mathbf{x})$:

$$I_u(\mathbf{x} - \mathbf{y}) = \langle I_u(\mathbf{x}, \mathbf{y}|\varepsilon) \rangle_\varepsilon$$

$$= \int d\sigma_o \exp\left\{ -\lambda \frac{\omega^4}{8} \iint\limits_0^u d\tau d\tau' P\left(\nu(\tau) - \nu(\tau') + (\mathbf{x} - \mathbf{y}) \frac{\tau - \tau'}{u} \right) \right\} .$$

The averaged Green function is

$$G(\mathbf{x}) = -\frac{1}{2} \int\limits_0^\infty \frac{du}{(2\pi i u)^{3/2}} \exp\left[\frac{i}{2} \left(\omega^2 u + \frac{\mathbf{x}^2}{u} \right) \right] \cdot I_u(\mathbf{x}) ,$$

where

$$I_u(\mathbf{x}) = C_0 \int\limits_{\nu(0)=\nu(u)=0} \delta\nu \exp\left\{ \frac{i}{2} \int\limits_0^u d\tau \dot{\nu}^2(\tau) \right. \tag{12.5}$$

$$\left. -\lambda \frac{\omega^4}{8} \iint\limits_0^u d\tau d\tau' P\left(\nu(\tau) - \nu(\tau') + \mathbf{x} \frac{\tau - \tau'}{u} \right) \right\} .$$

For further convenience we introduce the following notation:

$$r = |\mathbf{x}|, \quad u = \frac{r}{\omega} z, \quad \beta = r\omega, \quad \tau = \frac{z}{\omega^2} t, \quad \tau' = \frac{z}{\omega^2} s, \quad g = \frac{\lambda z^2}{4}$$

and change the variable of the PI:

$$\nu(\tau) = \frac{\sqrt{z}}{\omega} \rho(t) .$$

Then we have

$$G(\beta) = -\frac{\omega}{2\sqrt{\beta}} \int\limits_0^\infty \frac{dz}{(2\pi i z)^{3/2}} \exp\left[i\frac{\beta}{2} \left(z + \frac{1}{z} \right) \right] \cdot I(\beta, z) ,$$

where

$$I(\beta, z) = C_o \int\limits_{\rho(0)=\rho(\beta)} \delta\rho \exp\left\{\frac{i}{2}\int\limits_0^\beta dt\, \dot\rho^2(t)\right. \tag{12.6}$$

$$\left. -\frac{g}{2}\iint\limits_0^\beta dt\,ds\, P\left(\frac{\sqrt{z}}{\omega}(\rho(t)-\rho(s)) + \mathbf{n}\frac{t-s}{\omega}\right)\right\},$$

where

$$\mathbf{n} = \frac{\mathbf{x}}{|\mathbf{x}|}, \qquad (\mathbf{nn}) = 1.$$

12.2 Calculation of the PI by the GER Method

In order to apply the formulae of chapter 9 let us introduce into (12.6) symmetrical limits by redefining

$$2T = \beta, \quad t \to t-T, \quad s \to s-T, \quad \rho(t) \to \rho(t-T).$$

Then we rewrite (12.6):

$$I_T(z) = C_0 \int\limits_{\rho(-T)=\rho(T)=0} \delta\rho \cdot \exp\left\{\frac{i}{2}\int\limits_{-T}^T dt\, \dot\rho^2(t)\right.$$

$$\left. -\frac{g}{2}\iint\limits_{-T}^T dt\,ds\, P\left(\frac{\sqrt{z}}{\omega}[\rho(t)-\rho(s)] + \mathbf{n}\frac{t-s}{\omega}\right)\right\}.$$

Let us introduce the operator

$$D_o^{-1}(t-s) = i\frac{\partial^2}{\partial t^2}\delta(t-s).$$

Note that it differs from (9.13) by the factor $-i$. The Green function $D_o(t,s)$ corresponding to this operator satisfies some periodic conditions and reads

$$D_o(t,s) = -\frac{i}{2}|t-s| - \frac{ts}{2T} \to -\frac{i}{2}|t-s|.$$

Its Fourier transform is

$$\tilde{D}_o(p) = \frac{i}{p^2}.$$

Then we rewrite

$$I_T(z) = C_o \int\limits_{\rho(-T)=\rho(T)=0} \delta\rho \exp\left\{-\frac{1}{2}\iint\limits_{-T}^T dt\,ds(\rho(t)\mathbf{D}_o^{-1}(t-s)\rho(s)) - gW[\rho]\right\},$$

$$C_o = (\det \mathbf{D}_o)^{-1/2} .$$

The free "kinetic" term is diagonal:

$$(\rho(t)\mathbf{D}_o^{-1}(t-s)\rho(s)) = (b_i(t)\delta_{ij} D_o^{-1}(t-s)b_j(s)) .$$

The interaction is given by

$$gW[\rho] = \frac{g}{2} \iint\limits_{-T}^{T} dtds\, P\left(\frac{\sqrt{z}}{\omega}(\rho(t) - \rho(s)) + \mathbf{n}\frac{t-s}{\omega}\right)$$

$$= \frac{g}{2} \iint\limits_{-T}^{T} dtds \int \frac{d\mathbf{k}}{\pi^{3/2}} \exp\left\{-\mathbf{k}^2 + i\frac{\mathbf{k}}{l\omega}\left(\sqrt{z}(\rho(t) - \rho(s)) + \mathbf{n}(t-s)\right)\right\} .$$

Comparing it with (9.17) we find that the measure $d\mathcal{K}$ of momentum integration now becomes

$$d\mathcal{K}_{\mathbf{n}}(\mathbf{k}, t-s) = \frac{d\mathbf{k}}{\pi^{3/2}} \exp\left\{-\mathbf{k}^2 + \frac{i}{l\omega}(\mathbf{kn})(t-s)\right\} ,$$

Following the GER method, we define the new measure

$$d\sigma = C\delta\rho\exp\left\{-\frac{1}{2} \iint\limits_{-T}^{T} dtds(\rho(t)\mathbf{D}^{-1}(t-s)\rho(s))\right\} ,$$

where

$$(\rho(t)\mathbf{D}^{-1}(t-s)\rho(s)) = (\rho_i(t)D_{ij}^{-1}(t-s)\rho_j(s)) .$$

Notice that, the operator D_{ij}^{-1} has nondiagonal elements owing to the presence of the vector $\mathbf{n}$ in $W[\rho]$.

In the following we will use the notations

$$\int d\sigma \exp\left\{i\frac{\sqrt{z}}{l\omega}\mathbf{k}\,(\rho(t) - \rho(s))\right\} = \exp\left\{-\frac{z}{(l\omega)^2}(\mathbf{k}F(t-s)\mathbf{k})\right\} ,$$

$$(\mathbf{k}F(t-s)\mathbf{k}) = (k_i F_{ij}(t-s)k_j) ,$$

$$\mathbf{F}(t-s) = \mathbf{D}(0) - \mathbf{D}(t-s) = \int\limits_{0}^{\infty} \frac{dp}{\pi}\, [1 - \cos p(t-s)]\, \tilde{\mathbf{D}}(p^2) , \qquad (12.7)$$

$$g\hat{W}[b] = \frac{g}{2} \iint\limits_{-T}^{T} dtds \int \frac{d\mathbf{k}}{\pi^{3/2}} \exp\left\{-\left(\mathbf{k}[\mathbf{I} + \frac{z}{(l\omega)^2}\mathbf{F}(t-s)]\mathbf{k}\right)\right\}$$

$$\exp\left\{i\frac{\sqrt{z}}{\omega l}\mathbf{k}\,(\mathbf{b}(t) - \mathbf{b}(s)) + i\mathbf{kn}\frac{t-s}{\omega l}\right\} ,$$

$$(|\mathbf{q}\rangle\langle\mathbf{q}|)_{ij} = q_i q_j \, ,$$

$$\Phi(t-s) = \frac{gz^3}{(\omega l)^4} \int \frac{d\mathbf{q}}{\pi^{3/2}} |\mathbf{q}\rangle\langle\mathbf{q}|$$

$$\times \exp\left\{-\left(\mathbf{q}[\mathbf{I} + \frac{z}{(l\omega)^2}\ vecF(t-s)]\mathbf{q}\right) + i q n\frac{t-s}{\omega l}\right\}$$

$$= \Phi_0(t-s) + |\mathbf{n}\rangle\langle\mathbf{n}|\,\Phi_1(t-s) \, ,$$

$$\Phi_{ij}(t-s) = \delta_{ij}\cdot\Phi_0(t-s) + n_i n_j \Phi_1(t-s) \, .$$

Then we get

$$w_{ij}(t-s) = g\frac{\delta^2 \hat{W}[\mathbf{b}]}{\delta b_i(t)\delta b_j(s)}|_{\mathbf{b}=0} = -\left[\tilde{\Phi}_{ij}(0) - \Phi_{ij}(t-s)\right] \, .$$

Following all the steps described in chapter 9 we finally obtain

$$I_T(z) = e^{-2T E_o(z)} \cdot J_T(z) \, , \tag{12.8}$$

$$J_T(z) = C \int \delta\rho \exp\left\{-\frac{1}{2}\iint\limits_{-T}^{T} dt ds (\rho D^{-1}\rho) - g : \tilde{W}[\rho] : \right\} \, ,$$

where the leading-order term (or the zeroth approximation of the GER method) is

$$E_o(z) = \frac{3}{2\pi} \int\limits_0^\infty dp \left[\ln\left(1 + \frac{i}{p^2}\Sigma(p)\right)\right]$$

$$-\frac{g}{2}\int\limits_{-\infty}^{\infty} dt \int \frac{d\mathbf{q}}{\pi^{3/2}} \exp\left\{-\left(\mathbf{q}[\mathbf{I} + \frac{z}{(l\omega)^2}F(t)]\mathbf{q}\right) + i q n\frac{t}{\omega l}\right\} \, .$$

The interaction functional in the new representation is

$$\tilde{W}[\rho] = -\frac{g}{2}\iint\limits_{-T}^{T} dt ds \int d\mathcal{K}_{\mathbf{n}}(\mathbf{q}, t-s) \tag{12.9}$$

$$\times \exp\left\{-\left(\mathbf{q}[\mathbf{I} + \frac{z}{(l\omega)^2}F(t-s)]\mathbf{q}\right) + i q n\frac{t-s}{\omega l}\right\} : e_2^{i\sqrt{\frac{z}{l\omega}}[\mathbf{q}(\rho(t)-\rho(s))]} :$$

where $: e_2^z := e^z - 1 - z - \frac{z^2}{2}.$

The requirement of the "correct" form for the interaction functional is hold if we put

$$\tilde{\mathbf{D}}(p) = \tilde{\mathbf{D}}_0(p) + \tilde{\mathbf{D}}_0(p)\Sigma(p)\tilde{\mathbf{D}}(p) \, ,$$

or

$$\tilde{\mathbf{D}}(p) = \frac{\tilde{\mathbf{D}}_0(p)}{I + \tilde{\mathbf{D}}_0(p)\Sigma(p)} = \frac{iI}{p^2 + i\Sigma(p)} \, .$$

Then (9.23) and (9.25) become

$$\Sigma(p) = -\tilde{w}(p) = \int_{-\infty}^{\infty} dt\,[1 - \cos(pt)]\,\Phi(t) \tag{12.10}$$

and

$$\mathbf{F}(t) = i \int_{0}^{\infty} \frac{dp}{\pi} \frac{1 - \cos p(t)}{p^2 + i\Sigma(p)} . \tag{12.11}$$

12.3 The Green Function for Large Distances

The initial (12.5) and the new (12.8) representations are equivalent. The next step is to solve (12.10) and (12.11) which allows to calculate the function $E_o(z)$. The explicit form of the interaction functional (12.9) allows to calculate the highest corrections. In principle, these calculations are similar to those in the polaron problem except that now all functionals are complex. Nevertheless, all transformations of the GER method applied here are valid. In the future we plan to solve these equations and investigate the behaviour of the Green function $G(\mathbf{x})$ for different values of the parameters λ and l.

So the main problem is to solve the integral equations (12.10) and (12.11). However this represents a laborous task and one can by-pass this difficulty considering the large distance's behaviour of the Green function $G(\beta)$.

We now consider wave propagation for large distances $\beta \to \infty$. Then by analogy with the polaron problem, where the similar asymptotics have been studied, we can expect that the following behaviour of the PI occurs

$$I(\beta, z) \sim \frac{1}{\beta^{O(1)}} \exp\left\{-\beta E(z; \lambda, \omega l)\right\} .$$

Consequently,

$$G(\beta) \sim \frac{1}{\beta^{O(1)}} \int_{0}^{\infty} \frac{dz}{z^{3/2}} \exp\left\{\beta\left[\frac{i}{2}\left(z + \frac{1}{z}\right) - E(z; \lambda, \omega l)\right]\right\}$$

$$\sim \frac{1}{\beta^{O(1)}} \int_{0}^{\infty} \frac{dz}{z^{3/2}} \exp\left\{\beta S(z)\right\} , \tag{12.12}$$

where

$$S(z) = \frac{i}{2}\left(z + \frac{1}{z}\right) - E(z; \lambda, \omega l) .$$

The main contribution to the PI in (12.12) for large β can be obtained by using the saddle-point method:

$$S(z) = S(z_0) - \frac{1}{2}S''(z_0)(z - z_0)^2 + O((z - z_0)^3) \, ,$$

with the conditions

$$S'(z_0) = 0 \, , \qquad S''(z_0) > 0 \, .$$

Finally, one gets

$$G(\beta) \sim \frac{1}{\beta^{O(1)}} \exp\left\{\beta S(z_0)\right\} \, .$$

13. Bound States in QFT

In this chapter we shall obtain formulae which give the mass of a bound state of two scalar particles interacting with each other by means of a gauge field $U(1)$. The $\pi^+\pi^-$-atom is the typical physical example. We shall be interested in the mechanism of formation of bound states and in the estimation of the contribution of nonpotential interactions because calculations of relativistic corrections are one of the main problems to describe the bound states of elementary particles. In particular the computing of these corrections can help to determine the validity bounds of the potential approach. Plenty of papers are devoted to these problems (see, for example, [123, 124, 125, 126]).

13.1 The Mass of Bound States

Our approach is based on the investigation of the polarization loop operator of scalar particles in a quantum gauge field $U(1)$. We introduce the representations of this loop in a form of functional integral and investigate its asymptotic behaviour for large distances. Our scenario is the following. We solve the equation for the Green function of a scalar particle in an external gauge field $A_\mu(x)$:

$$[(i\partial_\mu + gA_\mu(x))^2 + m^2]G(x, y|A) = \delta(x - y), \tag{13.1}$$

where m is the mass of our scalar particle and g is a coupling constant.

The averaging over the gauge field $A_\mu(x)$ is defined by the relation

$$\langle \exp\{i\int dx J_\mu(x)A_\mu(x)\}\rangle_A = \exp\left\{-\frac{1}{2}\int\int dx_1 dx_2 J_\mu(x)D_{\mu\nu}(x - y)J_\nu(y)\right\},$$

where $J_\mu(x)$ is a real current. The propagator $D_{\mu\nu}(x - y)$ of the vector field $A_\mu(x)$ has the form

$$D_{\mu\nu}(x - y) = \langle A_\mu(x)A_\nu(y)\rangle_A = \delta_{\mu\nu}D(x - y) + \partial_\mu\partial_\nu D_1(x - y),$$

$$D(x) = \int \frac{d^4k}{(2\pi)^4}\tilde{D}(k^2)e^{ik(x-y)}.$$

The function $D_1(x - y)$ defines the gauge of the field $A_\mu(x)$ and physical characteristics do not depend on this function.

The function $D(x - y)$ defines the character of the interaction which is realized by this vector gauge field $A_\mu(x)$. In the case of the electromagnetic field we have

$$D(x) = \frac{1}{x^2(2\pi)^2} \qquad (13.2)$$

In some rough approximation one can hope that (13.1) reproduces the situation for non-Abelian gauge fields at large distances. In this case the true behaviour of the function $D(x)$ is not known. So far as we can investigate the bound states we can consider arbitrary functions $D(x)$.

Under gauge transformations

$$A_\mu(x) \to A_\mu(x) + \partial_\mu \lambda(x)$$

the Green function acts as follows:

$$G(x, y|A + \partial\lambda) = e^{ig\lambda(x)} G(x, y|A) e^{-ig\lambda(x)}.$$

The main object of our interest is the gauge-invariant loop function of two scalar particles with different masses m_1 and m_2:

$$\Pi(x - y) = \langle G_{m_1}(x, y|A) G_{m_2}(x, y|A) \rangle_A . \qquad (13.3)$$

When $(x - y)^2 \to \infty$ this function $\Pi(x - y)$ looks like

$$\Pi(x - y) \simeq \exp\{-M\sqrt{(x - y)^2}\},$$

where M is the mass of the bound state of our two scalar particles. Thus our aim is to find

$$M = - \lim_{x^2 \to \infty} \frac{1}{\sqrt{x^2}} \ln \Pi(x) . \qquad (13.4)$$

The following situations can be realized:

(i) if $M < \infty$ and $M \neq m_1 + m_2$, then a bound state with a mass M arises;

(ii) if $M = m_1 + m_2$, then the interaction is so weak that the bound state can not arise and the scalar particles exist as two independent states.

The solution of (13.1) can be represented in the form of the following functional integral (analogous representation for the Green function is obtained in papers [125, 127] and described also in section 8.4):

$$G(x, y|A) = \int_0^\infty d\alpha \exp\{-\alpha m^2 + \alpha(\partial_\mu - ig A_\mu(x))^2\} \delta(x - y)$$

$$= \int_0^\infty d\alpha \int d\nu \exp\left\{-\int_0^1 d\beta \nu_\mu^2(\beta)\right\}$$

$$\times \exp\left\{2ig\sqrt{\alpha}\int_0^1 d\beta\nu_\mu(\beta)A_\mu(x - 2\sqrt{\alpha}\int_\beta^1 d\beta'\nu(\beta'))\right\}$$

$$\times\delta\left(x - y - 2\sqrt{\alpha}\int_0^1 d\beta\nu(\beta)\right)$$

which leads to

$$G(x,y|A) = \int d\Sigma \exp\left\{ig\int_0^\alpha d\beta \dot{z}_\mu(\beta)A_\mu(z(\beta))\right\},$$

$$d\Sigma = d\mu_\alpha d\sigma_B,$$

$$d\mu_\alpha = \frac{d\alpha}{8\pi^2\alpha^2}\exp\left\{-\frac{1}{2}\left(m^2\alpha + \frac{(x-y)^2}{\alpha}\right)\right\}, \qquad 0 < \alpha < \infty$$

$$d\sigma_B = C\delta B \exp\left\{-\frac{1}{2}\int_0^\alpha dB\dot{B}_\mu^2(\beta)\right\}, \qquad \int d\sigma_B = 1,$$

$$z_\mu(\beta) = x_\mu\frac{\beta}{\alpha} + y_\mu\left(1 - \frac{\beta}{\alpha}\right) + B_\mu(\beta),$$

$$B_\mu(0) = B_\mu(\alpha) = 0.$$

The loop function $\Pi(x)$ (13.3) has the form

$$\Pi(x - y) = \iint d\Sigma_1 d\Sigma_2 \exp\{-W_{11} + 2W_{12} - W_{22}\},$$

$$W_{ij} = \frac{g^2}{2}\int_0^{\alpha_i} d\beta_1 \int_0^{\alpha_j} d\beta_2 \left(\dot{z}_\mu^{(i)}(\beta_1)\dot{z}_\mu^{(j)}(\beta_2)\right)D(z^{(i)}(\beta_1) - z^{(j)}(\beta_2)).$$

The term with D_1 disappears because

$$\dot{z}_\mu(\beta)\partial_\mu F(z(\beta)) = \frac{\partial}{\partial\beta}F(z(\beta)).$$

Thus the function $\Pi(x - y)$ does not depend on a gauge. For simplicity let us put $y = 0$. We shall study the asymptotic behaviour of the function $\Pi(x)$ at $x \to \infty$. To this end let us introduce the following variables:

$$\alpha_j = x\xi_j, \qquad \beta_j = \tau_j\xi_j, \qquad B_j(\tau) = \sqrt{\xi_j}A_j(\tau).$$

One can obtain

$$\Pi(x) = \int\limits_0^\infty \int\limits_0^\infty \frac{d\xi_1 d\xi_2}{(8\pi^2 \xi_1 \xi_2 x)^2} \exp\left\{-x \sum_{j=1}^2 \frac{1}{2}\left(m_j^2 \xi_j + \frac{1}{\xi_j}\right)\right\} J_x(\xi_1, \xi_2), \quad (13.5)$$

$$J_x(\mu_1, \mu_2) = C_1 C_2 \int \int \delta B_1 \delta B_2$$

$$\cdot \exp\left\{-\int\limits_0^x d\tau \sum_{j=1}^2 \frac{\xi_j}{2}\left(\dot{A}^j(\tau)\right)^2 - W_{11} + 2W_{12} - W_{22}\right\}, \quad (13.6)$$

$$W_{ij} = \frac{g^2}{2}\int\limits_0^x \int\limits_0^x d\tau_1 d\tau_2 \left(\dot{z}^{(i)}(\tau_1)\dot{z}^{(j)}(\tau_2)\right)D(z^{(i)}(\tau_1) - z^{(j)}(\tau_2)),$$

$$z_\mu^{(j)}(\tau) = n_\mu \tau + \sqrt{\xi_j}A_\mu^{(j)}(\tau), \quad n_\mu = \frac{x_\mu}{x},$$

$$A_\mu^{(j)}(0) = A_\mu^{(j)}(x) = 0.$$

The functional integral for $J_x(\mu_1, \mu_2)$ looks like the Feynman path integral in nonrelativistic quantum mechanics for the four-dimensional motion of particles $B_\mu^{(1)}(\tau)$ and $B_\mu^{(2)}(\tau)$ with "masses" μ_1 and μ_2. The interaction of these particles is defined by the nonlocal functional $-W_{11} + 2W_{12} - W_{22}$ which contains potential and nonpotential interactions. In the multiplier

$$(\dot{z}^i(\tau_1)\dot{z}^j(\tau_2)) = 1 + (n(\dot{B}^i(\tau_1) + \dot{B}^j(\tau_2))) + (\dot{B}^i(\tau_1)\dot{B}^j(\tau_2))$$

there are terms with the "velocities" $\dot{B}_\mu^{(i)}(\tau)$ of these particles. These terms define the so-called interaction with a radiation field and Lamb shift.

The asymptotic behaviour of the functional $\Pi(x)$ at $x \to \infty$ is determined by a saddle point of the integral in the representation (13.5). As the asymptotic of the function $J_x(\mu_1, \mu_2)$ looks like

$$J_x(\mu_1, \mu_2) \sim \exp\{-xE(\mu_1, \mu_2)\}, \quad (13.7)$$

then the mass M in (13.4) is

$$M = \min_{(\mu_1, \mu_2)} \left[\frac{1}{2}(m_1^2/\mu_1 + 1/\mu_1) + \frac{1}{2}(m_2^2/\mu_2 + 1/\mu_2) + E(\mu_1, \mu_2)\right] \quad (13.8)$$

and in the case $m_1 = m_2 = m$

$$M = \min_\mu[m^2\mu + 1/\mu + E(\mu, \mu)]. \quad (13.9)$$

Formulae (13.6)–(13.9) define the mass of our bound state completely. The main problem is to compute the functional integral (13.6). It is not a simple problem. We have performed the first preliminary variational estimation of this integral with the interaction term W_{12} in the case of the electromagnetic interaction (13.2). These preliminary calculations have shown that the contribution of the nonpotential interactions is essential. We will not describe these calculations here as they have a very preliminary character.

13.2 The Nonrelativistic Limit

In this section we obtain the nonrelativistic limit $c \to \infty$ for the loop function $\Pi(x)$ in (13.5). To this end let us restore the parameters $\hbar$ and c in (13.1):

$$[(i\partial_\mu + \frac{g}{c\hbar} A_\mu(x))^2 + \frac{c^2}{\hbar^2}m^2]G(x,y|A) = \delta(x-y),$$

where

$$x_\mu = (x_4, \mathbf{x}) = (ct, \mathbf{x}), \quad x = \sqrt{x^2} = \sqrt{c^2t^2 + \mathbf{x}^2} \to ct, \quad c \to \infty$$

$$n_\mu = (1, \mathbf{x}/ct) \to (1, 0),$$

$$\langle A_\mu(\quad \nu(y)) = \hbar c D_{\mu\nu}(x-y) = \hbar c[\delta_{\mu\nu} D(x-y) + \partial_\mu \partial_\nu D_1(x-y)],$$

$$D(x) = \int \frac{d^4k}{(2\pi)^4} \tilde{D}(k^2)e^{ikx} = \int\limits_{-\infty}^{\infty} \frac{d\mathbf{k}}{(2\pi)^3} \int \frac{dv}{2c\pi} \tilde{D}\left(\mathbf{k}^2 + \frac{v^2}{c^2}\right) e^{i(vt+\mathbf{k}\mathbf{x})}.$$

Our task is to introduce the parameter c in explicit form into $\Pi(x)$ (13.5) and then to find the limit $c \to \infty$ in this expression for $\Pi(x)$. Let us introduce in (13.5) new variables $\tau_j \to \tau_j c$, $\mu_j \to \mu_j c$. We then obtain

$$\Pi(t) = \frac{1}{(8\pi^2t)^2} \int\limits_0^\infty\!\!\!\int limits_0^\infty d\mu_1 d\mu_2 \exp\left\{-t\sum_{j=1}^{2} \frac{c^2}{2}\left(\frac{m_j^2}{\mu_j} + \mu_j\right)\right\} J_t(\mu_1,\mu_2) \quad (13.10)$$

$$J_t(\mu_1,\mu_2) = C_1 C_2 \iint \delta B_1 \delta B_2 \exp\left\{-\int_0^t d\tau \sum_{j=1}^{2} \frac{\mu_j}{2}(\dot{B}_j(\tau))^2\right\}$$

$$\times \exp\left\{-\frac{g^2}{2}\int_0^t\!\!\int_0^t d\tau_1 d\tau_2 \sum_{i,j}^{2}(-)^{i+j}((n + \frac{1}{c}\dot{B}_i(\tau_1))(n + \frac{1}{c}\dot{B}_j(\tau_2)))\right.$$

$$\times \int \frac{d\mathbf{k}}{(2\pi)^3} \int \frac{dv}{2\pi} \tilde{D}(\mathbf{k}^2 + \frac{v^2}{c^2}) \exp\left\{iv((\tau_1 - \tau_2) + \frac{1}{c}(B_{i4}(\tau_1) - B_{j4}(\tau_2)))\right.$$

$$\left.\left. +i\mathbf{k}(\mathbf{B}_i(\tau_1) - \mathbf{B}_j(\tau_2))\}\}\right\}.$$

We shall consider the term W_{12} only, because the terms W_{11} and W_{22} contribute to the renormalization masses of scalar particles and have a purely relativistic origin.

Now we are able to go to the limit $c \to \infty$ in the functional integral. We can put all terms with $\frac{1}{c}$ equal zero. The δ-function $\delta(\tau_1 - \tau_2)$ arises in the last term in (13.10). The fourth components $B_{j4}(\tau)$ disappear in the interaction function and we can integrate over these components. Thus we get

$$J_t(\mu_1, \mu_2) = C_1 C_2 \iint \delta \mathbf{B}_1 \delta \mathbf{B}_2 \exp\left\{ - \int_0^t d\tau \sum_{j=1}^2 \frac{\mu_j}{2} (\dot{\mathbf{B}}^{(j)}(\tau))^2 \right\}$$

$$\times \exp\left\{ -g^2 V(0) t + g^2 \int_0^t d\tau V(\mathbf{B}^{(1)}(\tau) - \mathbf{B}^{(2)}(\tau)) \right\}, \qquad (13.11)$$

where

$$V(\mathbf{r}) = \int \frac{d\mathbf{k}}{(2\pi)^3} \tilde{D}(\mathbf{k}^2) e^{i\mathbf{kr}} \qquad (13.12)$$

is a nonrelativistic potential. The term $g^2 V(0)$ is the contribution of the gauge field to the mass renormalization of the particle given by $\langle A_\mu^2 \rangle$ in the nonrelativistic limit. One can see that the representation (13.11) coincides with Feynman PI in the quantum mechanics [40] for the situation where there are two particles $\mathbf{B}^{(1)}$ and $\mathbf{B}^{(2)}$ with masses μ_1 and μ_2 and the interaction between these particles is described by the potential $V(\mathbf{B}^{(1)} - \mathbf{B}^{(2)})$. We should stress that these two particles are not the real scalar particles with masses m_1 and m_2. These "particles" $\mathbf{B}^{(1)}$ and $\mathbf{B}^{(2)}$ can be considered some auxiliary particles in our representation of the loop function $\Pi(x)$.

Let us introduce the standard variables

$$\mathbf{B}^{(1)}(\xi) = \mathbf{R}(\xi) + \frac{\mu_2}{\mu_1 + \mu_2} \mathbf{r}(\xi),$$

$$\mathbf{B}^{(2)}(\xi) = \mathbf{R}(\xi) - \frac{\mu_1}{\mu_1 + \mu_2} \mathbf{r}(\xi)$$

and integrate over $\mathbf{R}(\xi)$. We obtain

$$J_t(\mu_1, \mu_2) = C \int \delta \mathbf{r} \exp\left\{ - \int_0^t d\xi \left[\frac{\mu}{2} \dot{\mathbf{r}}^2(\xi) - g^2 V(\mathbf{r}(\xi)) \right] \right\}, \qquad \mu = \frac{\mu_1 \mu_2}{\mu_1 + \mu_2}.$$

When $t \to \infty$ the function $J_t(\mu_1, \mu_2)$ becomes

$$J_t(\mu_1, \mu_2) \to \exp\{-t E(\mu)\},$$

where $E(\mu)$ is the energy of the bound state of our two auxiliary nonrelativistic particles in the potential $V(\mathbf{r})$. The asymptotic behaviour of the function $\Pi(x)$ is $\Pi(x) = \Pi(ct) \to \exp(-Mt)$, where

$$
\begin{aligned}
M &= \min_{\mu_1, \mu_2} \left[\frac{c^2}{2} \left(\frac{m_1^2}{\mu_1} + \mu_1 \right) + \frac{c^2}{2} \left(\frac{m_2^2}{\mu_2} + \mu_2 \right) + E\left(\frac{\mu_1 \mu_2}{\mu_1 + \mu_2} \right) \right] \\
&= (m_1 + m_2) c^2 + E\left(\frac{m_1 m_2}{m_1 + m_2} \right) + O\left(\frac{1}{c^2} \right).
\end{aligned}
$$

Thus in the nonrelativitic limit the mass of the bound state of two scalar particles is the sum of their masses plus the bound state energy which is defined by the nonrelativistic potential interaction.

References

1. *Proceed. Int. Conference "Path Integrals from meV to MeV, Tutzing-92"*, eds. H. Grabert and et al., (World Scientific, Singapore, 1993)
2. L.S. Schulman, *"Techniques and Application of Path Integration "*, (Wiley, N.Y., 1981)
3. L.H. Ryder, *Quantum Field Theory*, (Cambridge Univ. Press, Cambrige, 1985)
4. P. Ramond, *Field Theory, A Modern Primer*, (Benjamin/Cummings, Massach., 1981)
5. *Proceed. Int. Workshop on Variational Calculations in Quantum Field Theory*, eds. L. Polley and D.E.L. Pottinger, (World Scientific, Singapore, 1987).
6. W. Fischer, H. Leschke and P. Müller, J. Phys., **A 25** (1992) 3835
7. G.V. Efimov and G. Ganbold, NATO ASI Series, **B255** (1991) 133
8. P.A.M. Dirac, *The Principles of Quantum Mechanics* , (Clarendon Press, Oxford, 1930)
9. J. Schwinger, Phys. Rev., **73** (1948) 416
10. S.N. Gupta, Proc. Phys. Soc., **A 63** (1950) 681
11. K. Bleuler, Helv. Phys. Acta, **23** (1950) 567
12. T. Kugo and I. Ojima, Phys. Lett., **B 73** (1978) 459
13. R.P. Feynman, Rev. Mod. Phys., **20** (1948) 367
14. R.P. Feynman, Phys. Rev., **80** (1950) 440; **84** (1951) 108
15. N. Wiener, Ann. Math., **22** (1920) 66
16. P.A.M. Dirac, Phys. Z. Sowjetunion, **3** (1933) 1
17. M. Kac, in *Proceed. Second Berkeley Simposium on Probability and Statistics*, ed. J. Neyman, (Univ. Calif. Press, Berkeley, 1951)
18. S.F. Edwards and R.E. Peierls, *Field equations in functional forms*, in Proceed. Roy. Soc., **224** (1954) 24
19. I.M. Gelfand and R.A. Minlos, Doklady Akad. Nauk Sov. Union, **97** (1954) 209
20. N.N. Bogoliubov, Doklady Akad. Nauk Sov. Union, **99** (1954) 225
21. J. Glimm and A. Jaffe, Comm. Math. Phys., **22** (1971) 253
22. J. Glimm and A. Jaffe, *Field Theory Models in Statistical Mechanics and QFT*, (Gordon Breach, N.Y., 1971)
23. L.D. Faddeev and V.N. Popov, Phys. Lett., **B 25** (1967) 29
24. De Witt, Phys. Rev., **160** (1967) 1113
25. L.D. Faddeev and A.A. Slavnov, *Gauge fields, Introduction to Quantum Theory*, (Benjamin/Cummings, Massachusets, 1980)
26. J.G. Taylor, Nucl. Phys., **B 33** (1971) 436
27. B.W. Lee and J. Zinn-Justin, Phys. Rev., **D 5** (1972) 3137
28. G. t'Hooft and M. Veltman, Nucl. Phys., **B44** (1972) 189
29. E.S. Fradkin and I.V. Tyutin, Phys. Lett., **B 30** (1969) 562; Phys. Rev., **D 2** (1970) 2841
30. D. Peak and A. Inomata, J. Math. Phys., **10** (1969) 1422

31. I.H. Duru and H. Kleinert, Phys. Lett., **B 84** (1979) 185
32. H. Kleinert, *"Path Integrals in Quantum Mechanics, Statistics and Polymer Physics"*, (World Scientific, Singapore, 1990)
33. N.N. Bogolubov and D.V. Shirkov, *An Introduction to the Theory of Quantized Fields*, (Wiley, N.Y., 1959)
34. *Path Integrals and their Applications in Quantum, Statistical and Solid State Physics*, eds. G.J. Papadopoulos and J.T. Devreese, (Plenum Press, N.Y., 1978)
35. B. Simon, *Functional Integration and Quantum Physics*, (Academic Press, N.Y., 1979)
36. *Functional Integration. Theory and Application*, eds. E.P. Antoine and E. Tirapegui, (Plenum Press, N.Y., 1980)
37. J. Glimm and A. Jaffe, *Quantum Physics: A Functional Integral Point of View*, (Springer-Verlag, N.Y., 1981)
38. G. Roepstorff, *"Path Integral Approach to Quantum Physics"*, (Springer Verlag, Berlin, 1994)
39. S. Albeverio, S. Paycha and S. Scarlatti, Prog. Phys., **13** (1989) 230
40. R.P. Feynman, Phys. Rev., **97** (1955) 660
41. R.P. Feynman and A.R. Hibbs, *Quantum Mechanics and Path Integrals*, (McGraw-Hill Book Company, N.Y., 1965)
42. R.P. Feynman, *Statistical Mechanics*, (Benjamin, N.Y., 1972)
43. G.V. Efimov and G. Ganbold, Int. J. Mod. Phys., **A 5** (1990) 531
44. G.V. Efimov and G. Ganbold, preprint Dublin Ins. Adv.Stud., DIAS-STP/90-22 (1990)
45. L.D. Landau, Phys. Z. Sowjetunion, **3** (1933) 644
46. H. Fröhlich, Proceed. Roy. Soc., **A 160** (1937) 230
47. H. Fröhlich, H. Peltzer and S. Zienau, Philos. Mag., **41** (1950) 221
48. H. Fröhlich, Advan. Phys., **3** (1954) 325
49. J.T. Marshall and L.R. Mills, Phys. Rev., **B 2** (1970) 3143
50. I.D. Feranchuk and I.I. Komarov, phys. stat. sol., **(b) 15** (1982) 1965
51. C. Alexandrou and R. Rosenfelder, Phys. Rep., **215** (1992) 1
52. N.N. Bogolubov Jr. and A.V. Soldatov, Mod. Phys. Lett., **B7** (1993) 1773
53. H.G. Dosch and U. Lisenfeld, Phys. Lett., **B 219** (1989) 493
54. A.H. Castro Neto and A.O. Caldeira, Phys. Rev. Lett., **67** (1991) 1960
55. *"Polarons and Excitons"*, eds. C.G. Kuper and G.D. Whitfield, (Oliver and Boyd, London, 1963)
56. *Polarons in ionic crystals and polar semiconductors*, ed. J.T. Devreese, (North-Holland, Amsterdam, 1972)
57. *Proceed. IV Int. Conf. on Electronic Properties of 2D-Systems, New Hampshire, August 1981*, Surf. Sci., **113** (1982) n. 1-3
58. *Physics of polarons and excitons in polar semiconductors and ionic crystals*, eds. J.T. Devreese and F.M. Peeters, (Plenum, N.Y., 1984)
59. T.K. Mitra, A. Chatterjee and S. Mukhopadhyay, Phys. Rep., **153** (1987) 91
60. N.N. Bogoliubov jr. and V.N. Plechko, Riv. Nuo. Cim., **9** (1988) 1
61. B. Gerlach and H. Löwen, Rev. Mod. Phys., **63** (1991) 63
62. M. Mikkor, K. Kanazawa and F.C. Brown, Phys. Rev., **162** (1967) 848
63. G. Ascarelli, Phys. Rev. Lett., **20** (1968) 44
64. S. Pekar, Zh. Eks. Teor. Fiz., **19** (1954) 796
65. S. Miyake, Jour. Phys. Soc. Jap., **38** (1975) 181
66. G. Höhler and A. Múllensiefen, Z. Phys., **157** (1959) 159
67. J. Röseler, phys. stat. sol., **25** (1968) 311
68. E.N. Lieb and K. Yamazaki, Phys. Rev., **111** (1958) 728
69. D.M. Larsen, Phys. Rev., **172** (1968) 967
70. M. Saitoh, Jour. Phys. Soc. Jap., **49** (1980) 878

71. J. Adamowski, B. Gerlach and H.Leschke, in *"Functional Integration, Theory and Applications"*, eds. J.P. Antoine and E. Tirapegui, (Plenum, N.Y., 1980)
72. B. Gerlach, H. Löwen and H. Schliffke, Phys. Rev., **B 36** (1987) 6320
73. M. Horst, V. Merkt and J.P. Kottaus, Phys. Rev. Lett., **50** (1983) 754
74. S.A. Jackson and P.M. Platzman, Phys. Rev., **B 24** (1981) 499
75. G. Ganbold and G.V. Efimov, Phys. Rev., **B 50** (1994) 3733
76. T.D. Lee, F. Low and D. Pines, Phys. Rev., **90** (1953) 297
77. J.M. Luttinger and C.-Y. Lu, Phys. Rev., **B 21** (1980) 4251
78. W. Becker, B. Gerlach and H. Schliffke, Phys. Rev., **B 28** (1983) 5735
79. F. Brosens and J.T. Devreese, phys. stat. sol., **(b) 145** (1988) 517
80. C. Alexandrou, W.Fleischer and R.Rosenfelder, Phys. Rev. Lett., **65** (1990) 2615
81. S. Das Sarma, Phys. Rev., **B 27** (1983) 2590
82. F.M.Peeters, Wu Xiaoguang and J.T. Devreese, Phys. Rev., **B 33** (1986) 3926
83. M.A. Smondyrev, Physica, **A 171** (1991) 191
84. J. Adamowski, B. Gerlach and H.Leschke, Phys. Lett., **A 79** (1980) 249
85. G.V. Efimov and G. Ganbold, phys.stat.sol., **(b) 168** (1991) 165
86. O.V. Selyugin and M.A. Smondyrev, phys. stat. sol., **(b) 155** (1989) 155
87. W. Fischer, H. Leschke and P. Müller, *"How to get a lower bound on ..."*, in *Proceed. Int. Conference on Path Integrals in Physics, Bangkok, Thailand, 7-12 January 1993*
88. S.W. Tjablikov, Soviet J. Eks. Teor. Fiz., **21** (1951) 16
89. A.V. Tulub, Soviet J. Eks. Teor. Fiz., **41** (1961) 1828
90. O. Hipolito, Sol. Stat. Commun., **32** (1979) 515
91. Wu Xiaoguang, F.M.Peeters, and J.T. Devreese, Phys. Rev., **B 31** (1985) 3420
92. S. Das Sarma and B.A. Mason, Ann. Phys., **163** (1985) 78
93. D.M. Larsen, Phys. Rev., **B 35** (1987) 4435
94. L.D. Landau and S. Pekar, Soviet J. Eks. Teor. Fiz., **16** (1946) 341
95. N.N. Bogoliubov and S.W. Tjablikov, Soviet J. Eks. Teor. Fiz., **19** (1949) 256
96. G.R. Allcock, Adv. Phys., **5** (1956) 412
97. J. Appel, Sol. Stat. Phys., **21** (1968) 193
98. D.Matz and B.C. Burkey, Phys. Rev., **B 3** (1971) 3487
99. H. Spohn, Ann. Phys., **175** (1987) 278
100. T.D. Schultz, Phys. Rev., **116** (1959) 526
101. P. Sheng and L.D. Dow, Phys. Rev., **B 4** (1971) 1343
102. W.J. Huybrecht, Sol. Stat. Commun., **28** (1978) 95
103. B. Simon and R.B. Griffiths, Comm. Math. Phys., **33** (1973) 145
104. J. Glimm and A. Jaffe, Phys. Rev., **D 10** (1975) 536
105. O.A. McBryan and J. Rosen, Comm. Math. Phys., **51** (1976) 97
106. S. Coleman and E. Weinberg, Phys. Rev., **D 7** (1973) 1888
107. S.-J. Chang, Phys. Rev., **D 13** (1976) 2778; **D 16** (1977) 1979
108. S.D. Drell, M. Weinstein and S. Yankielowicz, Phys. Rev., **D 14** (1976) 487
109. *Proceed. Int. Works. "Variational Calculations in Quantum Field Theory"*, eds. L. Polley and D.Pottinger, (World Scientific, Singapore, 1988)
110. S.-J. Chang, Phys. Rev., **D 12** (1975) 1071
111. P.M. Stevenson, Phys. Rev., **D 30** (1984) 1712; **D 32** (1985) 1389
112. M. Funke, U. Kaulfuss and H. Kummel, Phys. Rev., **D 35** (1987) 631
113. L. Polley and U. Ritschel, Phys. Lett., **B 221** (1989) 44
114. G.V. Efimov and G. Ganbold, Commun. JINR, E2-91-437 (1991)
115. G.V. Efimov and G. Ganbold, preprint JINR, E2-92-191 (1992)
116. G.V. Efimov and G. Ganbold, Mod. Phys. Lett., **A7** (1992) 2189
117. N. Garcia and A.Z. Genack, Phys. Rev. Lett., **66** (1991) 1850;
118. J.P. Bouchaud, Europhys. Lett., **11** (1990) 505

119. S.M. Flatte, D.R. Bernstein and R. Dashen, Phys. Fluids **26** (1983) 1701
120. P.L. Chow, J. Math. Phys., **13** (1972) 1224
121. R. Dashen, J. Math. Phys., **20** (1979) 894
122. J.L. Codona et al., J. Math. Phys., **27** (1986) 171
123. T. Appelquist, M.Dine,and I.Muzinich, Phys. Rev., **D 17** (1978) 2074
124. M.B. Voloshin, Nucl.Phys., **B 154** (1979) 365; H. Leutwyler, Phys. Lett., **B 98** (1981) 447
125. H.G. Dosch and Yu.A. Simonov, Phys. Lett., **B 205** (1988) 339; Yu.A. Simonov, Nucl. Phys., **B 324** (1989) 67
126. D. Gromes, Phys. Rep., **200** (1991) 186
127. M.B. Halpern and P. Sejanovic, Phys. Rev., **D 15** (1977) 1655

Part III

Oscillator Representation in
Quantum Mechanics

14. The Oscillator in Quantum Mechanics

One of the basic problems of nonrelativistic quantum mechanics is to find the energy spectrum and eigenfunctions of a microsystem described by the Schrödinger equation with an appropriate potential. Exact solutions of this equation are found [1]–[4] for a limited class of potentials such as the harmonic oscillator, the Coulomb potential, and some others. Most quantum systems are described by potentials for which the Schrödinger equation cannot be solved analytically. Thus, the solution of the Schrödinger equation with a sufficiently arbitrary potential represents the main mathematical task. With this aim many approximate analytical and numerical methods have been worked out. Great progress in the development of computer techniques and effective algorithms for the numerical solution of differential equations enables us to obtain numerical solutions for the energy spectrum and wave functions with quite a high accuracy, although practical calculations are usually very laborious and require powerful computers.

Nevertheless, the development of analytical methods is very important because only analytical methods permit us to investigate qualitative features of quantum physical systems and indicate effective ways for improvement of numerical algorithms.

Among numerous analytical methods we would like to decide upon two popular approaches for the investigation of this problem: the perturbation method and the variatonal method (see for example, [1]–[4], [5]). Our aim is to join these two methods in one approach which will be called the *oscillator representation method*.

Approximate analytical perturbative methods imply the procedure in which the total Hamiltonian is divided into two parts: $H = H_0 + H_I$, the solution of the zeroth approximation $H_0 \Psi^{(0)} = E^{(0)} \Psi^{(0)}$ is assumed to be obtained analytically and perturbation corrections to the zeroth approximation $E^{(0)}$ and $\Psi^{(0)}$ can be calculated. The main physical and mathematical point is that the Hamiltonian H_0 in an appropriate representation of the Schrödinger equation should be chosen in such a way as to catch the main dynamic properties of a quantum system and to give the possibility of calculating all physical characteristics of the system under consideration analytically. The interaction Hamiltonian H_I should give small corrections to the zeroth approximation and these corrections can be calculated.

Here we mention the standard perturbation Reley–Schrödinger theory [1]–[4], the quasi-classical or WKB method [1]-[4], and the $1/N$–expansion [6], [7]. This method works quite well if the interaction described by H_I is weak and does not change the physical character of the system described by H_0. We will not enter into the details of these methods and refer readers to the vast literature (see, for example, [1]–[7]).

The variational methods permit one to carry out highly precise calculations, and for complicated quantum Hamiltonians (for example, three–body Coulomb systems) these methods are realized as a set of program packages for computers. In the framework of this method the energy spectra and other parameters of quantum mechanical systems as atoms, molecules, nucleons and nuclei have been successfully calculated to a high accuracy.

Nevertheless, it should be noted that the variational method has some disadvantages: there are no regular rules of how to choose test basis functions, and criteria of the accuracy of approximations are absent as well as the possibility of calculations of the corrections to the basic variational approximation in general.

In this part of our book, the oscillator representation method developed in the previous parts and the papers [8], [9] will be applied to quantum mechanics problems. In the previous parts, the oscillator representation method based on the ideas and methods of quantum field theory has been proposed to investigate strong coupling regime of quantum field models and functional integrals over the Gaussian measure. Now we want to use it to calculate the energy spectrum of quantum mechanical systems described by the nonrelativistic Schrödinger equation.

For our aims it is important to focus attention on the following result of the QFT formalism, namely that the total Hamiltonian $H = H_0 + H_I$ of a quantum field system is usually represented in *the correct form*, i.e. all field operators in the Hamiltonian are written in the normal product and the interaction Hamiltonian contains field oparators in powers more than 2. For example, in the simplest case of one-component scalar field the correct form of the total Hamiltonian in the formalism of creation and annihilation operators a_k^+ and a_k is

$$H = H_0 + H_I, \qquad H_0 \sim \int dk \omega_k a_k^+ a_k,$$

$$H_I \sim \sum_{n>2} \sum_{m=0}^{n} \int dk_1 \cdots \int dk_n$$

$$\times f_{n,m}(k_1, \cdots, k_m; k_{m+1}, \cdots, k_n) a_{k_1}^+ \cdots a_{k_m}^+ a_{k_{m+1}} \cdots a_{k_n},$$

The normal ordering of the Hamiltonian results in that the main quantum contributions to the ground state or vacuum of the system are taken into account.

In reality, for example, in quantum field theory of scalar fields, noninteracting particles are described by a Hamiltonian representing a set of oscillators and the interaction is described by the product of field operators of the type ϕ^4. A Hamiltonian like that contains diverging contributions related to the divergence of the scalar field propagator at zero

$$\langle \phi(x)\phi(x)\rangle = D(0) \sim \hbar \int \frac{d^4 k}{m^2 + k^2} \sim \infty \,.$$

Note that this term is the highest divergence in renormalizable theory of the type ϕ^4. This singularity can be removed by renormalizing the mass m and vacuum energy. This renormalization procedure is equivalent to the postulate that field operators in the interaction Hamiltonian have to be written in terms of normal products over particle creation and annihilation operators. Thus, one can say that *the normal ordering of the Hamiltonian means essentially that the main quantum contributions to the ground state or vacuum of the system are taken into account.*

Our idea is to represent the Hamiltonian for a quantum mechanical system in this *correct form*, so that we hope the main quantum contributions to be taken into account.

However, in the course of materialization of this idea we face the following trouble. The most remarkable difference between quantum field theory and quantum mechanics is that quantized fields in QFT are an infinite set of oscillators, and any interactions of fields do not change the oscillator nature of these quantized fields. In the same time, in quantum mechanics most of the potentials and, therefore, the corresponding wave functions are quite far from the oscillator behaviour.

The application of the oscillator repesentation method imply that a wave function, being a bound ground state of a quantum system with an attractive potential, is expanded over the oscillator basis. In most cases, the asymptotic behaviour of a true wave function for short and large distances does not coincide with the Gaussian asymptotic behaviour of the oscillator wave functions. As a result, the expansion of these wave functions over the oscillator basis, although being methematically correct, means the series do not converge sufficiently fast for practical purposes.

Therefore, before applying the oscillator representation method, we have to modify the variables in the starting Schrödinger equation so that the modified equation should have solutions with the Gaussian asymptotic behaviour at large and short distances. For example, in the Coulomb system such a modification is performed by going over to the four–dimensional space where the wave function of the Coulomb system becomes an oscillator one. In an early paper [10], Schrödinger noted the existence of such a transformation which transforms the three-dimensional Coulomb system into an oscillator one in the four-dimensional space. Kustaanheimo and Stiefel [11] gave the explicit form of this transformation and used it to solve the classical Kepler problem.

In the case of other potentials, these transformations effectively result in that modified equations having the Gaussian asymptotics of solutions are written in a space of any dimension. It should be emphasized that these transformations are not canonical ones. It means that a quantum system after the Kustaanheimo-Stiefel-type transformation becomes another quantum system with a diffirent set of quantum numbers and corresponding wave functions. However this new quantum system contains a subset of wave functions which coincide with the wave functions of the initial system and we are be able to pick out these desirable quantum numbers and wave functions. Finally, one can say that these transformations should be considered as a successful mathematical technical method.

Thus, the first step is to find such transformations of vatiables in the initial Schrödinger equation leading to the modified Hamiltonian whose wave functions have the Gaaussian asymptotics.

The next step is to devide the modified Hamiltonian in the form $H = H_0 + H_I$. As the wave functions have Gaussian asymptotics, we can introduce the canonical variables coordinate q and momenta p in the representation of the creation and annihilation operators a^+ and a:

$$q = \frac{1}{\sqrt{2}} \left(a + a^+ \right), \qquad p = \frac{1}{i\sqrt{2}} \left(a - a^+ \right).$$

The conception of normal products introduced into nonrelativistic quantum mechanics is actually not new (see for example, [12]); however, the question is on what principles the realization of this idea should be based. Undoubtedly, we want the zeroth Hamiltonian H_0 to concentrate the main quntum contributions in some optimal way. We claim that this optimal way is the representation of the total Hamiltonian in the correct form.

The general scheme to do this is the following. Let the modified Hamiltonian with the Gaussian asymptotic of wave functions be given. First of all, we have to rewrite the modified Hamiltonian in the representation of the creation and annihilation operators. Let us extract the pure oscillator part with some unknown, at this stage, frequency ω and write this Hamiltonian in the form $H_0 = \omega a^+ a$. The rest of the interaction Hamiltonian,

$$H_I = H - H_0 = H - \omega a^+ a,$$

should be represented in terms of normal products over the operators a^+ and a. Then we have to require that this interaction Hamiltonian does not contain terms quadratic in the canonical variables. This requirement leads to the equation which determines the oscillator frequency ω. This equation is called the *oscillator representation condition* (ORC). As a result, the total Hamiltonian is written in the representation where the main quantum contributions to the ground state of the system are taken into account.

It turns out that the oscillator representation condition coincides with the variation equation arising after averaging of the total Hamiltonian over a

Gaussian test function. Thus, one can say that the oscillator representation method is the next step in developing of the variational approach in Quantum Mechanics.

15. The Oscillator Representation in R^d

In this chapter we formulate the Oscillator Representation for the Hamiltonian describing a quantum system in the space R^d. In other words the representation of this Hamiltonian in the correct form will be done.

First of all, let us list the well known formulae for the quantum oscillator in the space R^d $(d = 1, 2, ...)$. The Hamiltonian is

$$H = \sum_{j=0}^{d} \frac{1}{2}\left(p_j^2 + \omega^2 q_j^2\right) = \frac{1}{2}(p^2 + \omega^2 q^2) , \qquad [q_j, p_i] = i\delta_{ij}. \qquad (15.1)$$

In the representation where q_j is diagonal, the operator p_j is

$$p_j = \frac{1}{i}\frac{\partial}{\partial q_j}$$

. The oscillator canonical variables (p, q) can be written in the form

$$q_j = \frac{a_j + a_j^+}{\sqrt{2\omega}} = \frac{Q_j}{\sqrt{\omega}} , \qquad p_j = \sqrt{\frac{\omega}{2}}\frac{a_j - a_j^+}{2i} = \sqrt{\omega}P_j , \qquad (15.2)$$

$$a_j = \frac{1}{\sqrt{2}}\left(\sqrt{\omega}q_j + \frac{i}{\sqrt{\leq mega}}p_j\right) = \frac{1}{\sqrt{2}}\left(Q_j + iP_j\right) = \frac{1}{\sqrt{2}}\left(Q_j + \frac{\partial}{\partial Q_j}\right) ,$$

$$a_j^+ = \frac{1}{\sqrt{2}}\left(\sqrt{\omega}q_j - \frac{i}{\sqrt{\omega}}p_j\right) = \frac{1}{\sqrt{2}}\left(Q_j - iP_j\right) = \frac{1}{\sqrt{2}}\left(Q_j - \frac{\partial}{\partial Q_j}\right) ,$$

$$[a_i, a_j^+] = \delta_{ij} .$$

The operators a_j and a_j^+ are called annihilation and creation operators.

The Hamiltonian (15.1) becomes

$$H = \frac{\omega}{2}\sum_{j=1}^{d}\left(a_j a_j^+ + a_j^+ a_j\right) = \omega(a^+ a) + \frac{d\omega}{2} = H_0 + \varepsilon_0 ,$$

$$H_0 = \omega(a^+ a)\sum_{j=1}^{d} a_j^+ a_j , \qquad \varepsilon_0 = \frac{d\omega}{2} . \qquad (15.3)$$

The operator

$$H_0 = \omega(a^+ a)$$

written in normal form is called the free oscillator Hamiltonian. The parameter

$$\varepsilon_0 = d\omega/2$$

is the energy of the ground state or vacuum of the Hamiltonian (15.1). The ground state wave function is

$$|0\rangle = \prod_{j=1}^{d} \frac{\omega^{1/4}}{\pi^{1/4}} e^{-\frac{\omega}{2}q_j^2} = \left(\frac{\omega}{\pi}\right)^{d/4} e^{-\frac{\omega}{2}q^2} , \qquad (15.4)$$

and satisfies the conditions

$$a_j|0\rangle = 0 , \qquad (15.5)$$

$$\langle 0|0\rangle = \int_{-\infty}^{\infty} d^d q \left(\frac{\omega}{\pi}\right)^{d/2} e^{-\omega q^2} = 1 .$$

All excited states are defined as

$$|n\rangle = \frac{1}{\sqrt{n!}} a_{j_1}^+ \ldots a_{j_n}^+ |0\rangle , \qquad H_0|n\rangle = n\omega|n\rangle .$$

Radial excitations with zero angular moment are defined in R^d as

$$|n_r\rangle = C_{n_r}(a_j^+ a_j^+)^{n_r}|0\rangle , \qquad C_{n_r}^{-2} = 2^{2n_r} n_r! \frac{\Gamma(d/2+n_r)}{\Gamma(d/2)} \qquad (15.6)$$

and obey the relations

$$\langle n_r|m_r\rangle = \delta_{n_r m_r} , \qquad H_0|n_r\rangle = 2n_r\omega|n_r\rangle .$$

15.1 Hamiltonians in the Oscillator Representation

Let us formulate *the oscillator representation* (OR) for the Hamiltonian

$$H = \frac{p^2}{2} + W(q^2) \qquad (15.7)$$

in the space R^d. The potential $W(q^2)$ is supposed to permit the existence of a bound state. The problem is to calculate the energy of the ground state of the Hamiltonian H. Let us pick out the pure oscillator

$$H = \frac{1}{2}(p^2 + \omega^2 q^2) + [W(q^2) - \frac{\omega^2}{2}q^2] , \qquad (15.8)$$

where ω is a parameter. The oscillator canonical variables (p, q) can be written in the form (15.2) and the vacuum is defined by (15.4, 15.5).

Let us make use of the idea of the normal product. Let us substitute the representation (15.2) into the potential $W(q)$ and go over to the normal ordering of the creation and annihilation operators:

$$
\begin{aligned}
W(q^2) &= \int \left(\frac{dk}{2\pi}\right)^d \tilde{W}(k^2) e^{ikq} = \int \left(\frac{dk}{2\pi}\right)^d \tilde{W}(k^2) \exp\left(ik\frac{a+a^+}{\sqrt{2\omega}}\right) \\
&= \int \left(\frac{dk}{2\pi}\right)^d \tilde{W}(k^2) \exp\left(-\frac{k^2}{4\omega}\right) \exp\left(ik\frac{a^+}{\sqrt{2\omega}}\right) \exp\left(ik\frac{a}{\sqrt{2\omega}}\right) \\
&= \int \left(\frac{dk}{2\pi}\right)^d \tilde{W}(k^2) \exp\left(-\frac{k^2}{4\omega}\right) : \exp(ikq) :,
\end{aligned}
$$

where $:*:$ is the symbol for the normal ordering, $(kq) = \sum k_j q_j$ and

$$
\widetilde{W}_d(k^2) = \int (d\rho)^d W(\rho^2) e^{i(k\rho)} .
$$

Now let us substitute the representation (15.2) into (15.8) and go to the normal product over the operators a_j^+ and a_j. One gets

$$
\frac{1}{2}(p^2 + \omega^2 q^2) = \omega \sum_j a_j^+ a_j + \frac{d}{2}\omega = \omega(a^+ a) + \frac{d}{2}\omega, \tag{15.9}
$$

$$
\begin{aligned}
W(q) - \frac{\omega^2}{2}q^2 &= \int \left(\frac{dk}{2\pi}\right)^d \widetilde{W}_d(k^2) \exp\left(-\frac{k^2}{4\omega}\right) : e^{i(kq)} : \\
&\quad - \frac{\omega^2}{2}\left(:q^2: + \frac{d}{2\omega}\right) .
\end{aligned}
$$

Our requirement is that the interaction Hamiltonian in the form of the normal product should not contain any terms quadratic in q. We call this requirement *the oscillator representation condition*.

We have required that the interaction part of the Hamiltonian should not contain the term with $:q^2:$ because this term is postulated to be completely included in the free oscillator part. This requirement gives the equation for the frequency ω:

$$
\omega^2 + \int \left(\frac{dk}{2\pi}\right)^d \widetilde{W}_d(k^2) \exp\left(-\frac{k^2}{4\omega}\right)\frac{k^2}{d} = 0 . \tag{15.10}
$$

Using these formulas we can rewrite the Hamiltonian (15.8) in the form

$$
H = H_0 + H_I + \varepsilon_0 \tag{15.11}
$$

with

$$H_0 = \omega(a^+ a) ,$$

$$H_I = \int \left(\frac{dk}{2\pi}\right)^d \widetilde{W}_d(k^2) \exp\left(-\frac{k^2}{4\omega}\right) : e_2^{i(kq)} :$$

$$= \int \left(\frac{d\rho}{\sqrt{\pi}}\right)^d e^{-\rho^2} W(\frac{\rho^2}{\omega}) : \exp(-Q^2 + 2(\rho Q)) - 1 + Q^2(1 - \frac{2\rho^2}{d}) : ,$$

$$\varepsilon_0 = \frac{d\omega}{4} + \int \left(\frac{dk}{2\pi}\right)^d \widetilde{W}_d(k^2) \exp\left(-\frac{k^2}{4\omega}\right) ,$$

$$e_2^z = \exp_2(z) = e^z - 1 - z - \frac{z^2}{2} .$$

One can see that (15.10) defines the minimum of ε_0 in (15.11), i.e.

$$
\begin{aligned}
\varepsilon_0 &= \min_\omega \left\{ \frac{d\omega}{4} + \int \left(\frac{dk}{2\pi}\right)^d \widetilde{W}_d(k^2) \exp(-\frac{k^2}{4\omega}) \right\} \\
&= \min_\omega \left\{ \frac{d\omega}{4} + \int \left(\frac{d\rho}{\sqrt{\pi}}\right)^d e^{-\rho^2} W\left(\frac{\rho^2}{\omega}\right) \right\} \\
&= \min_\omega \left\{ \frac{d\omega}{4} + \int_0^\infty \frac{du\, u^{d/2-1} e^{-u}}{\Gamma(d/2)} W\left(\frac{u}{\omega}\right) \right\} \\
&= \int_0^\infty \frac{du\, u^{d/2-1} e^{-u}}{\Gamma(d/2)} \frac{d}{du}\left[uW\left(\frac{u}{\omega}\right) \right] ,
\end{aligned}
$$
(15.12)

where ω is the solution of (15.10), which can be rewritten in the form

$$\omega = \int_0^\infty \frac{du\, u^{d/2} e^{-u}}{\Gamma(d/2+1)} \frac{d}{du} W\left(\frac{u}{\omega}\right) .$$
(15.13)

Thus the frequency ω is determined as a function of the input parameters defining the potential $W(q^2)$.

The interaction Hamiltonian in the normal form (15.11) does not contain cactus-type diagrams in any perturbation order. Their contributions are included in the characteristics of the ground state, i.e., into the oscillator frequency ω. Thus, the contributions of all cactus-type diagrams are effectively summed and define the main quantum corrections to the formation of the ground state. The highest quantum corrections can be calculated by perturbation expansion over H_I.

15.2 The Second Correction

The ground state energy is the eigenvalue of the equation

$$H\Psi_0 = E_0\Psi_0 \qquad \text{or} \qquad (H_0 + H_I + \varepsilon_0)\Psi_0 = E_0\Psi_0$$

so that in the zeroth approximation we have

$$E_0^{(0)} = \varepsilon_0 \quad \text{and} \quad \Psi_0^{(0)} = |0>$$

The first correction equals zero:

$$\varepsilon_1 = \langle 0|H_I|0\rangle = 0.$$

The second correction is defined by the standard formula

$$\varepsilon_2 = -\langle 0|H_I \frac{1}{H_0} H_I|0\rangle = -\frac{1}{\omega} \int \left(\frac{dk_1}{2\pi}\right)^d \int \left(\frac{dk_2}{2\pi}\right)^d \qquad (15.14)$$

$$\times \widetilde{W}_d(k_1^2)\widetilde{W}_d(k_2^2) \exp\left(-\frac{k_1^2 + k_2^2}{4\omega}\right) S\left(\frac{(k_1 k_2)}{2\omega}\right) ,$$

$$S(z) = \int_0^1 \frac{dt}{t}\left[\cosh(tz) - 1 - \frac{1}{2}z^2 t^2\right] = \sum_{n=2}^{\infty} \frac{z^{2n}}{2n(2n)!} ,$$

$$\varepsilon_2 = -\frac{1}{2\omega} \sum_{n=2}^{\infty} B\left(\frac{d}{2}, \frac{d}{2} + n\right) A_n^2 ,$$

where $B(x, y)$ is the beta function and

$$A_n = \frac{(4\omega)^{d/2}}{n!} \int \left(\frac{dq}{2\pi}\right)^d \widetilde{W}_d(4\omega q^2)(q^2)^n e^{-q^2} \qquad (15.15)$$

$$= \left(\frac{\omega}{\pi}\right)^{d/2} \int_0^{\infty} \frac{du\, u^{d/2+n-1}e^{-u}}{n!\Gamma(d/2)} \widetilde{W}_d(4\omega u)$$

$$= \left(-\frac{\partial}{\partial \alpha}\right)^n \int_0^{\infty} \frac{du\, u^{d/2-1}e^{-u}}{n!\Gamma(d/2)} W\left(\frac{\alpha u}{\omega}\right)|_{\alpha=1} .$$

Thus the background energy of the Hamiltonian is

$$E_0 = \varepsilon_0 + \varepsilon_2 + \ldots = \varepsilon_0\left(1 + \frac{\varepsilon_2}{\varepsilon_0} + \ldots\right) . \qquad (15.16)$$

The accuracy of the oscillator representation can be estimated as

$$\delta \sim \left|\frac{\varepsilon_2}{\varepsilon_0}\right| . \qquad (15.17)$$

15.3 The General Case

The oscillator representation introduced above can be generalized in the following way. The Hamiltonian is (15.7). Let us introduce a constant ε_0 and a potential $W_0(q, \xi)$ depending on some parameters ξ and rewrite the Hamiltonian in the form

$$H = \frac{p^2}{2} + W(q^2) = H_0 + H_I + \varepsilon_0 , \qquad (15.18)$$

$$H_0 = \frac{p^2}{2} + W_0(q^2, \xi) - \epsilon(\xi),$$

$$H_I = (W(q^2) - \varepsilon_0) - (W_0(q^2, \xi) - \epsilon(\xi)).$$

We suppose that the Schrödinger equation

$$\left(\frac{p^2}{2} + W_0(q^2, \xi)\right)\Psi_0^\xi = \epsilon(\xi)\Psi_0^\xi$$

can be solved, i.e. the wave function

$$\Psi_0^\xi = \Psi_0(q, \xi)$$

can be found in an explicit form. Naturally the potential $W_0(q^2, \xi)$ should be chosen to be "close" to the potential $W(q^2)$ as far as possible. Now let us make the variational estimation of the initial Hamiltonian over the wave function $\Psi_0(q, \xi)$:

$$\varepsilon_0 = \min_\xi \langle \Psi_0^\xi | H | \Psi_0^\xi \rangle = \min_\xi \langle \Psi_0^\xi | \frac{p^2}{2} + W(q^2) | \Psi_0^\xi \rangle, \qquad (15.19)$$

where the minimization is performed over the parameters ξ. As a result these parameters ξ and the wave function $\Psi_0(q, \xi)$ become fixed. Thus we can consider the operators H_0 and H_I in (15.18) as the free and interaction Hamiltonians respectively and develop perturbation calculations over H_I. The energy of the ground state E_0 can be calculated by the perturbation method and looks like (15.16), because the first perturbation correction equals zero, i.e.

$$\varepsilon_1 = \langle \Psi_0^\xi | H_I | \Psi_0^\xi \rangle = 0.$$

It should be noted that ε_0 is the upper estimation of the ground state energy

$$E_0 < \varepsilon_0.$$

If the potential $W_0(q^2, \xi)$ is chosen as an oscillator with an unknown frequency, we get the oscillator representation introduced above. It should be stressed that the oscillator representation has an indubitable advantage because calculations become simpler in the creation and annihilation operator formalism.

As an example, let us consider the Hamiltonian having a direct connection with the three-body problem (see chapter 20)

$$H = \frac{p^2}{2} + W(u) = -\frac{1}{2}\frac{d^2}{du^2} - \frac{B(u)}{\cosh^2 u} \ , \tag{15.20}$$

where $-\infty < u < \infty$ and $B(u)$ is a positive bounded function. We have

$$H = H_0 + H_I + \varepsilon_0 \ ,$$

$$H_0 = -\frac{1}{2}\frac{d^2}{du^2} - \frac{\theta(\theta + 1)}{2\cosh^2 u} + \frac{\theta^2}{2} ,$$

$$H_I = -\frac{2W(u) - \theta(\theta + 1)}{2\cosh^2 u} - \frac{\theta^2}{2} - \varepsilon_0 \ ,$$

where θ is a constant. The equation

$$H_0\Psi_0 = \left(-\frac{1}{2}\frac{d^2}{du^2} - \frac{\theta(\theta + 1)}{2\cosh^2 u} \right)\Psi_0 = \varepsilon(\theta)\Psi_0$$

has the solution

$$\Psi_0(u) = \frac{1}{N(\theta)[\cosh u]^\theta} \ , \qquad \varepsilon(\theta) = -\frac{\theta^2}{2} \ .$$

The constant $N(\theta)$ is determined by the normalization condition and equals

$$N^2(\theta) = B(\frac{1}{2}, \theta) = \frac{\Gamma(\frac{1}{2})\Gamma(\theta)}{\Gamma(\frac{1}{2} + \theta)} \ .$$

The next steps should be done according to the rules formulated above.

16. The Oscillator Representation in the Space R^3

Our starting point is the radial Schrödinger equation in 3 dimensions:

$$\left[-\frac{1}{2r}\left(\frac{d}{dr}\right)^2 r + \frac{l(l+1)}{2r^2} + V(r)\right]\psi_{nl}(r) = E_{nl}\psi_{nl}(r) . \qquad (16.1)$$

We shall consider the potentials of Coulomb or Yukawa type, i.e. the potentials that decrease for large distances

$$V(r) \to 0 \qquad \text{for} \qquad r \to \infty \qquad (16.2)$$

and the anharmonic potentials of confinement type for which

$$V(r) \to r^{2\sigma}, \qquad (\sigma > 0) \quad \text{for} \quad r \to \infty. \qquad (16.3)$$

These potentials can have a repulsive region at short distances.

Our aim is to calculate the energy spectrum E_{nl} and to find the wave functions $\psi_{nl}(r)$ by using the oscillator representation method. This means that the wave functions $\psi_{nl}(r)$ should be expanded over the oscillator basis. This expansion can be done but it will not be effective for the simple reason that the asymptotic behaviours of the true and oscillator wave functions disagree for large $r \to \infty$ and short $r \to 0$ distances. Thus we can not apply the oscillator representation method directly but we have to transform the Schrödinger equation (16.1) in such a way that the true wave function should provide the necessary asymptotic behaviour for large and small distances. For this aim we will use the well-known technique of changing the independent coordinate (see, for example, [13], [14]) which has been applied to show the equivalence between solutions for different power-law potentials in the spaces of different dimensions. For example, there exists equivalence between the Coulomb potential in 3 dimensions and the oscillator potential in 4 dimensions (see [11], [14]).

Our idea consists in the following. We want to change the variable $r = r(s)$ and identify the transformed equation with a Schrödinger equation in the space with another dimension. The transition in the radial Schrödinger equation to the highest dimensions from the general point of view was considered earlier (see, for example, [15]). Thus calculation of the function $\psi_{0l}(r)$ would be equivalent to calculation of the ground state wave function of a modified

Hamiltonian in another dimension. Moreover the wave functions in this auxiliary space should have the oscillator Gaussian asymptotic behaviour. The radial excitation wave functions

$$\psi_{nl}(r) = |n_r\rangle$$

will be equivalent to the highest oscillator states. Let us perform all necessary transformations.

16.1 Large Distances

Let us consider large distances. Usually we can find the asymptotic behaviour of the wave function $\psi(r)$ for $r \to \infty$ analytically. Let this asymptotic be

$$\psi(r) \sim \exp(-a(r)) \ .$$

For example, for the class of potentials (16.3) one can get

$$a(r) \sim r^{1+\sigma}. \tag{16.4}$$

The case $\sigma = 0$ corresponds to Coulomb– and Yukawa–type potentials (16.2). The new variable $r = r(s)$ should be introduced so that

$$a(r(s)) \sim s^2 \quad \text{for} \quad s \to \infty$$

and

$$\psi(r(s)) \sim \exp(-s^2).$$

The substitutions

$$r = r(s) \ \text{and} \ \psi(r) = \sqrt{r'(s)}\phi(s)$$

in (16.1) give

$$\left[-\frac{1}{2}\left(\frac{d}{ds}\right)^2 + \left\{\frac{1}{8}\left(\frac{r''(s)}{r'(s)}\right)^2 - \frac{1}{4}\left(\frac{r''(s)}{r'(s)}\right)' + \frac{l(l+1)}{2}\right.\right.$$

$$\left.\left. \times \left(\frac{r'(s)}{r(s)}\right)^2\right\} + (r'(s))^2\left(V(r(s)) - E\right)\right]\phi(s) = 0. \tag{16.5}$$

This equation in the case

$$r^{1+\sigma} = s^2 \quad \text{or} \quad r = s^{\frac{2}{1+\sigma}} = s^{2\rho}, \qquad \rho = \frac{1}{1+\sigma}$$

after some transformations becomes

$$\left[-\frac{1}{2}\left(\frac{d}{ds}\right)^2 + \frac{L(L+1)}{2s^2} + W(s^2; E)\right]\phi(s) = 0 \ , \tag{16.6}$$

where

$$W(s^2; E) = \left(\frac{2}{1+\sigma} s^{\left(\frac{1-\sigma}{1+\sigma}\right)}\right)^2 (V(s^{\frac{2}{1+\sigma}}) - E)$$
$$= 4\rho^2 s^{2(2\rho-1)}[V(s^{2\rho}) - E],$$
$$L = L(l, \rho) = \rho(2l+1) - 1.$$

Now let us list the well-known formulae in the space R^d:

$$\Delta_d = \sum_{j=0}^{d} \left(\frac{\partial}{\partial x_j}\right)^2, \quad r^2 = \sum_{j=0}^{d} (x_j)^2, \qquad (16.7)$$

$$\Delta_d f(r) = \left[\left(\frac{d}{dr}\right)^2 + \frac{d-1}{r}\frac{d}{dr}\right] f(r)$$

$$= r^{-(L+1)}\left[\left(\frac{d}{dr}\right)^2 - \frac{L(L+1)}{r^2}\right] r^{(L+1)} f(r),$$

$$L = L_d = \frac{d-3}{2}.$$

Comparing (16.6) with (16.7) one can say that the operator in (16.6) can be identified with the radial part of Δ_d in the space R^d with

$$d = d(l, \rho) = 2L(l, \rho) + 3 = 2\rho(2l+1) + 2. \qquad (16.8)$$

In the particular case $\sigma = 0$ ($\rho = 1$) corresponding to Coulomb type potentials we have $d = 4 + 4l$.

Thus the equation (16.6) takes the form

$$\left[-\frac{1}{2}\Delta_d + W(s^2; E)\right]\phi(s) = 0 \qquad (16.9)$$

and we have to look for a solution depending on the radial variable s only.

16.1.1 Small Distances

The Hamiltonian in (16.6) contains the orbital repulsive part and this repulsion is absorbed by d dimensions in (16.9), but the potential itself can be repulsive at short distances. Therefore the wave function $\psi(s)$ decreases when $s \to 0$. Let us represent it as

$$\Psi(s) = s^{(L_D+1)}\Phi(s),$$

where the parameter L_D secures the decreasing of the wave function at short distances. According to (16.7) the following chain of transformations is valid:

$$\Delta_d \rightarrow \left(\frac{d}{ds}\right)^2 - \frac{L_d(L_d+1)}{s^2}$$

$$\rightarrow \left[\left(\frac{d}{ds}\right)^2 - \frac{L_D(L_D+1)}{s^2}\right] + \left[\frac{L_D(L_D+1)}{s^2} - \frac{L_d(L_d+1)}{s^2}\right]$$

$$\rightarrow \Delta_D + \frac{(D-d)(D+d-4)}{4} \cdot \frac{1}{s^2}.$$

These correlations are valid if the operators Δ_d and Δ_D act on functions depending on the radius s only.

As a result the equation (16.9) can be rewritten in the form

$$\left[-\frac{1}{2}\Delta_D + W_{lD,\rho}(s^2, E)\right]\Phi(s) = 0, \tag{16.10}$$

$$W_I(s^2, E) = W_{lD,\rho}(s^2, E) = -\frac{K(l, \rho, D)}{2s^2} + W(s^2, E),$$

$$K(l, \rho, D) = \frac{(D - d(l, \rho))(D + d(l, \rho) - 4)}{4},$$

where $s_j \in R^D$ and D is an additional parameter. We can see that the potential $W_{lD,\sigma}(s, E)$ contains an additional attractive term which can compensate the repulsive part of the potential $W(s, E)$. Let us stress once more that the wave function is assumed to depend on the radial variable s only.

In the case of Coulomb– or Yukawa–type potentials (16.2) when $\rho = 1$ and the equation (16.10) looks like

$$\left[-\frac{1}{2}\Delta_D + W_I(s^2, E)\right]\Phi(s) = 0, \tag{16.11}$$

$$W_I(s^2; E) = -\frac{K(l, D)}{2s^2} + 4s^2(V(s^2) - E),$$

$$K(l, D) = \frac{(D - 4l - 4)(D + 4l)}{4}.$$

16.2 Formulation of the Problem

We would like to stress that the energy E enters into the Schrödinger equations (16.10) and (16.11) as a parameter. Thus, our problem is formulated in the following way. We have the Hamiltonian H in D dimensions

$$H(E) = -\frac{1}{2}\Delta_D + W_I(s^2, E) = \frac{p^2}{2} + W_I(s^2, E) \tag{16.12}$$

and we have to solve the Schrödinger equation

$$H(E)\Phi = \left[\frac{p^2}{2} + W_I(s^2, E)\right]\Phi = \varepsilon(E) \, Phi \tag{16.13}$$

for radial excitations only. The desired energy spectrum $E_{n\ell}$ of the initial problem in (16.1) is contained in the radial excitation spectrum $\varepsilon^{[n]}$ of the Hamiltonian (16.12):

$$H(E)\Phi^{[n]}(s) = \varepsilon^{[n]}(E)\Phi^{[n]}(s) , \qquad (n = 0, 1, 2, ...) \qquad (16.14)$$

and it is defined by the equation

$$\varepsilon^{[n]}(E) = 0. \qquad (16.15)$$

We shall solve the equation (16.14) by the oscillator representation method. Formulas (16.11) give for the Hamiltonian (16.12)

$$H = H_0(E) + H_I(E) + \varepsilon_0(E) = H_0 + H_I + \varepsilon_0 , \qquad (16.16)$$

where

$$
\begin{aligned}
H_0 &= \omega(a^+ a) , \\
H_I &= \int \left(\frac{dk}{2\pi}\right)^D \widetilde{W}_I(k^2, E) e^{-k^2/(4\omega)} : e_2^{ikq} : , \\
\varepsilon_0(E; \omega, D) &= \frac{D\omega}{4} + \int_0^\infty \frac{du\, u^{D/2-1} e^{-u}}{\Gamma(D/2)} \cdot W_I\left(\frac{u}{\omega}, E\right).
\end{aligned}
$$

The equation defining the oscillator representation

$$\frac{\partial}{\partial\omega}\varepsilon_0(E; \omega, D) = 0 \qquad (16.17)$$

determines the parameter

$$\omega = \omega(E, D) ,$$

as the function of the energy E, D and other parameters defining the potential $V(r)$ in (16.1).

The ground state energy $\varepsilon(E, D)$ of the Hamiltonian H in (16.16) will be calculated by perturbation expansion over the interaction Hamiltonian H_I, and in the N–th approximation order has the form

$$\varepsilon_{(N)}(E, D) = \varepsilon_0(E, D) + \varepsilon_2(E, D) + ... + \varepsilon_N(E, D) .$$

The ground state energy E of the initial problem in the N-th perturbation order of the oscillator representation method is defined by (16.17 and

$$\varepsilon_{(N)}(E, D) = \varepsilon_0(E, D) + \varepsilon_2(E, D) + ... + \varepsilon_N(E, D) = 0 .$$

This equation determines the energy $E_{(N)}(D)$ in the N-th perturbation order as a function of D and other parameters defining the potential. The parameter D can be defined by the condition

$$E_{(N)} = \min_{\{D\}} E_{(N)}(D) . \qquad (16.18)$$

16.2.1 The Ground State Energy in the Zeroth and Second Approximations

In fact we shall use the oscillator representation method in the zeroth and second approximations only. Here we give formulae simplifying these calculations for the ground state energy. The function $\varepsilon(E;\omega, D)$ in (16.16) depends on two parameters D and ω. We shall consider problems for the solution of which a greater number of auxiliary parameters should be introduced. Let us denote these parameters by $\{\alpha_j\}$ and the auxiliary ground state energy

$$\varepsilon_{(2)}(E;\omega, \alpha_j) = \varepsilon_0(E;\omega, \alpha_j) + \varepsilon_2(E;\omega, \alpha_j) .$$

According to (16.16) the function $\varepsilon_0(E;\omega, \alpha_j)$ has the form

$$\varepsilon_0(E;\omega, \alpha_j) = A(\omega, \alpha_j) - E \cdot B(\omega, \alpha_j) ,$$

where $A(\omega, \alpha_j)$ and $B(\omega, \alpha_j)$ are known functions.

In the zeroth approximation the equation defining the oscillator representation (16.17) and the equation

$$\varepsilon_0(E_0;\omega, \alpha_j) = A(\omega, \alpha_j) - E_0 B(\omega, \alpha_j) = 0.$$

(16.18) give

$$\frac{\partial}{\partial \omega} \left(\frac{A(\omega, \alpha_j)}{B(\omega, \alpha_j)} \right) = 0 . \tag{16.19}$$

The equations

$$\frac{\partial}{\partial \alpha_k} E_0(\omega, \alpha_j) = \frac{\partial}{\partial \alpha_k} \left(\frac{A(\omega, \alpha_j)}{B(\omega, \alpha_j)} \right) = 0 \qquad \text{for all } k$$

define the parameters $\{\alpha_j\}$ as functions of E_0:

$$\alpha_j = \alpha_j(E) . \tag{16.20}$$

As a result in the zeroth approximation the energy E_0 is determined by the minimum

$$E_0 = \min_{\{\omega, \alpha_j\}} \frac{A(\omega, \alpha_j)}{B(\omega, \alpha_j)} = \frac{A(\omega_0, \alpha_j^0)}{B(\omega_0, \alpha_j^0)} , \tag{16.21}$$

where the parameters ω_0 and α_j^0 defines the minimum.

In the second approximation we have to solve the equation (16.17) and

$$\varepsilon_{(2)}(E;\omega, \alpha_j) = \varepsilon_0(E;\omega, \alpha_j) + \varepsilon_2(E;\omega, \alpha_j) = 0 . \tag{16.22}$$

We expect that the second correction is small so that in the second approximation the energy

$$E_{(2)} = E_0 + E_2$$

and

$$\varepsilon_0(E_{(2)}; \omega_0, \alpha_j^0) = A(\omega_0, \alpha_j^0) - E_{(2)}B(\omega_0, \alpha_j^0)$$
$$= -E_2 B(\omega_0, \alpha_j^0) + O(E_2^2) ,$$

where $\omega_0 = \omega(E_0)$. Thus the second correction is

$$E_2 = \frac{\varepsilon_2(E_0; \omega_0, \alpha_j^0)}{B(\omega_0, \alpha_j^0)} + O(E_2^2) .$$

Finally we get

$$E_{(2)} = E_0 + E_2 \qquad (16.23)$$
$$= \min_{\{\omega, \alpha_j\}} \frac{A(\alpha_j)}{B(\alpha_j)} + \frac{\varepsilon_2(E_0; \omega_0, \alpha_j(E_0))}{B(\omega_0, \alpha_j(E_0))} + E_0 O\left(\left|\frac{E_2}{E_0}\right|^2\right)$$
$$= \frac{A(\omega_0, \alpha_j^0) + \varepsilon_2(E_0; \omega_0, \alpha_j^0)}{B(\omega_0, \alpha_j^0)} + E_0 O\left(\left|\frac{E_2}{E_0}\right|^2\right) .$$

Using this formula calculations become more simple in comparison with (16.18) and (16.19).

16.2.2 Radial Excitations

Radial excitations in the oscillator representation are defined in (15.6). We shall apply the oscillator representation to the Hamiltonian in the form (16.16), and then we get the Schrödinger equation (16.14). The desired energies E_n ($n = 0, 1, ...$) of the initial equation (16.1) for the ground and radial excited states are defined by equation (16.15), and therefore, we should find the functions $\varepsilon^{[n]}(E)$ for the ground and radial excited states. For the state $|n\rangle$ ($n = 0, 1, ...$) the matrix element

$$\langle n|H_I|n\rangle = A^{[n]}(\omega, \alpha_j) - E_0 B^{[n]}(\omega, \alpha_j) \neq 0 .$$

The energy $\varepsilon^{[n]}$ in the lowest approximation is

$$\varepsilon_1^{[n]}(E) = \langle n|H|n\rangle = \varepsilon_0(E) + 2n\omega(E) + \langle n|H_I|n\rangle$$
$$= A_1^{(n)}(\omega, \alpha_j) - E B_1^{(n)}(\omega, \alpha_j) , \qquad (16.24)$$

where

$$A_1^{(n)}(\omega, \alpha_j) = \varepsilon_0(E) + 2n\omega(E) + A^{[n]}(\omega, \alpha_j) ,$$
$$B_1^{(n)}(\omega, \alpha_j) = B(\omega, \alpha_j) + B^{[n]}(\omega, pha_j) .$$

Two equations,

$$\frac{\partial}{\partial \omega} A(\omega, \alpha_j) - E \frac{\partial}{\partial \omega} B(\omega, \alpha_j) = 0 , \tag{16.25}$$

$$A_1^{(n)}(\omega, \alpha_j) - E B_1^{(n)}(\omega, \alpha_j) = 0 ,$$

determine the functions $\omega(\alpha_j)$ and $E(\alpha_j)$. The energy of the n-th excited state in the first approximation of the oscillator representation is determined as

$$E_1^{[n]} = \min_{\{\alpha_j\}} \frac{A_1^{[n]}(\omega(\alpha_j), \alpha_j)}{B_1^{[n]}(\omega(\alpha_j), \alpha_j)} .$$

In the second approximation the energy is defined as

$$\varepsilon_{(2)}^{[n]}(E) = \varepsilon_0(E) + 2n\omega(E) + \langle n|H_I|n\rangle \tag{16.26}$$

$$- \langle n|(H_I - \langle n|H_I|n\rangle) \cdot \frac{1}{H_0 - 2n\omega(E)} \cdot (H_I - \langle n|H_I|n\rangle)|n\rangle .$$

The wave function in the second approximation is

$$\Phi_2^{[n]} = \{1 - \frac{1}{H_0 - 2n\omega(E)} \cdot (H_I - \langle n|H_I|n\rangle)\}|n\rangle .$$

16.2.3 The Upper Estimate

We see that the energy of the initial problem in the zeroth approximation is defined by the equation

$$\varepsilon_0(E_0) = 0 , \tag{16.27}$$

and the exact value of this energy is defined by the equation

$$\varepsilon(E) = \sum_{n=0}^{\infty} \varepsilon_n(E) = 0 . \tag{16.28}$$

Let us show that

$$E \le E_0 . \tag{16.29}$$

In other words the energy in the zeroth perturbation order gives the upper variational estimate for the exact energy. Let us consider the function $\varepsilon_0(E)$ (16.16). Taking into account (16.17) and (16.11) we get

$$\frac{\partial}{\partial E} \varepsilon_0(E, D(E), \omega(E)) = -\langle 0|4s^2|0\rangle < 0 , \tag{16.30}$$

i.e. $\varepsilon_0(E)$ is a decreasing function of the parameter E. From the point of view of the Schrödinger equation (16.13) $\varepsilon_0(E)$ is the upper variational estimate for the background energy of the Hamiltonian (16.12) so that for any E

$$\varepsilon(E) \le \varepsilon_0(E) .$$

Since E_0 is defined by (16.20) so that for any $E > E_0$, $\varepsilon(E) \le \varepsilon(E_0) < 0$ and the solution of equation (16.21) can be only for $E \le E_0$ and inequality (16.22) is proved.

16.2.4 The Parameter D and the Oscillator Basis

Here we want to make the remark concerning the connection of the space R^D, for which D can be a noninteger, and the algebra of the creation and annihilation operators implying the number D to be an integer.

The initial Schrödinger equation (16.1) can be written in the form

$$\int d^3 r \Psi(\mathbf{r}) \left[-\frac{1}{2}\Delta + V(r) - E \right] \Psi(\mathbf{r}) = 0 .$$

If

$$\Psi(\mathbf{r}) = r\psi_{nl}(r)Y_{nl}(\theta, \phi)$$

then this equation for the wave function of the l-th orbital excitation looks like

$$\int\limits_0^\infty dr \left(r\psi_{nl}(r) \right) \left[-\frac{1}{2}\left(\frac{d}{dr}\right)^2 + \frac{l(l+1)}{2r^2} + \left(V(r) - E \right) \right] \left(r\psi_{nl}(r) \right) = 0 .$$

The wave function $\psi_{nl}(r)$ depends on one radial variable r only. After the substitutions

$$r = s^{2\rho} \quad \text{and} \quad r\psi_{nl}(r) = s^a \Phi(s), \tag{16.31}$$

where ρ and a are parameters, this equation becomes after some transformations

$$\int\limits_0^\infty ds\, s^{D-1} \;\; \Phi(s) \left[-\frac{1}{2}\left(\left(\frac{d}{ds}\right)^2 + \frac{D-1}{s}\cdot\frac{d}{ds} \right) \right.$$

$$\left. + \;\; W_I(s^2, E) \right] \Phi(s) = 0 , \tag{16.32}$$

with

$$W_I(s^2, E) \;\; = \;\; \bar{W}(s^2; l, \rho, D; E) = -\frac{K(l, \rho, D)}{2s^2}$$

$$+ \;\; 4\rho^2(s^2)^{2\rho-1}\left(V(s^{2\rho}) - E \right) ,$$

where

$$D = 2a - 2\rho + 2 , \qquad K(l, \rho, D) = \frac{1}{4}\left((D-2)^2 - 4\rho^2(2l+1)^2 \right) .$$

One can see that in the case when the function $\Phi(\mathbf{s}) = \Phi(s)$ depends on the s^2 only this equation can be identified with the equation in .the space R^D with

$$D = 2a - 2\rho + 2$$

on a wave function $\Phi(s)$ depending on the radius s only. The equation can be rewritten as

$$\int d^D s \Phi(s)\left[-\frac{1}{2}\Delta_D + W_I(s^2, E) - \varepsilon(E)\right]\Phi(s) = 0 ,$$

where the function

$$\varepsilon(E) = \varepsilon(l, \rho, D; E)$$

should be considered the eigenvalue of the Schrödinger equation in D-dimensions

$$\left[-\frac{1}{2}\Delta_D + W_I(s^2, E)\right]\Phi(s) = \varepsilon(E)\Phi(s). \tag{16.33}$$

The desired energy E is defined by the equation

$$\varepsilon(E) = \varepsilon(l, \rho, D; E) = 0.$$

The parameters ρ and D are arbitrary and can be chosen in an appropriate way. They can be considered additional variational parameters which can be found, for example, by the minimization of the energy in the zeroth approximation:

$$\varepsilon_0(E) = \min_{\{\rho, D\}} \varepsilon(l, \rho, D; E).$$

On the other hand, the parameter ρ can be connected with the behaviour of the wave function $\psi_{nl}(r)$ at large distances to get the Gaussian asymptotics. For example, for potentials (16.3) we can choose the parameter $\rho = 1/(1+\sigma)$ so that one can get

$$\Psi(r) \sim \exp(-r^{1+\sigma}) \sim \exp(-r^{1/\rho}) \sim \exp(-s^2) \sim \Phi(s).$$

The parameter a or D can be connected with the behaviour of the wave function at short distances. If the potential $V(r)$ has no repulsive character for $r \to 0$ then we choose

$$K(l, \rho, D) = 0$$

and

$$D = 2 + 2\rho(2l + 1).$$

If the potential $V(r)$ has a repulsive character for $r \to 0$ then D is a parameter which should be chosen to dump the repulsive behaviour of the potential at short distances. For example, it can be found by minimization of the background energy in the zeroth approximation. This implies that D can be any positive number. In other words, the dimension D of the space R^D can be considered an additional parameter which can be chosen to improve the zeroth approximation.

One can see that radial quantum number n does not enter into the Schrödinger equation (16.33) in an explicit form. The orbital quantum number l enters into (16.33) but it is absorbed by the "dimension" parameter D. From the point of view of the space R^D the functions

$$\Phi_n(s) = s^{2\rho-a}\psi_{nl}(s^{2\rho}) \qquad \text{or} \qquad \psi_{nl}(r) = r^{\frac{D-2\rho-2}{4\rho}}\Phi_n(r^{1/(2\rho)})$$

for any n and for a fixed l are eigenfunctions of the basic series of radial excitations in the space R^D with radial quantum number n and zeroth orbital momenta.

Thus the solution of the equation in 3-dimensions for the l-th orbital excitation is equivalent to the solution of the Schrödinger equation in the space R^D for states with zeroth angular moment.

As a result the initial Schrödinger equation is represented in the form (16.33) in which the wave function of the ground state $\Phi(s)$ has

– Gaussian asymptotics for large distances $\Phi(s) \sim \exp(-s^2)$,
– a maximum at the point $s = 0$.

The oscillator representation method consists in looking for the solution of the Schrödinger equation (16.33) in the form

$$\Phi_n(s) = \exp(-\frac{\omega}{2}s^2)\sum_m c_{nm}P_m^{(D)}(s^2\omega) \, ,$$

where $\{P_m^{(D)}(t)\}$ is the class of orthogonal polynomials which are orthogonal on the interval $0 < t < \infty$ with a weight function

$$\rho_D(t) = t^{\frac{D}{2}-1}\exp(-t),$$

i.e.

$$\int_0^\infty dt \, t^{\frac{D}{2}-1}e^{-t}P_n^{(D)}(t)P_m^{(D)}(t) = \delta_{nm} \, .$$

These orthogonal polynomials can be constructed by using the formalism of creation and annihilation operators a_j and a_j^+ in the space R^D (see chapter 15). We have

$$|0\rangle \sim \exp\left(-\frac{\omega}{2}s^2\right).$$

All radial excitations can be written in the form

$$\Phi_n \sim (a^+a^+)^n|0\rangle \sim P_n^{(D)}(\omega s^2)e^{-\frac{\omega}{2}s^2}$$
$$\sim P_n^{(D)}(\omega r^{1/\rho})\exp(-\frac{\omega}{2}r^{1/\rho}) \qquad (16.34)$$

where $P_n^{(D)}(t)$ is a polynomial of the n-th order. The parameter D in this representation can be considered to be any positive number. These polynomials satisfy the orthogonal condition

$$\left(\Psi_n, \Psi_m\right) \sim \langle 0|(aa)^n(a^+a^+)^m|0\rangle$$

$$\sim \int_0^\infty ds \; s^{D-1} \exp(-s^2) P_n^{(D)}(s^2) P_m^{(D)}(s^2)$$

$$\sim \int_0^\infty dt \; t^{D/2-1} \exp(-t) P_n^{(D)}(t) P_m^{(D)}(t) \sim \delta_{nm} \; .$$

This condition can be imposed for any positive D. Thus the algebra of creation and annihilation operators is nothing other than the mathematical method of performing any calculations connected with orthonormal polynomials. Moreover the explicit form of these polynomials is given by (16.34).

17. Anharmonic Potentials

Anharmonic oscillator models have played an important role in the evolution of many branches of quantum physics. In spite of their seeming simplicity it is not an easy problem to find the spectrum and eigenfunctions of an anharmonic interaction. There is a vast literature where different analytical and numerical methods are worked out to solve this problem (see, for example, [16]–[22]). From an other point of view the anharmonic potential is a good touchstone to test any new method.

Bender and Wu [17] have made a valuable contribution to the investigation of the anharmonic oscillator, which is of particular interest to field theoreticians because it can be regarded as a field theory in one dimension. The main hope is that the unusual and unexpected properties of this non-linear model may give some indication of the analytical structure of a more realistic field theory. Nevertheless the technique developed turned out to be quite complicated even for this simple case.

The standard way of attacking this problem is to invoke perturbation theory. Perturbation series for any physical characteristics are asymptotical ones, i.e. they have a zeroth radius of covergence. Summation methods should be applied to calculate high order corrections. As a result we have quite a cumbersome process. A thorough discussion of these difficulties has been give by Stevenson [18].

The quasi–classical approximation has been applied to the three-dimensional anharmonic oscillator [19]. The problem of the calculation of the energy levels is reduced to the solution of a very cumbersome transcendental equation, invoking the complete elliptic integrally. However, its accuracy drastically worsens for the low lying energy levels and moderate anharmonicity.

Another known approach for treating systems with strong interaction is a modified perturbation theory [20]. The accuracy of the modified perturbation theory with the principle of minimal sensitivity has been carefully analyzed [21] for the anharmonic oscillator.

The $1/N$–expansion for the anharmonic oscillator was used in [22]. In [7] the $1/N$–expansion was applied to calculate the spectrum of the anharmonic oscillator.

In this chapter we would like to present our contribution to these numerous investigations. We shall demonstrate the oscillator representation method for calculating the bound state energies of anharmonic oscillators [23].

17.1 Anharmonic Potentials in R^1

Here we demonstrate the oscillator representation method in the calculation of the bound state energy of a one-dimensional anharmonic oscillator. The Hamiltonian is

$$H = \frac{p^2}{2m} + \frac{m\nu^2}{2}q^2 + \lambda q^4. \tag{17.1}$$

In the case of symmetric potentials $V(q^2)$ the ground state wave function depends on q^2 only, i.e.

$$\Psi = \Psi(q^2), \qquad \Psi'(0) = 0,$$

so that we can write

$$\int_{-\infty}^{\infty} dq\, \Psi(q^2)\left[-\frac{1}{2}\cdot\frac{d^2}{dq^2} + V(q^2) - mE\right]\Psi(q^2) = 0$$

or

$$\int_{0}^{\infty} dq\, \Psi(q^2)\left[-\frac{1}{2}\cdot\frac{d^2}{dq^2} + V(q^2) - mE\right]\Psi(q^2) = 0, \tag{17.2}$$

$$V(q^2) = \frac{m^2\nu^2}{2}q^2 + m\lambda q^4.$$

Thus we can consider the wave equation on the positive semiaxis $0 \le q \le \infty$.

We are going to apply the oscillator representation to this Hamiltonian so that we should coordinate the Gaussian asymptotic behaviour of functions in OR with the true one. For large q this asymptotic is defined by the anharmonic term λq^4 and the wave function is proportional to

$$\Psi(q^2) \sim \exp(-q^3) \qquad \text{for} \quad q \to \infty.$$

However, it is clear that for small λ the true wave function is close to the Gaussian wave function then to the anharmonic one. Thus we can expect that the behaviour

$$\Psi(q^2) \sim \exp(-q^\alpha) \qquad \text{for} \quad q \to \infty,$$

where $2 \le \alpha \le 3$ is a parameter, could be an acceptable approximation. Let us introduce the new variable

$$q = s^{2/\alpha} = s^{2\rho}, \qquad \rho = \frac{1}{\alpha}.$$

After some transformations the integral in (17.2) can be written as

$$\int_0^\infty ds\, s^{1-2\rho}\, \Psi(s) \left[-\frac{1}{2}\left[\frac{d^2}{ds^2} + \frac{1-2\rho}{s}\frac{d}{ds} \right] + W(s^2, E) \right] \Psi(s) = 0 ,$$

where

$$W(s^2, E) = 4\rho^2 \left[\frac{m^2\nu^2}{2} \cdot (s^2)^{4\rho-1} + m\lambda(s^2)^{6\rho-1} - mE(s^2)^{2\rho-1} \right] . \quad (17.3)$$

Now we can identify the operator

$$\frac{d^2}{ds^2} + \frac{1-2\rho}{s}\frac{d}{ds} = \frac{d^2}{ds^2} + \frac{d-1}{s}\frac{d}{ds} \longrightarrow \Delta_d, \qquad d = 2 - 2\rho$$

and the measure

$$ds\, s^{1-\rho} = ds\, s^{d-1} \longrightarrow (ds)^d$$

with the "Laplacian" Δ_d and the "measure" $(ds)^d$ in an auxiliary space R^d if these operators act on a function depending on the radius only. Relation (16.7) can be used and (17.3) becomes

$$\int (ds)^d \Psi(s) \left[-\frac{1}{2}\Delta_d + W(s^2, E) \right] \Psi(s) = 0 . \quad (17.4)$$

The wave function $\Psi(s)$ in (17.4) can be considered to be the wave function of the ground state satisfying the Schrödinger equation

$$[-\frac{1}{2}\Delta_d + W(s^2, E)]\Psi(s) = H\Psi(s) = \varepsilon(E)\Psi(s) ,$$

$$H = \frac{p^2}{2} + W(s^2, E) , \quad (17.5)$$

and the desired energy E is determined by the equation

$$\varepsilon(E) = 0 . \quad (17.6)$$

Now we can apply the oscillator representation method to the Hamiltonian (17.5). According to (16.16) the Schrödinger equation and the Hamiltonian (17.5) in the oscillator representation look like

$$(H_0 + H_I + \varepsilon_0)\Psi = \varepsilon(E)\Psi ,$$

$$H = \frac{p^2}{2} + W(s^2, E) = H_0 + H_I + \varepsilon_0 , \quad (17.7)$$

where H_0 and H_I are given by (15.11) and ε_0 (15.12) is

$$\varepsilon_0(E) = \min_{\{\omega, \rho\}} \varepsilon_0(E; \omega, \rho) ,$$

with

$$\varepsilon_0(E;\omega,\rho) = \frac{d\omega}{4} + \int_0^\infty \frac{du\,u^{\frac{d}{2}-1}e^{-u}}{\Gamma(\frac{d}{2})} W\left(\frac{u}{\omega},E\right)$$

$$= A(\omega,\rho) - E \cdot B(\omega,\rho) ,$$

$$A(\omega,\rho) = \frac{1-\rho}{2}\omega + \frac{4\rho^2 m\omega}{\Gamma(1-\rho)}\left[\frac{m\nu^2}{2x^2}\cdot\Gamma(3\rho) + \frac{\lambda}{x^3}\cdot\Gamma(5\rho)\right] ,$$

$$B(\omega,\rho) = \frac{4\rho^2 m\omega}{x} \cdot \frac{\Gamma(\rho)}{\Gamma(1-\rho)} .$$

where $x = \omega^{2\rho}$. The functions $\omega(E)$ and $\rho(E)$ are defined by the equations

$$\frac{\partial}{\partial\omega}\varepsilon_0(E,\omega,\rho) = 0, \qquad \frac{\partial}{\partial\rho}\varepsilon_0(E,\omega,\rho) = 0.$$

The ground state energy $\varepsilon_2(E)$ in the second perturbation order is defined by (15.14, 15.15) and after some calculations is

$$\varepsilon_2(E;\omega,\rho) = -\frac{8\rho^2 m^2\omega}{\Gamma(1-\rho)} \cdot \sum_{n=2}^\infty \left(\frac{R_n}{n!}\right)^2 \cdot \frac{\Gamma(n+1-\rho)}{\Gamma(n+2-2\rho)} , \quad (17.8)$$

where

$$R_n = \frac{m\nu^2}{6x^2} \cdot \frac{\Gamma(n+1-4\rho)\Gamma(1+3\rho)}{\Gamma(1-4\rho)} + \frac{\lambda}{5x^3}$$

$$\times \frac{\Gamma(n+1-6\rho)\Gamma(1+5\tau)}{\Gamma(1-6\rho)} - \frac{E}{x} \cdot \frac{\Gamma(n+1-2\rho)\Gamma(1+\rho)}{\Gamma(1-2\rho)} .$$

The ground state energy E in the zeroth perturbation order is defined by (16.21)

$$E_0 = \min_{\{\omega,\rho\}} \frac{A(\omega,\rho)}{B(\omega,\rho)} = \frac{A(\omega_0,\rho_0)}{B(\omega_0,\rho_0)}, \quad (17.9)$$

where ω_0 and ρ_0 define the minimum. The energy in the second perturbation order equals

$$E_{(2)} = E_0 + E_2, \quad (17.10)$$

where

$$E_0 = \min_{\{x,\rho\}} \left[\frac{\Gamma(2-\rho)}{8m\rho\Gamma(1+\rho)} \cdot x \right.$$

$$+ \frac{m\nu^2}{6} \cdot \frac{\Gamma(1+3\rho)}{\Gamma(1+\rho)} \cdot \frac{1}{x} + \frac{\lambda}{5} \cdot \frac{\Gamma(1+5\rho)}{\Gamma(1+\rho)} \cdot \frac{1}{x^2}\right],$$

$$E_2 = \frac{\varepsilon_2(E_0;\omega_0,\rho_0)}{B(\omega_0,\rho_0)}$$

$$= -\frac{2m}{\Gamma(\rho)} \sum_{n=2}^\infty \left(\frac{R_n}{n!}\right)^2 \cdot \frac{\Gamma(n+1-\rho)}{\Gamma(n+2-2\rho)}$$

In Table 17.1 the numerical results for the background energy are given for the case $m = \frac{1}{2}$ and $m\nu = 1$ in the zeroth and second approximations.

The accuracy of the zeroth approximation can be defined as

$$\delta = \frac{|E^{(0)} - E^{(2)}|}{E^{(0)}} \cdot 100\%$$

and from Table 17.1 one can see that this is less than 1 per cent, i.e. the perturbation series converges fairly fast.

If $\nu = 0$ and $m = 1$ then the ground state energy equals

$$E = c\lambda^{\frac{1}{3}}.$$

The constant c in the zeroth perturbation order is defined by (17.9) and equals

$$c_0 = \min_{\{\rho\}} \frac{3}{4\Gamma(1+\rho)} \cdot \left[\frac{\Gamma(5\rho)\Gamma^2(2-\rho)}{4\rho} \right]^{1/3} = .66933.... .$$

The second approximation is done by (17.10). The result is

$$c_{(2)} = c_0 + c_2 = .66846..., \qquad c_2 = -.00087... .$$

The exact numerical value is $c = .667986...$ (see [16]).

Table 17.1. Results of the calculation the ground state energy of a one-dimension anharmonic oscillator for the case $m = \frac{1}{2}$ and $m\nu = 1$ as a function of the parameter λ. $E^{(0)}$ and $E^{(2)}$ are the energies of the zeroth and second approximations, $E^{\text{ex.}}$ is the exact value in [24]

λ	ρ	$E^{(0)}$	$E^{(2)}$	$E^{\text{ex.}}$
0.02	2.02	1.015	1.015	
0.1	2.07	1.065	1.065	
0.2	2.12	1.119	1.118	1.118
0.5	2.18	1.243	1.242	
1.0	2.23	1.394	1.393	
1.5	2.25	1.511	1.510	
2.0	2.27	1.610	1.609	1.608
5.0	2.31	2.022	2.020	
10.0	2.32	2.454	2.452	
20.0	2.34	3.016	3.014	3.010
100.0	2.36	5.009	5.008	

17.2 Anharmonic Potentials in R^3

In this section we consider the three-dimensional anharmonic potential. The Schrödinger equation is

$$\left[-\frac{1}{2mr}\left(\frac{d}{dr}\right)^2 r + \frac{l(l+1)}{2mr^2} + \frac{m\nu^2}{2}r^2 + \lambda r^4\right]\psi(r) = E\psi(r) . \quad (17.11)$$

According to (16.31) we make the transformation $r = s^{2\rho}$ and get the representation

$$H(E)\Phi(s) = \varepsilon(E)\Phi(s) ,$$

$$H(E) = \frac{1}{2}p^2 + W(s^2, E) ,$$

$$W(s^2, E) = 4\rho^2 m\left[\frac{m\nu^2}{2}(s^2)^{4\rho-1} + \lambda(s^2)^{6\rho-1} - E(s^2)^{2\rho-1}\right] \quad (17.12)$$

where $s \in R^d$ with

$$d = 2\rho(2l+1) + 2.$$

Now we can apply the oscillator representation method. The Hamiltonian is

$$H = \frac{p^2}{2} + W(s^2, E) = H_0 + H_I + \varepsilon_0, \quad (17.13)$$

where H_0 and H_I are given by (15.12) and ε_0 according to (17.12) equals

$$\varepsilon_0(E) = \min_{\{\omega,\rho\}} \varepsilon_0(E; \omega, rho) ,$$

with

$$\varepsilon_0(E; \omega, \rho) = \frac{d\omega}{4} + \int_0^\infty \frac{du\, u^{\frac{d}{2}-1}e^{-u}}{\Gamma(\frac{d}{2})} \cdot W\left(\frac{u}{\omega}, E\right)$$

$$= A(l, \omega, \rho) - E \cdot B(l, \omega, \rho) ,$$

$$A(l, \omega, \rho) = \frac{\rho(2l+1)+1}{2} \cdot \omega + \frac{4\rho^2 m\omega}{\Gamma(\rho(2l+1)+1)}$$

$$\times \left[\frac{m\nu^2}{2x^2} \cdot \Gamma(\rho(2l+5)) + \frac{\lambda}{x^3} \cdot \Gamma(\rho(2l+7))\right] ,$$

$$B(l, \omega, \rho) = \frac{4\rho^2 m\omega}{x} \cdot \frac{\Gamma(\rho(2l+3))}{\Gamma(\rho(2l+1)+1)} .$$

where $x = \omega^{2\rho}$.

The ground state energy E_{0l} in the zeroth perturbation order is

$$E_{0l} = \min_{\{\omega,\rho\}} \frac{A(l,\omega,\rho)}{B(l,\omega,\rho)} = \min_{\{x,\rho\}} \left[\frac{\Gamma(\rho(2l+1)+2)}{\Gamma(\rho(2l+3))} \cdot \frac{x}{8m\rho^2} \right. \tag{17.14}$$

$$+ \left. \frac{m\nu^2}{2x} \cdot \frac{\Gamma(\rho(2l+5))}{\Gamma(\rho(2l+3))} + \frac{\lambda}{x^2} \cdot \frac{\Gamma(\rho(2l+7))}{\Gamma(\rho(2l+3))} \right] .$$

The ground state energy of the anharmonic oscillator has been extensively studied numerically and exact result [7] for E_{00} in the case $\lambda = 1$, $m = 1/2$ and $\nu = 2$ is

$$E_{00}^{\text{ex}} = 4.64881.... .$$

Formula (17.14) gives for this case

$$E_{00} = 4.6511.... .$$

One can see that the oscillator representation method in the zeroth approximation gives quite an acceptable accuracy.

17.2.1 Power–Law Potentials

In this section the oscillator representation method will be applied to the calculation of the ground, orbital and radial excitation energy spectrum of three-dimensional power–law potentials

$$V(r) = \lambda r^\nu . \tag{17.15}$$

The Schrödinger equation is

$$\left[-\frac{1}{2mr}\left(\frac{d}{dr}\right)^2 r + \frac{l(l+1)}{2mr^2} + \lambda r^\nu \right]\psi(r) = E\psi(r) . \tag{17.16}$$

The transformation $r = s^{2\rho}$ leads to the representation

$$H(E)\Phi(s) = \varepsilon(E)\Phi(s) ,$$
$$H(E) = \frac{1}{2}p^2 + W(s^2, E) ,$$
$$W(s^2, E) = 4\rho^2 m \cdot \left[\lambda(s^2)^{\rho(2+\nu)-1} - E(s^2)^{2\rho-1} \right] , \tag{17.17}$$

where $s \in R^d$ with

$$d = 2\rho(2l+1) + 2.$$

The Hamiltonian in the oscillator representation is

$$H = \frac{p^2}{2} + W(s^2, E) = H_0 + H_I + \varepsilon_0, \tag{17.18}$$

where H_0 and H_I are given by (17.11) and ε_0 according to (17.12) equals

$$\varepsilon_0(E) = \min_{\{\omega,\rho\}} \varepsilon_0(E; \omega, rho) ,$$

with

$$\varepsilon_0(E;\omega,\rho) = \frac{d\omega}{4} + \int_0^\infty \frac{du\, u^{\frac{d}{2}-1} e e^{-u}}{\Gamma(\frac{d}{2})} \cdot W\left(\frac{u}{\omega}, E\right)$$

$$= A(l,\omega,\rho) - E \cdot B(l,\omega,\rho) ,$$

$$A(l,\omega,\rho) = \frac{\rho(2l+1)+1}{2} \cdot \omega + \frac{4\rho^2 m\lambda\omega}{y^{2+\nu}}$$

$$\times \frac{\Gamma(\rho(2l+3+\nu))}{\Gamma(\rho(2l+1)+1)} ,$$

$$B(l,\omega,\rho) = \frac{4\rho^2 m\omega}{y^2} \cdot \frac{\Gamma(\rho(2l+3))}{\Gamma(\rho(2l+1)+1)} ,$$

where $y = \omega^\rho$.

The ground state energy E_{0l} in the zeroth perturbation order is

$$E_{0l} = \min_{\{\omega,\rho\}} \frac{A(l,\omega,\rho)}{B(l,\omega,\rho)} = \min_{\{y,\rho\}} \left[\frac{\Gamma(\rho(2l+1)+2)}{\Gamma(\rho(2l+3))} \right.$$

$$\times \left. \frac{y^2}{8m\rho^2} + \frac{\lambda}{y^\nu} \cdot \frac{\Gamma(\rho(2l+3+\nu))}{\Gamma(\rho(2l+3))} \right] .$$

$$= \min_{\{\rho\}} \frac{2+\nu}{\Gamma(\rho(2l+3))} \cdot \left[\frac{\Gamma(\rho(2l+1)+2)}{8\nu m\rho^2} \right]^{\frac{\nu}{2+\nu}} \qquad (17.19)$$

$$\times \left[\frac{\lambda}{2} \cdot \Gamma(\rho(2l+3+\nu)) \right]^{\frac{2}{2+\nu}} .$$

Now let us obtain the formula for radial excitations. According to (16.24) we have in the first approximation

$$\varepsilon_1^{(n)} = \varepsilon_0 + 2n\omega + \langle n|H_I|n\rangle ,$$

where

$$\langle n|H_I|n\rangle = \int \left(\frac{dk}{2\pi}\right)^d \cdot \widetilde{W}_d(k^2) \cdot \exp\left(-\frac{k^2}{4\omega}\right) \langle n| : e_2^{ikq} : |n\rangle .$$

Let us define the polynomials

$$e_n(t,d) = \langle n| : e_2^{ikq} : |n\rangle = \sum_{m=2}^{2n} c_m(n,d) t^m, \qquad t = -\frac{k^2}{4\omega} ,$$

$$c_m(n,d) = \frac{n!\Gamma(\frac{d}{2})}{\Gamma(\frac{d}{2}+n)\Gamma(d\,\text{over}2+m)} \cdot \sum_p^{[\frac{m}{2}]} \frac{2^{m-2p}\Gamma(\frac{d}{2}+n+p)}{(n-m+p)!(m-2p)!(p!)^2} ,$$

where

$$p = \max(0, m - n).$$

The coefficients for $n = 1$ and $n = 2$ are

$$c_2(1, d) = \frac{2}{d} \; ;$$

$$c_2(2, d) = \frac{4(d+8)}{d(d+2)}, \qquad c_3(2, d) = \frac{16}{d(d+2)}, \qquad c_4(2, d) = \frac{2}{d(d+2)}.$$

The three first polynomials are

$$e_0(t) = 0 \; ,$$

$$e_1(t) = \frac{2}{d}t^2 \; ,$$

$$e_2(t) = \frac{4}{d(d+2)} \cdot \left[(d+8)t^2 + 4t^3 + \frac{1}{2}t^4 \right] \; .$$

Table 17.2. Results of the calculation of the ground state energy power-law potentials for $n = \ell = 0$ and $2m = 1$ of various ν. E_{00} is the zeroth approximation of the oscillator representation. The results of the numerical [25] and $1/N$- expansion methods [7], are also shown.

ν	$1/N$	Num.	E_{00}
−1.5	−0.29888	−0.29609	−0.29703
−1.25	−0.22035	−0.22029	−0.22027
−1.0	−0.25	−0.25	−0.25
0.15	1.32795	1.32795	1.3279
0.5	1.83341	1.83339	1.8335
0.75	2.10815	2.10814	2.1082
1.5	2.70806	2.70809	2.7081
2.0	3.0	3.0	3.0
3.0	3.45111	3.45056	3.4511
4.0	3.80139	3.79967	3.8024
5.0	4.09146	4.08916	4.0962
]6.0	4.33801	4.33860	4.3524
7.0	4.54690	4.55866	4.5815
8.0	4.71772	4.75587	4.7901
10.0	4.92220	5.09786	5.1607

If the potential has the form

$$W(s^2) = \sum_k w_k(s^2)^{\sigma_k}$$

Table 17.3. Results of the calculation of the energy spectrum for a wide class of potentials in the zeroth approximation. The numerical ones are taken from [7], [25] (in parentheses).

		$V(r)$			
		$-\frac{2^{1.7}}{r^{0.2}}$	$-\frac{2^{.8}}{r^{0.8}}$	$2^{3.5}r$	$\ln r$
		$2m = 1$	$2m = 1$	$2m = 1$	$m = 1$
$n = 0$	$l = 0$	−2.686	−1.2186	9.353	1.045
		(−2.686)	(−1.218)	(9.35243)	(1.0443)
	$l = 1$	−2.345	−0.5004	13.445	1.641
		(−2.345)	(−0.500)		(1.643)
	$l = 2$	−2.156	−0.2947	16.993	2.014
		(−2.156)	(−0.295)		(2.015)
	$l = 3$	−2.029	−0.2019	20.204	2.284
		(−2.029)	(−0.202)		(2.286)
$n = 1$	$l = 0$	−2.253	−0.462	16.355	1.848
		(−2.253)	(−0.462)	(16.3518)	(1.8474)
	$l = 1$	−2.101	−0.281	19.540	2.151
		(−2.101)	(−0.281)		(2.151)
	$l = 2$	−1.990	−0.195	22.521	2.388
		(−1.990)	(−0.195)		(2.388)
	$l = 3$	−1.905	−0.146	25.330	2.580
$n = 2$	$l = 0$	−2.044	−0.265	22.084	2.290
		(−2.044)	(−0.265)	(22.08224)	(2.290)
	$l = 1$	−1.951	−0.187	24.833	2.491
		(−1.951)	(−0.187)		(2.491)
	$l = 2$	−1.875	−0.142	27.478	2.663
		(−1.875)	(−0.142)		(2.663)
	$l = 3$	−1.812	−0.113	30.021	2.811

then one gets

$$\langle n|H_I|n\rangle = \int \left(\frac{dk}{2\pi}\right)^d \tilde{W}_d(k^2)\exp(-\frac{k^2}{4\omega})\langle n| : e_2^{ikq} : |n\rangle$$

$$= \sum_k \frac{w_k}{\omega^{\sigma_k}} \cdot C_k(d, \sigma_k) ,$$

$$C_k(d, \sigma_k) = \frac{\Gamma(\frac{d}{2} + \sigma_k)}{\Gamma(\frac{d}{2} + 1)} \cdot \sum_{s=2}^{2n} c_s(n, d) \cdot \frac{\Gamma(\sigma_k + 1)}{\Gamma(\sigma_k + 1 - s)} .$$

the next step is to solve the equations (16.25). The result is

$$E_{nl} = \lambda \min_{\rho} \frac{\Gamma(\rho(2l+3+\nu))}{\Gamma(\rho(2l+3))} \cdot \frac{F(\nu)}{F(0)}$$

$$\times \left[\frac{F(0)}{8\rho^2 m\lambda J} \cdot \frac{\Gamma(\rho(2l+1)+1)}{\Gamma(\rho(2l+3+\nu))}\right]^{\frac{\nu}{2+\nu}} ,$$

with

$$\begin{aligned} F(\nu) &= 4n[\rho(2+\nu)-1] + [\rho(2l+1)+1] \\ &\times [\rho(2+\nu) + C_n(d, \rho(2+\nu)-1)] , \\ J &= \rho\nu + [\rho(2+\nu)-1]C_n(d, 2\rho-1) \\ &- (2\rho-1)C_n(d, \rho(2+\nu)-1) , \end{aligned}$$

where $d = 2\rho(2l+1)+2$. The numerical results are shown in Tables 17.2 and 17.3.

17.2.2 The Logarithmic Potential

Now we consider the logarithmic potential

$$V(r) = \lambda \ln(r) . \tag{17.20}$$

This is one of the potentials which has been used in heavy quarkonum spectroscopy [25]. The standard calculations according to (16.16) give the following result for the energy E_{nl} in the lowest approximation:

$$E_{nl} = \lambda \min_{\rho} \left\{ \frac{\partial}{\partial\sigma} \ln\left[(\frac{d}{2}+4n)\sigma + \frac{d}{2}(1+C_n(d,\sigma))\right] + \psi(\frac{d}{2}+\sigma) \right. \tag{17.21}$$

$$\left. + \frac{1}{2\rho} \ln\left[\frac{1}{(\sigma+1)^3 m\lambda} \cdot \frac{\Gamma(\frac{d}{2})}{\Gamma(\frac{d}{2}+\sigma)} \cdot \frac{(\frac{d}{2}+4n)\sigma + \frac{d}{2}(1+C_n(d,\sigma))}{1+C_n(d,\sigma)-\sigma\frac{\partial}{\partial\sigma}C_n(d,\sigma)}\right] \right\},$$

where

$$\frac{d}{2} = \rho(2l+1)+1 , \quad \sigma = 2\rho-1 , \quad \psi(x) = \frac{d}{dx}\ln\Gamma(x)$$

and the functions $C_n(d,\sigma)$ are defined in the previous section.

The numerical results are shown in the Table 17.3. One can see that the first approximation of OR method coincides with the exact values in four signs.

From Table 17.1–17.3 we can see that, the oscillator representation method gives the following possibilities:

-describing in a unified way one- and three-dimensional anharmonic oscillators;

- enaling the perturbation series to converge fairly fast, i.e. the corrections connected with the interaction Hamiltonian are small enough;

-defining the energy levels ground and orbital, radial excitation states for a wide class of potentials. The results of the zeroth approximation oscillator representation agreement with the exact values in four signs.

18. Coulomb–Type Potentials

The Coulomb potential and potentials connected with it play a fundamental role in atomic and molecular physics. The solutions of the Schrödinger equation for these potentials are well-known and were obtained with different methods (see, for example, [1]–[7]). In this chapter we demonstrate the oscillator representation method ([9], [26]) in the calculation of the energy eigenvalues of the Coulomb and the screened Coulomb potential.

18.1 The Coulomb Potential

The radial Schrödinger equation for the Coulomb potential is

$$\left[-\frac{1}{2}\left(\frac{\partial^2}{\partial r^2} + \frac{2}{r}\frac{\partial}{\partial r}\right) + \frac{l(l+1)}{2r^2} - \frac{m\alpha}{r}\right]\Psi_{nl}(r) = mE_{nl}\Psi_{nl}(r) , \qquad (18.1)$$

where l is the orbital quantum number and m is the mass of electron. Performing all the transformations of the chapter 16 for $r = s^2$ this radial equation is equivalent to

$$\left[-\frac{1}{2}\Delta_d + \frac{(-8mE)}{2}\cdot s^2 - 4m\alpha\right]\Phi = 0. \qquad (18.2)$$

This equation is defined in the space R^d with $d = 4 + 4l$. We get the pure oscillator Schrödinger equation, which in the oscillator representation is

$$\left[\frac{1}{2}p^2 + \frac{1}{2}\Omega^2 s^2\right]\Phi = \left[\Omega(a^+a) + \frac{d}{2}\Omega\right]\Phi = 4\alpha m\Phi , \qquad (18.3)$$

where $\Omega^2 = -8mE$.

The radial excited eigenfunctions are given by (15.6):

$$\Phi_{n_r} = |n_r\rangle = C_{n_r}(a^+a^+)^{n_r}|0\rangle , \qquad (18.4)$$

so that one gets

$$\Omega(2n_r + \frac{d}{2})\Phi_{n_r} = 4\alpha m\Phi_{n_r} ,$$

and we obtain the Coulomb spectrum

$$E_{n,l} = -\frac{\alpha^2 m}{2(1 + n_r + \ell)^2} = -\frac{\alpha^2 m}{2(1 + n)^2} \tag{18.5}$$

with $n = n_r + l$. Formula (18.4) gives the Coulomb wave functions

$$\Psi_{n,l}(r) \sim \Phi_{n_r}(s^2) = C_{n_r}\left[\frac{1}{2}\left(s_j - \frac{\partial}{\partial s_j}\right)\left(s_j - \frac{\partial}{\partial s_j}\right)\right]^{n_r} e^{-\frac{1}{2}s^2}, \tag{18.6}$$

where $s_j \in R^d$ with $d = 4 + 4\ell$ and $s^2 = r$.

18.2 The Screened Coulomb Potential

In this section, we calculate the bound-state energies of a hydrogen atom in a shielded Coulomb field for the ground as well as the orbital excited states as a function of classical Debye screening. The screened Coulomb potential (SCP) is known to adequately describe the effective interaction in many-body atomic phenomena. Since the Schrödinger equation for such a potential does not admit exact solutions, various approximation methods, both numerical [27]–[29] and analytical [30] have been developed. In [27]–[29], a numerical technique is used to calculate the eigenstate energy as a function of the screening length for the 1s-3d states.

Here, we consider the eigenstate energy as a function of the screening length for the ground and orbital excited states in the framework of the oscillator representation in the zeroth approximation.

The screened Coulomb (Debye) potential is given by

$$V(r) = -\frac{Ze^2}{r} \exp(-Ar), \tag{18.7}$$

where $1/A$ is the screening length. The Schrödinger equation can be written in the form:

$$\left[-\frac{1}{2} \cdot \frac{1}{r}\left(\frac{d}{dr}\right)^2 r + \frac{l(l+1)}{2r^2} + \frac{U}{2} - \frac{1}{2r}\exp(-\mu r)\right]\psi(r) = 0, \tag{18.8}$$

where the following dimensionless variables are introduced:

$$\mu = \frac{A}{2Zme^2}, \qquad E = -2mZ^2e^4U.$$

Our aim is to calculate the bound-state energies E or U as a function of the parameter μ for the ground and excited states. After the substitution $r = s^{2\rho}$ (18.8) takes the form

$$\left[-\frac{1}{2}\Delta_d + W(s^2, U)\right]\Phi(s) = 0, \tag{18.9}$$

where

$$W(s^2, U) = 2\rho^2 \left[-(s^2)^{\rho-1} e^{-\mu(s^2)^\rho} + U \cdot (s^2)^{2\rho-1} \right] ,$$
$$d = 2\rho(2l + 1) + 2 .$$

The Hamiltonian H in the oscillator representation looks the (15.19), for which

$$\varepsilon_0(U, \rho) = \min_\omega \varepsilon_0(U, \omega, \rho) ,$$

with

$$\varepsilon_0(U, \omega, \rho) = A(\omega, \rho) + U \cdot B(\omega, \rho) ,$$

where

$$
\begin{aligned}
A(\omega, \rho) &= \frac{\rho(2l+1)+1}{2} \cdot \omega - \frac{2\rho^2}{\omega^{\rho-1}} \\
&\quad \times \int_0^\infty \frac{du\, u^{\rho(2l+2)-1}}{\Gamma(\rho(2l+1)+1)} \exp\left(-u - \frac{\mu}{\omega^\rho} u^\rho \right) , \\
B(\omega, \rho) &= \frac{2\rho^2}{\omega^{2\rho-1}} \cdot \frac{\Gamma(\rho(2l+3))}{\Gamma(\rho(2l+1)+1)} .
\end{aligned}
$$

According to (16.17) the parameters ω and U as a function of ρ are defined by the equations

$$\varepsilon_0(U, \omega, \rho) = A(\omega, \rho) + U \cdot B(\omega, \rho) = 0 , \tag{18.10}$$
$$\omega \frac{d}{d\omega} \varepsilon_0(U, \omega, \rho) = \omega \frac{d}{d\omega} A(\omega, \rho) + U\omega \cdot \frac{d}{d\omega} B(\omega, \rho) = 0 .$$

Introducing the variable $x = \mu/\omega^\rho$ and after some simple analytical transformations one gets

$$U = U(\mu) = \max_{F(\rho, x)=\mu} \frac{\rho^2 I_+(\rho, x) I_-(\rho, x)}{\Gamma(\rho(2l+3))\Gamma(\rho(2l+1)+2)} , \tag{18.11}$$

with

$$I_\pm(\rho, x) = \int_0^\infty du\, u^{\rho(2l+2)-1} \exp(-u - xu^\rho)[1 \pm xu^\rho] ,$$

where the parameters ρ and x are connected by the correlation

$$\mu = F(\rho, x) = \frac{2\rho^2 x I_+(\rho, x)}{\Gamma(\rho(2l+1)+2)} . \tag{18.12}$$

This correlation defines the parameter U as a function of the parameters μ and l in the lowest approximation. The numerical results are shown in Table 18.1.

Table 18.1. The eigenstate energy as a function of the screening length for 1s and 2p states.

μ	1s		2p	
	$U^{(0)}$	[29]	$U^{(0)}$	[28]
0.001	0.24900	0.24900	0.061500	0.061500
0.01	0.24015	0.24015	0.05298	0.05298
0.0125	0.237743	0.23773	0.05075	0.05075
0.025	0.225908	0.22590	0.04037	0.040375
0.05	0.20354	0.20353	0.02326	0.02368
0.1	0.16342	0.16340	0.001999	0.00205
0.125	0.14547	0.14546		
0.25	0.07399	0.07395		
0.357	0.03357	0.03377		
0.5	0.004645	0.00515		

18.2.1 The Critical Screening Length

One of the characteristics of screened Coulomb potentials is the critical screening length, which is defined by the value of the parameter μ when the lowest bound state energy equals zero. Of special interest is the critical screening length for the ground state of the two-body system for SCP– or Yukawa–type potentials. This parameter has been computed by a variety of techniques. In [27], [28] numerical methods are used to compute the critical screening length for one-electron (n, l) eigenstates.

Table 18.2. The critical screening length μ_c for two-particle (n, l) eigenstates.

l	$n_r = 0$	
	$\mu_c^{[0]}$	[28]
0	0.579	0.59530
1	0.109023	0.11011
2	0.04531	0.04567
3	0.024698	0.02492
4	0.015522	0.015672
5	0.010654	0.01076
6	0.007764	0.00784
7	0.005909	0.00597

Let us compute the critical screening length for the two-body system. The critical screening length in the zeroth approximation of OR is defined by the condition

$$U(\mu_c) = 0 \ . \tag{18.13}$$

According to (18.9) this equation is equivalent to the equations

$$I_-(\rho, x) = 0, \qquad \frac{\partial}{\partial \rho} I_-(\rho, x) = 0 \ .$$

These equations define the parameters x and ρ as functions, on l and the critical value of μ_c is determined by (18.13). The numerical results are shown in Table 18.2.

19. The Relativized Schrödinger Equation

The standard Schrödinger equation describes the behaviour of nonrelativistec particles. Nonrelativistic potential models turned out to give a successful description not only of heavy quarkonia but also of ordinary hadrons. One can say that these models work much better then we would naively expect. Nevertheless the quark–quark systems are relativistic ones so that the calculation of relativistic corrections is one of the important problems of quark bound states. The relativistic character of the quark–quark interaction was studied in papers ([31]-[34]). The main point is that the complete quantum field theory of bound states formulated yet, so that we have differently motivated approaches like Bethe-Salpeter and Breit-Fermi equations and the so-called relativized Schrödinger equation which will be considered in this chapter.

There exists a vast literature where the bound states of quark–quark systems are studied in the framework of the nonrelativistic and relativized Schrödinger equation (see, for example, [32]). Here we do not attemp to discuss any physical aspects of quarkonia. Our aim is to attract attention to the OR method and to show its application to the spectroscopy of the relativized Schrödinger equation.

The relativized Schrödinger equation is based on the simplest idea (see, for example, [34]) for taking into account the kinematic relativistic corrections using the relativistic kinetic energy instead of the nonrelativistic one

$$\frac{p^2}{2m} \rightarrow \sqrt{p^2 + m^2} \ .$$

Thus we get the following relativized Schrödinger equation in the space R^3:

$$\left[\sqrt{p^2 + m^2} + V(r^2) \right] \Psi = E \Psi \ . \tag{19.1}$$

The usual solution of this equation is by numerical calculations on computers and by variational methods (see([31]–[33])).

We shall solve this equation by the OR method. Let the orbital moment is l; then the wave function is

$$\Psi(r, \theta, \phi) = Y_{l,m}(\theta, \phi)\Psi_{nl}(r) \ . \tag{19.2}$$

The radial Schrödinger equation becomes

$$\left[\sqrt{-\frac{1}{r}(\frac{d}{dr})^2 r + \frac{l(l+1)}{r^2} + m^2} + V(r^2)\right]\Psi_{nl} = E_{nl}\Psi_{nl} \ . \qquad (19.3)$$

Introducing the function

$$\Psi_{nl}(r) = r^l \Phi(r),$$

$$L_d = \frac{d-3}{2} = l, \qquad d = 3 + 2l \ ,$$

one gets

$$\left[\sqrt{-\frac{1}{r^{\frac{d-1}{2}}}(\frac{d}{dr})^2 r^{\frac{d-1}{2}} + \frac{L_d(L_d+1)}{r^2} + m^2} + V(r^2)\right]\Psi_{nl} = E_{nl}\Psi_{nl} \ .$$

This equation coincides with the radial Schrödinger equation in the space R^d:

$$\left[\sqrt{-\Delta_d + m^2} + V(r^2)\right]\Phi(r) = E\Phi(r) \ . \qquad (19.4)$$

Now we can apply the oscillator representation to the Hamiltonian

$$H = \sqrt{p^2 + m^2} + V(r^2)$$

in the space R^d. Let us rewrite this Hamiltonian in the form

$$H = \frac{1}{2\mu}(p^2 + \vartheta r^2) + \left[\sqrt{p^2 + m^2} - \frac{p^2}{2\mu}\right] + \left[V(r^2) - \frac{\vartheta}{2\mu}r^2\right] , \qquad (19.5)$$

where μ and ϑ are parameters, and introduce the oscillator canonical variables

$$r_j = \frac{Q_j}{\sqrt{\vartheta}} \ , \qquad Q_j = \frac{a_j + a_j^+}{\sqrt{2}} \ ; \qquad (19.6)$$

$$p_j = \sqrt{\vartheta}P_j \ , \qquad P_j = \frac{a_j - a_j^+}{i\sqrt{2}} \ ; \qquad (j = 1, ...d) \ .$$

The vacuum is defined by the standard way:

$$\langle 0|0\rangle = 1 \ , \qquad a_j|0\rangle = 0 \ ,$$

$$\langle 0|r_i r_j|0\rangle = \frac{\delta_{ij}}{2\vartheta} \ , \qquad]\langle 0|p_i p_j|0\rangle = \delta_{ij}\frac{\vartheta}{2} \ .$$

Let us substitute the representation (19.6) in (19.5), go over to the normal product of the operators a_i and a_i^+ in the Hamiltonian and require that the interaction Hamiltonian should not contain terms with $: p^2 :$ and $: r^2 :$. After some transformations one can get

$$H = H_0 + E_0 + H_I \ ,$$

with

$$H_0 = \omega a_j^+ a_j , \qquad \omega = \frac{\vartheta}{\mu} ;$$

$$H_I = \int \left(\frac{d\rho}{\sqrt{\pi}}\right)^d e^{-\rho^2} \left[\sqrt{\rho^2\vartheta + m^2} : \exp(-2p\rho - p^2) - 1 + p^2(1 - \frac{2}{d}\rho^2) : \right.$$

$$+ V(\frac{\rho^2}{\vartheta}) : \exp(-2Q\rho - Q^2) - 1 + Q^2(1 - \frac{2}{d}\rho) : \Bigg] \tag{19.7}$$

$$= \int \left(\frac{du}{\sqrt{\pi}}\right)^d \cdot \left[\tilde{K}(u^2) \exp(-\frac{u^2\vartheta}{4}) : e_2^{iup} : + \tilde{V}(u^2) \exp(-\frac{u^2}{4\vartheta}) : e_2^{iur} : \right] ,$$

where

$$\tilde{K}(u^2) = \int (d\rho)^d \sqrt{\rho^2 + m^2} e^{iu\rho} ;$$

$$\tilde{V}(u^2) = \int (d\rho)^d V(\rho^2) e^{iu\rho} ,$$

$$e_2^z = e^z - 1 - z - \frac{z^2}{2} ;$$

$$E_0 = \min_{\vartheta} \int \left(\frac{d\rho}{\sqrt{\pi}}\right)^d e^{-\rho^2} \cdot \left[\sqrt{\rho^2\vartheta + m^2} + V(\frac{\rho^2}{\vartheta}) \right]$$

$$= \min_{\vartheta} \int_0^\infty \frac{du\, u^{d/2-1} e^{-u}}{\Gamma(\frac{d}{2})} \cdot \left[\sqrt{u\vartheta + m^2} + V(\frac{u}{\vartheta}) \right] .$$

The parameters μ and ϑ are determined by a condition of the oscillator representation, i.e. the interaction Hamiltonian H_I should not contain the quadratic terms with $: p^2 :$ and $: r^2 :$. These equations are

$$\int_0^\infty du\, u^{d/2} e^{-u} \frac{d}{du} \cdot \left[\sqrt{u\vartheta + m^2} - V(\frac{u}{\vartheta}) \right] = 0 ,$$

$$\omega = \frac{2}{\Gamma(\frac{d}{2}+1)} \cdot \int_0^\infty du\, u^{d/2} e^{-u} \frac{d}{du} V(\frac{u}{\vartheta}) . \tag{19.8}$$

These formulae permit us to calculate the spectrum of the relativized Hamiltonian.

19.1 Examples

Here we would like to consider the Cornell potential for the quark mass $m = 0$, because in this case all calculations can be performed analytically. The Cornell potential gives the simplest interpolation from Coulomb behaviour at short distances to a linearly growing confined potential at large distances. The Cornell potential is

$$H = \sqrt{p^2} - \frac{\kappa}{r} + hr , \tag{19.9}$$

This potential was used to give a semirelativistic description of quark–quark bound states. Phenomenologically the first term in the potential is connected with a one–gluon exchange and describes small distances. The second term ensures the confinement of quarks.

Equations (19.8) for ϑ and ω can be easily solved:

$$\vartheta = \frac{h}{1 - \kappa_l} , \qquad \kappa_l = \frac{\kappa}{1 + l} ,$$

$$\omega = \sqrt{\frac{h}{1 - \kappa_l}} \cdot \frac{\Gamma(l + 2)}{\Gamma(l + 5/2)} . \tag{19.10}$$

One can see that

1. if $h = 0$, i.e. for the pure Coulomb potential the relativized Schrödinger equation has no solutions describing any bound states;

2. for $\kappa \geq 1$ there exists the "downfall on the center", i.e. there are no stable states in this system.

The Hamiltonian in the OR is

$$H = \omega a_j^+ a_j + H_I + \varepsilon_0 ,$$

where

$$
\begin{aligned}
H_I &= -\sqrt{\frac{h}{1 - \kappa_l}} \frac{\Gamma(2 + l)}{\pi^{2+l}} \int \frac{(du)^d}{u^{2(2+l)}} \exp\left(-\frac{u^2}{4}\right) \cdot \left[: e_2^{iuP} : \right. \\
&\quad + \left. \left(1 - \kappa_l + \frac{1}{2}\kappa_l u^2\right) : e_2^{iuQ} : \right] , \\
E_0 &= 2\sqrt{h(1 - \kappa_l)} \cdot \frac{\Gamma(l + 2)}{\Gamma(l + 3/2)} .
\end{aligned}
\tag{19.11}
$$

The second correction can be calculated. We give this formula for the case $\kappa = 0$. It is

$$
\begin{aligned}
E &= E_0 + E_2 = E_0(1 - \delta_l) = 2\sqrt{h}\frac{\Gamma(l + 2)}{\Gamma(l + 3/2)} \cdot (1 - \delta_l) , \\
\delta_l &= \frac{1}{\pi^2} \sum_{n=1}^{\infty} \frac{\Gamma(l + 2)\Gamma(2n + 1/2)}{\Gamma(2n + l + 1/2)} \cdot \frac{2^{4n-1}\Gamma^2(2n - 1/2)}{\Gamma(4n + 2)} .
\end{aligned}
$$

The numerical values of δ_l are

$$\delta_0 = .006, \qquad \delta_1 = .004, \qquad \delta_l < .003 \quad \text{for } l \geq 2.$$

For asymptotically large l we obtain

$$E = E_{0l} \approx 2\sqrt{hl} \;. \tag{19.12}$$

For the same linear potential in the nonrelativistic kinetic energy case in the Hamiltonian one can get

$$E = E_{0l} \approx \frac{3}{2}(hl)^{2/3} \;. \tag{19.13}$$

The phenomena of the "downfall on the centre" for $\kappa \geq 1$ leads to the sensitivity of the eigenvalues of the Hamiltonian on the mass m in the kinetic term. In particular if κ is close to 1, the nonrelativistic limit comes for quite large m. Let us demonstrate this statement. We have two Hamiltonians:

$$H_{\text{rel}} = \left(\sqrt{p^2 + m^2} - m \right) - \frac{\kappa}{r} + hr \;,$$

$$H = \frac{p^2}{2m} - \frac{\kappa}{r} + hr \;.$$

Our aim is to find the values of μ as a function of κ for which the ground state energies of these two Hamiltonian practically coincide. It is convenient to make the following substitutions:

$$r \to \frac{r}{\sqrt{h}}, \qquad \mu = \frac{m}{\sqrt{h}}, \qquad H \to H/\sqrt{h}$$

We shall consider the Hamiltonians

$$H_{\text{rel}} = \left(\sqrt{p^2 + \mu^2} - \mu \right) - \frac{\kappa}{r} + r \;,$$

$$H = \frac{p^2}{2\mu} - \frac{\kappa}{r} + r \;,$$

According to all the above one gets for the ground state energy for both cases

$$E_{00}^{\text{rel}} = \min_x \frac{2}{\sqrt{\pi}} \cdot \left\{ \int_0^\infty dt \sqrt{t} e^{-t} [\sqrt{\mu^2 + tx^2} - \mu] - \kappa x + \frac{1}{x} \right\} \;,$$

$$E_{00} = \min_{\{\rho,x\}} \left\{ \frac{\Gamma(2+\rho)}{\Gamma(3\rho)8\rho^2\mu} \cdot x^2 - \kappa x \cdot \frac{\Gamma(2\rho)}{\Gamma(3\rho)} + \frac{1}{x} \cdot \frac{\Gamma(4\rho)}{\Gamma(3\rho)} \right\} \;.$$

The numerical results are shown in Table 19.1. One can see that for $\kappa = 1/2$ the nonrelativistic approach be comes approximate for $\mu^2 = m^2/h \approx 100 \div 500$ only.

Table 19.1. Results of the calculations of the ground state energy for the relativized and nonraletivized Hamiltonian as a function of the parameters κ and μ.

μ	$\kappa = 0$		$\kappa = 0.1$		$\kappa = 0.5$		$\kappa = 0.9$	
	rel.	nonrel.	rel.	nonrel.	rel.	nonrel.	rel.	nonrel.
0.01	2.247	8.614	2.131	8.591	1.586	8.500	0.704	8.409
0.1	2.162	3.998	2.046	3.949	1.500	3.749	0.616	3.543
0.5	1.878	2.338	1.757	2.254	1.186	1.896	0.256	1.503
1.0	1.654	1.856	1.523	1.749	0.902	1.279	−0.134	0.730
2.0	1.396	1.473	1.246	1.337	0.517	0.702	−0.802	−0.114
5.0	1.068	1.085	0.877	0.896	−0.168	−0.120	−2.526	−1.706
10.0	0.857	0.861	0.616	0.614	−0.934	−0.968	−5.226	−3.885
40.0					−4.607	−4.925	−21.127	−16.158
100.0							−52.850	−40.483

20. Three-Body Coulomb Systems

One of the classic problems of quantum mechanics is the three-body Coulomb system. Fock was the first to suggest a method [35] to solve the Schrödinger equation for a helium atom as a three-body system with a Coulomb interaction. This method became one of the fundamental approaches to the solution of the Schrödinger equation for the three–body problem. The progress that was subsequently made led to the development of two widely used was of investigating this problem: the adiabatic representation method and the variatonal method (see, for example, ([36], [5])).

The variational methods enaling one to carry out highly precise calculations for three–body Coulomb systems have already been encoded and realized as a set of program packages for computers. In the framework of this method the nonadiabatic corrections to the energy levels of mesomolecules have also been successfully calculated, likewise the binding energy and other parameters of quantum mechanical systems, such as the atom, the molecule and nuclei. To achieve better convergence for arbitrary values of the masses and charges of the three-body systems, one uses expansions of the wave functions in terms of various basis states: the Hiller basis [37], the exponential expansion [38], expansion in terms of spheroidal coordinates [39] and so on.

The main purpose of these investigations is to obtain highly accurate numerical solutions of the Schrödinger equation for three–body Coulomb systems of given masses and charges. While advanced computer facilities have increased the accuracy of the calculation of bound states, attempts to describe analytical solutions have been unsuccessful so far.

The determination of eigenvalues of the Hamiltonian for the three-body Coulomb system to a high accuracy by means of numerical methods is certainly important. However, to understanding the formation dynamics of three–body bound states one also wishes to study the dependence of eigenvalues of the Hamiltonian on the masses and charges of particles qualitatively. Therefore, the development of analytical tools permitting the study of these dependences with an accuracy of a few per cent is justified.

In this chapter, we apply the OR method to calculate the ground state energy of a three-body system with the Coulomb interaction.

Our scenario is the following. First of all we transform the variables of the three–body Coulomb system in such a way as to get Gaussian asymptotic behaviour for the ground state wave function. As a result we obtain the modified Schrödinger equation with the Hamiltonian of oscillator type. Then this Hamiltonian is written in the oscillator representation and the equations defining the background energy in the zeroth approximation are obtained in the explicit form.

The results of our computations show that in the zeroth approximation the deviation from the exact value is less than 1 per cent, i.e. the perturbation series converge fairly fast. Beyond that, the oscillator representation method enables one to compute the ground state energy with arbitrary accuracy by taking into account higher perturbation orders. Here we shall consider the zeroth approximation only. As real physical applications of the oscillator representation method we determine the regiona stability of the unit charges and arbitrary masse of the three–body Coulomb systems, calculate bound–state energies of mesic molecules of light nuclei (He, Li, Be), find the dependence of these binding energies on the mass and charge and determine the stability boundary for mesic molecules with masses $m_N = 2Zm_p$, where m_p is the mass proton and Z is the charge of nuclei.

20.1 The Three–Body Hamiltonian for the Ground State

Let us consider a three–body Coulomb system in 3 dimensions. Let m_1, m_2, m_3 and Z_1e, $-Z_2e$, Z_3e be masses and charges of particles where Z_j are positive or negative so that $Z_i Z_j > 0$ always. The Hamiltonian for this system has the form

$$H = \frac{1}{2} \sum_{j=1}^{3} \frac{(\mathbf{p}_j)^2}{m_j} - \frac{Z_1 Z_2 e^2}{|\mathbf{r}_1 - \mathbf{r}_2|} + \frac{Z_1 Z_3 e^2}{|\mathbf{r}_1 - \mathbf{r}_3|} - \frac{Z_2 Z_3 e^2}{|\mathbf{r}_2 - \mathbf{r}_3|} . \qquad (20.1)$$

Let us introduce the Jacobi coordinates $\{\mathbf{x}, \mathbf{y}\}$:

$$\mathbf{r}_1 = a_1 \mathbf{x} + b_2 \mathbf{y} + \mathbf{R} ,$$
$$\mathbf{r}_2 = -b_1 \mathbf{y} + \mathbf{R} ,$$
$$\mathbf{r}_3 = b_2 \mathbf{y} - a_3 \mathbf{x} + \mathbf{R} .$$

It is convenient to work with the dimensionless variables

$$\mathbf{x} = \frac{1}{Me^2} \mathbf{r} ,$$
$$\mathbf{y} = \frac{1}{\sqrt{M \mu} e^2} \mathbf{s} .$$

The various symbols represent the following

$$a_3 = \frac{m_1}{m_1 + m_3} \; ; \qquad a_1 = \frac{m_3}{m_1 + m_3} \; ;$$

$$b_1 = \frac{m_1 + m_3}{m_1 + m_2 + m_3} \; ; \qquad b_2 = \frac{m_2}{m_1 + m_2 + m_3} \; ;$$

$$a_1 + a_3 = 1 \; ; \qquad b_1 + b_2 = 1 \; ;$$

$$M = \frac{m_1 m_3}{m_1 + m_3} \; ;$$

$$\mu = \frac{(m_1 + m_3) m_2}{m_1 + m_2 + m_3} \; .$$

After some simplifications the Hamiltonian (20.1) can be expressed by

$$H = M e^4 \left\{ \frac{1}{2} p_r^2 + \frac{1}{2} p_s^2 + \frac{Z_1 Z_3}{r} - \frac{Z_1 Z_2 c}{|s + c_1 r|} - \frac{Z_2 Z_3 c}{|s - c_3 r|} \right\} , \qquad (20.2)$$

where the term of the total kinetic energy is omitted and

$$c_1 = \frac{1}{m_1} \sqrt{\frac{m_1 m_2 m_3}{m_1 + m_2 + m_3}} \; ;$$

$$c_3 = \frac{1}{m_3} \sqrt{\frac{m_3 m_2 m_1}{m_1 + m_2 + m_3}} \; ;$$

$$c = c_1 + c_3 \; .$$

The energy of the ground state in the scale $\frac{1}{2} M e^4$ has the form

$$E = -\frac{1}{2} M e^4 U , \qquad (20.3)$$

so that the energy is defined by the dimensionless parameter U in this energy scale.

Our problem is to calculate the ground state energy E of the Hamiltonian (20.2). The wave function of the ground state $\Psi = \Psi(r, s, \phi)$ depends on three independent variables only: two radii $r = \sqrt{r^2}$, $s = \sqrt{s^2}$ and the angle ϕ between the vectors $\mathbf{r}$ and $\mathbf{s}$: $(\mathbf{rs}) = rs \cos \phi$. The Schrödinger equation has the form

$$h \Psi = 0 , \qquad (20.4)$$

where

$$h = -\frac{1}{2} \left(\frac{\partial^2}{\partial r^2} + \frac{2}{r} \frac{\partial}{\partial r} \right) - \frac{1}{2} \left(\frac{\partial^2}{\partial s^2} + \frac{2}{s} \frac{\partial}{\partial s} \right) \qquad (20.5)$$

$$- \frac{1}{2} \left(\frac{1}{r^2} + \frac{1}{s^2} \right) \left(\frac{\partial^2}{\partial \phi^2} + \text{ctg} \phi \cdot \frac{\partial}{\partial \phi} \right) + V_1,$$

$$V_1 = \frac{Z_1 Z_3}{r} + \frac{U}{2} - \frac{c Z_3 Z_2}{\sqrt{s^2 + c_3^2 r^2 - 2 c_3 rs \cdot \cos \phi}}$$

$$- \frac{c Z_2 Z_1}{\sqrt{s^2 + c_1^2 r^2 + 2 c_1 rs \cdot \cos \phi}} .$$

Let us consider the variables r, s, and ϕ in the Hamiltonian h of (20.5) as independent variables. Under this assumption the Hamiltonian h is non–Hermitian in the variable ϕ because the operator

$$\Delta_\phi = \frac{\partial^2}{\partial\phi^2} + \mathrm{ctg}\phi \cdot \frac{\partial}{\partial\phi} = \frac{1}{\sin\phi} \cdot \frac{\partial}{\partial\phi}(\sin\phi \cdot \frac{\partial}{\partial\phi})$$

is non–Hermitian in the interval $\phi \in [0, \pi]$. In order make the Hamiltonian also Hermitian in the variable ϕ, let us consider the following matrix element:

$$\left(\Psi, h\Phi\right) = \int d\mathbf{r} \int d\mathbf{s} \cdot \Psi^*(r, s, \phi) h\Phi(r, s, \phi)$$

$$= 8\pi^2 \int\limits_0^\infty dr r^2 \int\limits_0^\infty ds s^2 \int\limits_0^\pi d\phi \sin\phi \cdot \Psi^*(r, s, \phi) h\Phi(r, s, \phi) ,$$

where Ψ and Φ depend on r, s and ϕ only. The part of this integral depending on the angle ϕ can be written in the form

$$\int\limits_0^\pi \frac{d\phi}{\sin\phi} \Psi(\phi) \left(-\frac{1}{2}(\sin\phi\frac{\partial}{\partial\phi})^2 + \sin^2\phi \cdot V_1(\cos\phi) \right)\Phi(\phi)$$

$$= \int\limits_{-\infty}^\infty du\Psi(u) \left(-\frac{1}{2}\frac{d^2}{du^2} + \frac{1}{\cosh^2 u} \cdot V_1(\tanh u) \right)\Phi(u) ,$$

where the variable

$$u = \ln(\mathrm{tg}\frac{\phi}{2})$$

is introduced. Thus, the operator

$$(\sin\phi \cdot \frac{\partial}{\partial\phi})^2 = \frac{\partial^2}{\partial u^2}$$

is Hermitian.

The Hermitian form of the Hamiltonian h results:

$$h = -\frac{1}{2\cosh^2 u}\left[\left(\frac{\partial^2}{\partial r^2} + \frac{2}{r}\frac{\partial}{\partial r}\right) + \left(\frac{\partial^2}{\partial s^2} + \frac{2}{s}\frac{\partial}{\partial s}\right)\right]$$

$$- \frac{1}{2}\left(\frac{1}{r^2} + \frac{1}{s^2}\right) \cdot \frac{d^2}{du^2} + V , \qquad (20.6)$$

where

$$V = V(r, s, u)$$

$$= \frac{1}{\cosh^2 u}\left[\frac{Z_1 Z_3}{r} + \frac{U}{2} - \frac{cZ_3 Z_2}{\sqrt{s^2 + c_3^2 r^2 - 2c_3 rs \cdot \tanh u}}\right.$$

$$\left. - \frac{cZ_2 Z_1}{\sqrt{s^2 + c_1^2 r^2 + 2c_1 rs \cdot \tanh u}}\right]$$

and r, s, and u are the independent variables.

According to chapter 15 and 16, let us introduce new variables

$$r = Q^2 , \qquad\qquad s = q^2 .$$

Then, the kinetic part of the Hamiltonian (20.6), after these substitutions, becomes

$$\left[\left(\frac{\partial^2}{\partial r^2} + \frac{2}{r}\cdot\frac{\partial}{\partial r}\right) + \left(\frac{\partial^2}{\partial s^2} + \frac{2}{s}\cdot\frac{\partial}{\partial s}\right)\right]\Phi(r, s)$$

$$= \left[\frac{1}{4Q^2}\left(\frac{\partial^2}{\partial Q^2} + \frac{3}{Q}\cdot\frac{\partial}{\partial Q}\right) + \frac{1}{4q^2}\left(\frac{\partial^2}{\partial q^2} + \frac{3}{q}\cdot\frac{\partial}{\partial q}\right)\right]\Phi(Q, q) .$$

The operators

$$X_Q = \left(\frac{\partial^2}{\partial Q^2} + \frac{3}{Q}\frac{\partial}{\partial Q}\right),$$

$$X_q = \left(\frac{\partial^2}{\partial q^2} + \frac{3}{q}\frac{\partial}{\partial q}\right)$$

can be considered the radial parts of the four-dimensional Laplacians $\Box_Q$ and $\Box_q$. The wave function Ψ of our system depends on the variables Q, q, and u only. Therefore, the operators X_Q and X_q acting on the function $\Psi(Q, q, u)$ can be identified with the operators $\Box_Q$, $\Box_q$ and the variables Q_μ and q_μ can be considered vectors of 4 dimensions.

Thus, the Hermitian Hamiltonian of the three–body system for the ground state for the substitutions

$$0 < \phi < \pi, \qquad \phi \to u = \ln(\mathrm{tg}\frac{\phi}{2}), \qquad -\infty < u < \infty ,$$

$$0 < s < \infty, \qquad s \to q = \sqrt{s}, \qquad 0 < q < \infty , \qquad (20.7)$$

$$0 < r < \infty, \qquad r \to Q = \sqrt{r}, \qquad 0 < Q < \infty ,$$

looks like

$$h = \frac{1}{2\cosh^2 u}\left(q^2\cdot\frac{P_Q^2}{4} + Q^2\cdot\frac{P_q^2}{4}\right) + \left(\frac{Q^2}{q^2} + \frac{q^2}{Q^2}\right)P_u^2 + V \qquad (20.8)$$

with

$$V = \frac{1}{\cosh^2 u} \cdot \left[2q^2 Z_1 Z_3 + U Q^2 q^2 \right.$$
$$- 2c Q^2 q^2 Z_2 \left(\frac{Z_1}{\sqrt{q^4 + c_1^2 Q^4 - 2c_1 q^2 Q^2 \cdot \tanh u}} \right.$$
$$\left. \left. + \frac{Z_3}{\sqrt{q^4 + c_3^2 Q^4 + 2c_3 q^2 Q^2 \cdot \tanh u}} \right) \right] ,$$

where $Q = \sqrt{Q_\mu^2}$, $q = \sqrt{q_\mu^2}$ and Q_μ, q_μ are vectors of the auxiliary spaces R_4; $\Box_Q = -P_Q^2$, $\Box_q = -P_q^2$. The wave function of the ground state depends on three independent variables Q, q and u only, i.e. $\Psi = \Psi(Q, q, u)$.

The wave function in the oscillator representation should describe correctly the behaviour of the Coulomb system at small distances too. One can see that the potential in (19.6) contains a repulsive term over the variable r. The minimum of the potential is not at the point $r = 0$ and the wave function as a function of r has a maximum at the point $r_{max} > 0$. The repulsion at the point $r = 0$ can be taken into account in the oscillator representation method by the transition to a higher dimensional space, as has been done in chapter 15. Effectively, the repulsion potential turns into an attractive one when the radius r is considered as a radius of a vector in a d-dimensional space ($d > 4$). The parameter d is considered an additional variational parameter.

Finally the Hamiltonian of the three–body Coulomb system for the ground state looks is:

$$h = \frac{1}{\cosh^2 u} \left(q^2 \cdot \frac{P_Q^2}{4} + Q^2 \cdot \frac{P_q^2}{4} \right) + \left(\frac{Q^2}{q^2} + \frac{q^2}{Q^2} \right) \cdot P_u^2 + W, \qquad (20.9)$$

with

$$W = -\frac{d(d-4)}{16 \cdot \cosh^2 u} \cdot \frac{q^2}{Q^2} + V ,$$

where $q_\mu \in R^4$, $Q_j \in R^d$, $u \in R^1$ and the potential V is defined by (20.6).

The parameter U defining the background energy of the initial three–body system enters into the Hamiltonian h (20.9) so that our problem is formulated in the following way. We have to find the background state of the Hamiltonian h, i.e. we have to solve the Schrödinger equation

$$h\Psi = \varepsilon(U)\Psi \qquad (20.10)$$

and to find the energy ε which is a function of the parameter U,

$$\varepsilon = \varepsilon(U) .$$

According to the Schrödinger equation (20.4), this energy equals zero so that the equation

$$\varepsilon(U) = 0 \qquad (20.11)$$

defines the parameter U. Then, the background energy of the three–body Coulomb system is calculated according to (20.3).

20.2 The Hamiltonian in the Oscillator Representation

In this section, the oscillator representation method will be applied to calculate the function $\varepsilon(U)$, which is the background energy of the Hamiltonian (20.9). Let us rewrite the Hamiltonian h of (20.9) in the form

$$h = h_0 + h_I + \varepsilon_0 . \qquad (20.12)$$

Here h_0 is the oscillator part of the Hamiltonian defined as

$$
\begin{aligned}
h_0 &= \frac{1}{2\kappa_Q}\left[\frac{1}{2}(P_Q^2 + \Omega^2 Q^2) - \frac{d\Omega}{2}\right] \\
&+ \frac{1}{2\kappa_q}\left[\frac{1}{2}(P_q^2 + \omega^2 q^2) - 2\omega\right] \\
&+ \frac{1}{\kappa_u}\left[\frac{1}{2}P_u^2 - \frac{\theta(\theta+1)}{2\cosh^2(u)} + \frac{\theta^2}{2}\right] ,
\end{aligned} \qquad (20.13)
$$

where κ_Q, κ_q, κ_u, Ω, ω, θ are positive parameters. The solution of the Schrödinger equation

$$h_0 \Psi_0(Q, q, u) = 0$$

has the form

$$|0\rangle = \Psi_0(Q, q, u) = N\frac{1}{[\cosh u]^\theta}\exp\left\{-\frac{\Omega}{2}Q^2 - \frac{\omega}{2}q^2\right\} , \qquad (20.14)$$

where N is the constant defined by the normalization condition.

Taking into account (20.8, 9, 13) one gets

$$h_I + \varepsilon_0 = \frac{1}{2\kappa_Q}\left(\frac{d\Omega}{2} - \frac{\Omega^2 Q^2}{2}\right) + \frac{1}{2\kappa_q}\left(\frac{4\omega}{2} - \frac{\omega^2 q^2}{2}\right) \qquad (20.15)$$

$$
+ \frac{1}{2\kappa_u}\left(\frac{\theta(\theta+1)}{2[\cosh u]^2} - \frac{\theta^2}{2}\right) + \frac{P_Q^2}{4}\left(\frac{q^2}{[\cosh u]^2} - \frac{1}{\kappa_Q}\right)
$$

$$
+ \frac{P_q^2}{4}\left(\frac{Q^2}{[\cosh u]^2} - \frac{1}{\kappa_q}\right) + \frac{P_u^2}{2}\left(\frac{2Q^2}{q^2} + \frac{2q^2}{Q^2} - \frac{1}{\kappa_u}\right)
$$

$$
- \frac{d(d-4)}{16[\cosh u]^2}\frac{q^2}{Q^2} + V ,
$$

where the potential V is given in (20.8).

Now let us define the interaction Hamiltonian and the background energy in the zeroth approximation. It is convenient to introduce the definition

$$: A := A - < 0|A|0 >= A - \left(\Psi_0 A \Psi_0\right) ,$$

where A is an operator, so that

$$\begin{aligned}
: P_Q^2 &:= P_Q^2 - \left(\Psi_0 P_Q^2 \Psi_0\right) = P_Q^2 - \frac{d\Omega}{2} , \\
: Q^2 &:= Q^2 - \left(\Psi_0 Q^2 \Psi_0\right) = Q^2 - \frac{d}{2\Omega} , \\
: P_q^2 &:= P_q^2 - \left(\Psi_0 P_q^2 \Psi_0\right) = P_q^2 - 2\omega , \\
: q^2 &:= q^2 - \left(\Psi_0 q^2 \Psi_0\right) = q^2 - \frac{2}{\omega} , \\
: P_u^2 &:= P_u^2 - \left(\Psi_0 P_u^2 \Psi_0\right) = P_u^2 - \frac{\theta^2}{2\theta+1} .
\end{aligned} \qquad (20.16)$$

Let us substitute the representation (20.16) into (20.15) and require the oscillator representation condition to be fulfilled, i.e. the interaction Hamiltonian should not contain the quadratic terms with $: Q^2 :$, $: q^2 :$, and $: P_Q^2 :$, $: P_q^2 :$, $: P_u^2 :$. The coefficients before the operators $: P_Q^2 :$, $: P_q^2 :$ and $: P_u^2 :$ lead the parameters κ_Q, κ_q and κ_u to be

$$\begin{aligned}
\frac{1}{\kappa_Q} &= \left(\Psi_0 \frac{q^2}{[\cosh u]^2} \Psi_0\right) = \frac{4}{\omega} \frac{\theta}{2\theta+1} , \\
\frac{1}{\kappa_q} &= \left(\Psi_0 \frac{Q^2}{[\cosh u]^2} \Psi_0\right) = \frac{d}{\Omega} \frac{\theta}{2\theta+1} , \\
\frac{1}{\kappa_u} &= \left(\Psi_0 [\frac{2Q^2}{q^2} + \frac{2q^2}{Q^2}] \Psi_0\right) = \frac{d\omega}{\Omega} + \frac{8}{d-2} \frac{\Omega}{\omega} .
\end{aligned} \qquad (20.17)$$

Taking into account these formulae for the background energy in the zeroth approximation, we get

$$\begin{aligned}
\varepsilon_0(U) = \min_{\Omega,\omega,\theta,d} \left(\Psi_0 h \Psi_0\right) = \min_{\Omega,\omega,\theta,d} \Bigg\{ & \frac{2\theta}{2\theta+1} \cdot \\
\times \Bigg[\frac{d}{4}\left(\frac{\Omega}{\omega} + \frac{\omega}{\Omega}\right) + \frac{\theta}{2}\left(\frac{4}{d-2}\frac{\Omega}{\omega} + \frac{d\omega}{2\Omega}\right) & + \frac{4Z_1 Z_3}{\omega} + \frac{Ud}{\omega\Omega} \\
- \frac{d(d-4)}{4(d-2)}\frac{\Omega}{\omega} \Bigg] - \frac{c}{2}\frac{d(d+2)(d+4)\Gamma(2\theta)}{4^\theta \Gamma^2(\theta)} & f(\Omega,\omega,\theta,d) \Bigg\} , \quad (20.18)
\end{aligned}$$

where

$$f = \int\limits_0^1 dy\, y^{d/2}(1-y)^2 \int\limits_{-1}^1 d\tau (1-\tau^2)^\theta$$

$$\times \left[\frac{Z_2 Z_3}{\sqrt{(1-y)^2\Omega^2 + c_3^2 y^2 \omega^2 - 2c_3 y(1-y)\tau\omega\Omega}} \right.$$

$$+ \left. \frac{Z_2 Z_1}{\sqrt{(1-y)^2\Omega^2 + c_1^2 y^2 \omega^2 + 2c_1 y(1-y)\tau\omega\Omega}} \right].$$

The interaction Hamiltonian h has the form

$$
\begin{aligned}
h_I \;=\;& \frac{\theta}{2\theta+1}\Big(\frac{1}{2} : P_Q^2 :: q^2 : + \frac{1}{2} : P_q^2 :: Q^2 : +\theta : \frac{1}{q^2} :: Q^2 : \\
+\;& \theta : \frac{1}{Q^2} :: q^2 : +2U : q^2 :: Q^2 : - \frac{d(d-4)}{8} : \frac{1}{Q^2} :: q^2 : \Big) \\
+\;& : P_u^2 : \Big(: \frac{1}{q^2} :: Q^2 : + : \frac{1}{Q^2} :: q^2 : +\omega : Q^2 : + \frac{2\Omega}{d-2} : q^2 : \\
+\;& \frac{d}{2\Omega} : \frac{1}{q^2} : + \frac{2}{\omega} : \frac{1}{Q^2} : \Big) + \Big(2Z_1 Z_3 : q^2 : +U : Q^2 :: q^2 : \\
+\;& \frac{d}{2\Omega} U : q^2 : + \frac{2}{\omega} U : Q^2 : + \frac{1}{4} : P_Q^2 :: q^2 : + \frac{1}{4} : P_q^2 :: Q^2 : \\
+\;& \frac{1}{2\omega} : P_Q^2 : + \frac{d}{8\Omega} : P_q^2 : + \frac{d\Omega}{8} : q^2 : + \frac{\omega}{2} : Q^2 : - \frac{d(d-4)}{4\omega} : \frac{1}{Q^2} : \\
-\;& \frac{d(d-4)\Omega}{8(d-2)} : q^2 : \Big) : \frac{1}{[\cosh u]^2} : - : W : ,
\end{aligned}
\tag{20.19}
$$

where

$$
\begin{aligned}
W \;=\;& \frac{2cQ^2 q^2 Z_3 Z_2}{\cosh^2 u \sqrt{q^4 + c_3^2 Q^4 - 2c_3 q^2 Q^2 \cdot \tanh u}} \\
+\;& \frac{2cQ^2 q^2 Z_2 Z_1}{\cosh^2 u \sqrt{q^4 + c_1^2 Q^4 + 2c_1 q^2 Q^2 \cdot \tanh u}} .
\end{aligned}
$$

20.2.1 The Accuracy of the Zeroth Approximation

From the principal point of view, the oscillator representation method gives the possibility of calculating the background energy of every three-body Coulomb system with any accuracy if the highest perturbation orders over the interaction Hamiltonian h_I are taken into account. However, from the practical point of view, it is important to know what the accuracy of the zeroth approximation is. If this accuracy is high enough, then the zeroth approximation being practically analytic gives the opportunity of investigating qualitative and semi–quantitative dependences of the energy spectra on masses and charges of particles. In this section, we determine the accuracy of our method comparing the zeroth approximation for some well–known three–body systems with the exact results obtained by numerical methods. To this end, we calculate the background energies of the following Coulomb systems:

$$H^- = (pee), \quad H_2^+ = (ppe), \quad (e^+e^-e^-), \quad (pp\mu), \quad (dd\mu), \quad (dt\mu).$$

The background energies of these molecules have been calculated by many authors and different methods (see, for example, [36]-[39], [43]). The accuracy of these calculations reached a very high level with the perfection of computers. Therefore, these results can be considered exact.

Formula (20.18) in the zeroth approximation and the equation $\varepsilon_0(U) = 0$ define the parameter U as the function of charges and masses of particles. The energy is determined by (20.3).

We have used the following values of masses:

$$m_\mu = 206.77 m_e, \quad m_p = 1836.15 m_e, \quad m_d = 3670.48 m_e, \quad m_t = 5496.9 m_e ,$$

where m_e is the electron mass.

The results of our calculations are given in Tables 20.1 and 20.2. The exact values of the energies are taken from [40]. One can see that the accuracy of the zeroth approximation is less than one per cent.

Table 20.1. Results of the calculations of the parameter U. U^{OR} is the zeroth approximation of the oscillator representation, $U^{ex.}$ is the exact value [40]. $\Delta U = |U^{ex,} - U^{OR}|/U^{ex.} \cdot 100\%$ defines the accuracy of the zeroth approximation.

	ω	Ω	θ	d	U^{OR}	$U^{ex.}$	$\Delta U(\%)$
e^+e^-e	5.076	0.1034	0.1022	4.102	0.520	0.52399	0.762
μe^-e^-	4.833	0.1102	0.103	4.106	1.050	1.05011	0.011
pe^-e^-	4.835	0.1105	0.104	4.107	1.050	1.0547	0.446
ppe^-	0.0852	0.1113	0.102	4.101	1.190	1.1943	0.360

Table 20.2. Results of the calculations of the mesomolecular terms. E^{OR} is the zeroth approximation of the oscillator representation, $E^{ex.}$ is the exact value [40]. $\Delta E = |E^{ex} - E^{OR}|/E^{ex} \cdot 100\%$ defines the accuracy of the zeroth approximation.

	ω	Ω	θ	d	E^{OR} ev	$E^{ex.}$ ev	$\Delta E(\%)$
$pp\mu$	5.339	0.1048	0.1031	4.1195	-2781	-2782	0.036
$dd\mu$	8.750	0.895	0.100	4.104	-2987	-2988	0.034
$dt\mu$	0.0867	0.8325	0.100	4.128	-3028	-3029	0.033

20.2.2 The Stability of Three Unit-Charge Systems

In this section, the binding of the three–body Coulomb system with unit charges and various constituent masses will be discussed. The calculation of the stability region as a function of masses and charges of the particles is one of the basic objects in the Coulomb three–body problem (see, for examples, [40], [41]). The dependence of the background energy on the masses of particles has been considered in [42] and numerical calculations have been made in [40, 41] to establish the stability boundary.

Let us formulate the problem. We consider a three–body Coulomb system $(A^{\pm}, B^{\mp}, C^{\pm})$ with unit charges $\pm, \mp, \pm$ and various masses m_A, m_B, m_C. Without loss of generality, one can restrict oneself to $m_A \geq m_C$. We shall look for the stability threshold according to the decay

$$(A, B, C) \rightarrow (A, B) + C , \qquad (20.20)$$

where (A, B) is a two-body atom and C is a free particle.

It is more convenient to use the following variables instead of masses m_j (see,[40]):

$$\alpha_j = \frac{1}{m_j}/(\frac{1}{m_A} + \frac{1}{m_B} + \frac{1}{m_C}) , \qquad (j = A, B, C), \qquad (20.21)$$

$$\alpha_A + \alpha_B + \alpha_C = 1 .$$

Any three–body system can be represented by a point in an equilateral triangle for which the altitude equals 1 so that identity (20.21) can be interpreted as a sum of altitudes from a point at three sides of this triangle (see Fig. 20.1). Let us call it *the stability triangle*.

Our problem is to find on the stability triangle the boundary separating stable and unstable states of three–body systems. The stability of a three–body system (A, B, C) is defined relative to the decay (20.20). The binding energy can be calculated by the formula

$$\Delta E = -\frac{1}{2}e^4 \frac{M_A M_B}{M_A + M_B} \cdot (U - 1) , \qquad (20.22)$$

where the parameter U as a function of masses α_j is determined by (20.18). The condition

$$\Delta E = 0 \quad \text{or} \quad U = 1$$

defines the searched boundary. We shall call the masses for which this condition is fulfilled *the critical masses*.

First of all, let us place on the triangle the points corresponding to the well–known systems

$$H_2 = (ppe^-), \quad H^- = (pe^-e^-), \quad (e^+e^-e^-), \quad (pe^-\mu^-), \quad (pe^-e^+).$$

The system $(e^+e^-e^-)$ is studied in ([47], [48]) and it is stable. The molecules $(ppe^-) = H_2^+$ and $(pe^-e^-) = H^-$ are well–known hydrogen ions and they are stable too ([47], [49]). The system $(pe^-\mu^-)$ is unstable [50]. The problem is open whether the system (pe^-e^+) is stable or unstable (see, [40] for details).

In oscillator representation the critical masses will be calculated according to (20.22). We shall proceed in the following way. The molecules (pe^-e^+) and (pe^-p) are distinguished by the masses of the positron and proton. Let us consider the binding energy of the system (pe^-C^+) as a function of the mass M_C of the C-particle in the limits $m_e \leq M_C \leq m_p$. It turns out that for $M_C = m_e$ the parameter $U = .901$, i.e. the system (pe^-e^+) is unstable, for $M_C = 1.945m_e$ the parameter $U = 1.0$ and this system passes to the stable state. Thus, the mass of the C-particle

$$M_C = 1.945m_e$$

is a critical one for the system (pe^-C^+). Restrictions on the critical mass M_C for the system (pe^-C^+) were considered by many authors. In particular in [44] the upper restriction $M_C < 2.20m_e$ and in [45] the lower restriction $M_C > 1.51m_e$ for $m_p = \infty$ were obtained. The result of ref. [46] for $m_p = \infty$ claims that $M_C \geq 1.57m_e$. Reference [40] gives $M_C \geq 1.9m_e$.

The molecules (pe^-e^+) and $(e^+e^-e^+)$ are distinguished by the masses of the positron and proton, too. Let us consider the system $(A^+e^-e^+)$. When $M_A = m_p$ the system is unstable and for $M_A = m_e$ it is stable. Decreasing the mass of the A-particle to

$$M_A = 4.35m_e$$

one gets $U = 1$ and for $M_A \leq 4.35m_e$ this system becomes stable. In [40] for the system $(A^+B^-B^+)$ the lower boundary for the critical mass of the A-particle was obtained: $M_A > 4.19M_B$.

For the systems (pe^-C^+) and $(A^+e^-e^+)$ the critical masses of the particles A and C are found to equal

$$M_C = 1.945m_e \quad \text{and} \quad M_A = 4.35m_e.$$

These systems have been considered in refs.([40], [44]–[46]) where the restriction on the critical masses is established.

Now let us consider the molecule (pB^-e^-) which contains the molecule $(p\mu^-e^-)$ and the hydrogen atom (pe^-e^-). These systems differ in the masses of muon and electron. Our calculation gives the critical mass of the B-particle to be

$$M_B = 1.575m_e.$$

In ([40], [46]), for this molecule (pB^-e^-) the restriction on the critical mass is $M_B \geq 1.57m_e$.

In the system $(A^+A^-e^-)$ which changes from the ion H^- for $M_A = m_e$ to the proton–antiproton ion $(p\bar{p}e)$ for $M_A = m_p$ the critical mass equals

$$M_A = 2.45m_e.$$

Thus, for the systems (pB^-e^-) and $(A^+A^-e^-)$ the critical masses are

$$M_B = 1.575m_e \quad \text{and} \quad M_A = 2.450m_e.$$

The boundaries for these critical masses are obtained in ([40], [46]).

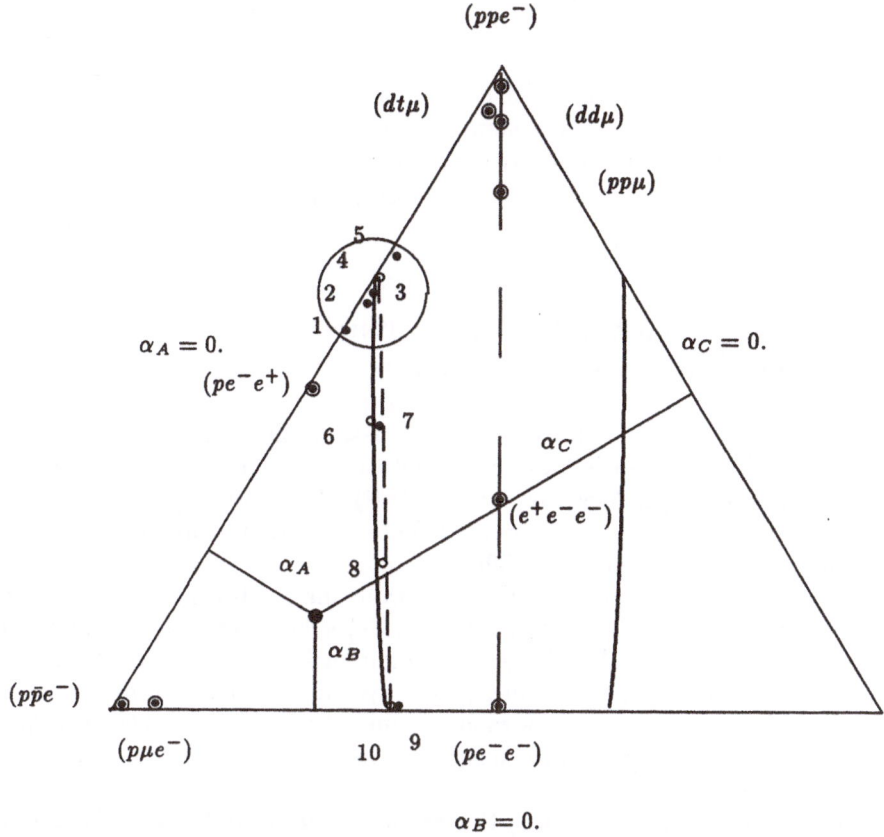

Fig. 20.1. The stability triangle. The systems (pe^-C^+) are denoted in the following way: the points indicate $1 - M_C = 1.51m_e$, $2 - M_C = 1.57m_e$, $3 - M_C = 1.9m_e$, $4 - M_C = 1.945m_e$, $5 - M_C = 2.2m_e$; the systems $(A^+e^-e^+)$ are denoted: $6 - M_A = 4.35m_e$, $7 - M_A = 4.19m_e$; the system $(A^+A^-e^-)$ is denoted: $8 - M_A = 2.45m_e$; the systems (pB^-e^-): $9 - M_B = 1.57m_e$, $10 - M_B = 1.575m_e$.

The stability boundary given by formula $U = 1$ in (20.22) is denoted by the circles on Fig. 20.1. Our results (four points) are shown in Table 20.3 and agree with the restrictions on critical masses obtained in refs.([40], [44]–[46]) Our curve can be approximated by equation (20.23). The stability boundary obtained in [40] is denoted by the dots in Fig. 20.1.

Table 20.3. Results for stability bound-aries (see Fig. 20.1).

α_A	0.000	0.103	0.225	0.388
α_B	0.662	0.440	0.229	0.002

$$\left(\frac{\alpha_A}{0.391}\right)^{0.89} + \left(\frac{\alpha_B}{0.662}\right)^{0.89} = 1 . \tag{20.23}$$

The vertical altitude on which every molecule is stable is the axis of charge symmetry of the three-body system. Therefore, the stability boundaries are symmetrical relative to this altitude. Thus the inside part of the stability triangle bounded by these lines is the stability region of three-body Coulomb systems.

20.3 Mesic Molecules of Light Nuclei in the Oscillator Representation

We study the ground state energies of mesic molecules of light nuclei in the framework of the oscillator representation method. This is one of the fundamental problems (see, for example [51]). Muon transfer from the mesic hydrogen to other nuclei is of considerable interest when studying mesic atom interaction processes refs. [51]–[59].

At present the transition of muons from mesic atoms of the hydrogen isotopes H (p, d, t) to light nuclei (He, Li, Be) is being investigated [54]–[59]. The ground state muonic hydrogen with light nuclei can lead to the formation of hydrogen-nuclei mesic molecules [52]. A system like that is accompanied by the decay (dissociation) of the molecule into the hydrogen nuclei (isotopes) and mesic molecules of the light nuclei, because there are no bound states for the lower term.

The molecular mechanism of the muon transfer from the ground state of the mesic hydrogen to helium is confirmed experimentally [53], [54]. Gershtein et al.[55] and Kravtsov et al. [56] have theoretically investigated the bound-state energies of the systems $(H\mu\,^3He)$ and $(H\mu\,^4He)$.

Muon transfer from the mesic hydrogen to lithium is of special interest (see, for example ref.[57]). The bound-state energies of mesic molecules of light nuclei (lithium) have first been calculated by using the Born-Oppenheimer approximation in ref.[58] and in ref.[59] these calculation were made on the basis of surface function expansions. In ref.[59] bound-state energies of the $(H\mu\,^7Be)$ system were calculated as well.

The main purpose of these investigations is mainly the construction of highly accurate numerical solutions of the Schrödinger equation for the three-body Coulomb system.

Research of the dependence of the binding energy of the mesic molecules $(H\mu N_Z)$ on the masses and charges of particles has both theoretical and experimental significance. It is one of the central points in understanding the molecular mechanism of the muon transfer from mesic hydrogen to other nuclei and the formation dynamics of the few-body Coulomb systems. It provides us, e.g., with the detailed information on the strong $H - N_Z$ interaction at very low energies (a few kev), information relevant to astrophysical questions, e.g., the pp-cycle in the sun.

Here we have calculated the bound-state energies of mesic molecules of light nuclei (He, Li, Be) by using the method of oscillator representation [60]. The OR method in the zeroth approximation provides an analytical study of the formation dynamics of the three-body Coulomb systems with the accuracy less then 1 per cent. As a consequence, the universal research of the dependence of bound state energy of the three-body Coulomb system on the masses and charges of particles becomes possible.

20.3.1 The Binding Energy of the Mesic Molecules of Light Nuclei

The binding energies ([55]-[61]) of the mesic molecules $(H\mu N_Z)$ are determined with respect to the ground state of the $(H\mu)$ atom

$$E_{bin} = E_{H\mu N_Z} - E_{H\mu} . \tag{20.24}$$

The details of calculation of the ground state energy of three-body Coulomb systems are given in the above. The ground state energy of the mesic molecules $(H\mu N_Z)$ can be represented in the form

$$E = -\frac{1}{2}\alpha^2 \frac{m_H m_N}{m_H + m_N} \cdot U ,$$

where m_H and m_N are the masses of the hydrogen isotopes and nuclei, respectively. According to (19.24), the binding energies of the mesic molecules $(H\mu N_Z)$ are written in the form

$$E_{bin} = -\frac{\alpha^2}{2} \frac{m_H m_\mu}{m_H + m_\mu} \cdot \left(\frac{m_N}{m_\mu} \frac{m_H + m_\mu}{m_H + m_N} U - 1 \right) , \tag{20.25}$$

where U is the energy parameter and m_μ is the muon mass. The energy parameter U is determined from the following equation as a function of masses and charges of particles

$$\varepsilon_0(U) = 0 , \tag{20.26}$$

where $\varepsilon_0(U)$ being the energy of oscillators in the zeroth approximation (20.18) of the OR, can be written in the form

$$
\varepsilon_0(U) = \min_{\Omega,\omega,\theta,d} \left\{ \frac{2\theta}{2\theta+1} \left[\frac{d}{4}\left(\frac{\Omega}{\omega} + \frac{\omega}{\Omega}\right) \right. \right.
$$

$$
+ \frac{\theta}{2}\left(\frac{d}{d-2}\frac{\Omega}{\omega} + \frac{d\omega}{2\Omega}\right) + \frac{4Z}{\omega} + \frac{Ud}{\omega\Omega} - \frac{d(d-4)}{4(d-2)}\frac{\Omega}{\omega}\right]
$$

$$
\left. \left. - \frac{c}{2}\frac{d(d+2)(d+4)\Gamma(2\theta)}{4^\theta\,\Gamma^2(\theta)} \cdot f(\Omega,\omega,\theta,d)\right\} . \right. \tag{20.27}
$$

where

$$
f = \int_0^1 dy\, y^{d/2}(1-y)^2 \int_{-1}^1 d\tau(1-\tau^2)^\theta
$$

$$
\times \left[\frac{Z}{\sqrt{(1-y)^2\Omega^2 + c_N^2 y^2\omega^2 - 2c_N y(1-y)\tau\omega\Omega}} \right.
$$

$$
\left. + \frac{Z}{\sqrt{(1-y)^2\Omega^2 + c_H^2 y^2\omega^2 + 2c_H y(1-y)\tau\omega\Omega}} \right]
$$

and the notation is used

$$
\lambda = \sqrt{\frac{m_H m_\mu m_N}{m_H + m_\mu + m_N}} \quad , \qquad M = \frac{m_H m_N}{m_N + m_H}
$$

$$
c_j = \frac{\lambda}{m_j} \quad , \qquad c = c_H + c_N \qquad j = H, N .
$$

The charge of light nuclei is denoted by Z.

For the calculation, we have used the following values of masses:

$$
m_\mu = 206.77 m_e, \quad m_p = 1836.15 m_e, \quad m_d = 3670.48 m_e, \quad m_t = 5496.9 m_e
$$

$$
m_{\,^3He} = 5495.92815 m_e, \quad m_{\,^4He} = 7294.3561 m_e, \quad m_{\,^6Le} = 10961.91216 m_e,
$$

$$
m_{\,^7Le} = 12786.40853 m_e, \quad m_{\,^7Be} = 12787.17851 m_e,
$$

where m_e is the electron mass. The results of calculations are given in Table 20.4.

Taking into account the values of the energy parameter U, which is represented in Table 20.4, we can determine the binding energy from (20.25) for each mesic molecule of light nuclei. The results our calculations are given in Tables 20.5 and 20.6. Table 20.5 represents the absolute values of the binding energies (in eV) of mesic molecules of Helium isotopes, determined in papers ([55], [56]), and our results. In Table 20.6 absolute values of the binding energies (in eV) of the $(H\mu\,^6Li)$, $(H\mu\,^7Li)$ and $(H\mu\,^7Be)$ systems are given.

From Table 20.5, one can see that the results of our calculations of the binding energies for the mesic molecules of Helium isotopes are in agreement with other theoretical calculations. But in Table 20.6 our results are not in good agreement with the other theoretical investigations. We want to note that the accuracy of the zeroth approximation for the $(H\mu\ ^7Be)$ system is not sufficient.

Table 20.4. The values of the parameters of our model.

system	Ω	ω	d	θ	U
$(p\mu\ ^3He)$	0.23719	0.10957	4.1001	0.10297	0.13863
$(d\mu\ ^3He)$	0.19336	0.10621	4.1000	0.10215	0.09128
$(t\mu\ ^3He)$	0.17423	0.10195	4.1028	0.10016	0.07448
$(p\mu\ ^4He)$	0.23029	0.10771	4.1000	0.10233	0.13035
$(d\mu\ ^4He)$	0.18513	0.10531	4.1052	0.10199	0.08245
$(t\mu\ ^4He)$	0.17210	0.10789	4.1000	0.10198	0.06542
$(p\mu\ ^6Li)$	0.21706	0.10740	4.1018	0.10236	0.11929
$(d\mu\ ^6Li)$	0.17335	0.10730	4.1045	0.10141	0.07191
$(t\mu\ ^6Li)$	0.16302	0.11148	4.1138	0.10010	0.05509
$(p\mu\ ^7Li)$	0.21394	0.10686	4.1000	0.10241	0.11698
$(d\mu\ ^7Li)$	0.17053	0.11041	4.1000	0.10377	0.06944
$(t\mu\ ^7Li)$	0.15821	0.11075	4.1040	0.10000	0.05252
$(p\mu\ ^7Be)$	0.21226	0.10777	4.1007	0.10226	0.11616
$(d\mu\ ^7Be)$	0.17113	0.10950	4.1020	0.10065	0.06900
$(t\mu\ ^7Be)$	0.15231	0.11059	4.1033	0.10001	0.05215

Table 20.5. The absolute values of the binding energies of mesic molecules of Helium isotopes (in eV).

system	[?]	[?]	[?]	OR
$(p\mu\ ^3He)$	67.70	69.0		67.4
$(d\mu\ ^3He)$	69.96	70.6	70.74	70.0
$(t\mu\ ^3He)$	71.91	72.3		73.8
$(p\mu\ ^4He)$	74.36	75.4		73.0
$(d\mu\ ^4He)$	77.96	78.4		76.0
$(t\mu\ ^4He)$	80.76	81.3		79.0

20.3.2 The Stability of Mesic Molecules of Light Nuclei

We consider the systems $(H\mu\ N_Z)$, where N_Z is the nuclei with charge Z and mass

$$m_N = 2Zm_p ,\qquad (20.28)$$

where m_p is the proton mass. We should calculate the energy parameter U as a function of Z for this system $(H\mu\ N_Z)$. The binding energy of the given system is determined also as a function of Z. The dependence of the binding

Table 20.6. The absolute values of the binding energies of the $(H\mu Li)$ and $(H\mu Be)$ systems(in eV).

system	[59]	[58]	OR
$(p\mu\ ^6Li)$	24.3	17.6	24.1
$(d\mu\ ^6Li)$	23.8	18.5	27.0
$(t\mu\ ^6Li)$	35.3	19.8	33.0
$(p\mu\ ^7Li)$	20.8	21.0	20.0
$(d\mu\ ^7Li)$	25.9	22.0	23.0
$(t\mu\ ^7Li)$	37.5	23.3	29.0
$(p\mu\ ^7Be)$	11.7		9.0
$(d\mu\ ^7Be)$	29.3		14.0
$(t\mu\ ^7Be)$			17.0

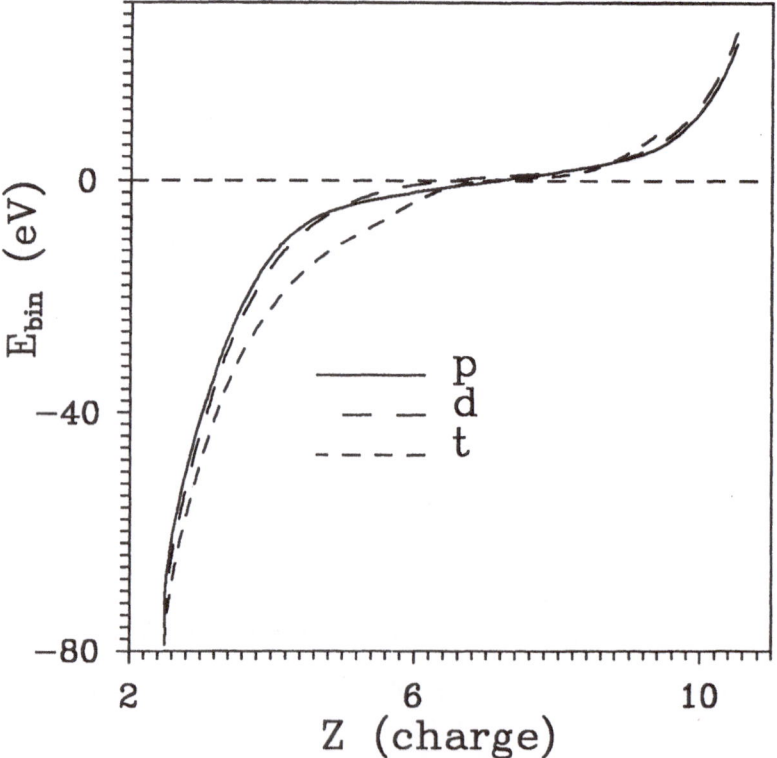

Fig. 20.2. The dependence of the binding energies of the system $(H\mu\ N_Z)$ on the charges Z of nuclei.

energy of the system $(H\mu\ N_Z)$ as a function of the variable Z is represented on Fig. 20.2.

The critical value of the variable Z_{cr} is determined from the condition

$$E_{bind}(Z_{cr}) = 0 \ . \tag{20.29}$$

From Fig. 20.2 one can see that the critical charges Z_{cr} for the mesic molecules $(p\mu\ N_Z)$, $(d\mu\ N_Z)$ and $(t\mu\ N_Z)$ are different and bounded in the interval

$$5,5 < Z_{cr} < 7,3 \ .$$

This lowest and upper estimations for the critical values of Z are qualitative. In principle, the oscillator representation method can improve the accuracy of calculation by taking into account higher perturbation corrections connected with the interaction Hamiltonian [2]. So, by taking into account higher perturbation orders one can be determine the exact values of the critical charges for any three-body Coulomb systems.

References

1. Landau L.D., and Lifschitz E.M., *Quantum Mechanics : Non − Relativistic Theory* (Pergamon Oxford, 1977)
2. Blokhintsev D.I.,*Quantum Mechanics* : M., atomizdat,(1981) (in Russian).
3. Messiah A., *Quantum Mechanics* (John Wiley and Sons, Inc., New York, 1961).
4. Fröman N., and Fröman P.O, *JWKB Approximation* (North-Holland, Amsterdam, 1965).
5. Kohn W., Phys. Rev., 71, p.902(1947);
   Bishop D.,CHeung L.: Phys. Rev., 16,p.640(1977);
   Alexander S.A.,Monkhorst H.J.: Phys. Rev.,38, p.26(1988).
6. Mlodinow L., and Papanicolaou N., Ann. Phys. (N.Y.) 128, p.314(1980):
   Bander C., Mlodinow L., and Papanicolaou N., Phys. Rev. A25, p.1305(1983);
   Ader J., Phys. Lett., 97A, p.178(1983);
   Yaffe L., Rev. Mod. Phys., 54, p.497(1982);
   Witten E., Nucl. Phys. B160, p.57(1979).
7. Sukhatme U., Imbo T., Phys. Rev. D28, p.418(1983);
   Imbo T., Pagnamenta A. and Sukhatme U., Phys. Rev. D29, p.418(1984).
8. Efimov G.V., Preprint IC/91/31, Miramare-Trieste (1991);
   Preprint IC/90/23, Miramare-Trieste (1990).
9. Dineykhan M., and Efimov G.V. The Oscillator Representation and the Stability of the Three-Body Coulomb Systems. Few-Body Systems 16, p.59(1994); Yad. Fiz. 57 ,p.38(1994) (in Russian).
10. Schrödinger E.: Proc.R.Irish Acad.46,p.183(1941).
11. Kustaanheimo P.and Stiefel E.: J. Reine Angew.Math. 218,p.204(1965).
12. Caswell W.E., Ann. Phys. 123, p.153(1979);
    Feranchuk I.D. and Komarov L.I., Phys.Lett. A88, p.211(1982).
13. Duru I.H. and Kleinert H., Fortsch. der Phys. 30,p.401(1982).
14. Mlodinow L.D., Papanicolaow N., Ann. Phys., 131, p.1(1981);
    Sergeev A.V., Yad. Fiz. 50 ,p.945(1989) (in Russian).
15. Johnson R., J.Math.Phys. 21,p.2640(1980);
    Papp E., Phys. Rev. A38, p.5910(1988).
16. Hioe F.T., Don Mae Millen and Montrall E.W., Phys. Rep. C43, p.307(1978).
17. Bender C.M., and Wu T.T., Phys. Rev. 184, p.1231(1969); Phys. Rev., D7, p.1620(1973).
18. Stevenson P.M., Nucl. Phys., B231, p.65(1984).
19. Seetharaman M., Raghavan S., and Vason S., J. Phys. A15, p.1537(1982); J. Phys. A17, p.2493(1984); J. Phys. A18, p.1041(1985).
20. Yukalov V.I. , Theor. Math. Phys., vol.28, p.652(1976) (in russian);
    Killingbeck J., J. Phys. A14, p.1005(1981).
21. Stevenson P.M., Phys. Rev. D23, p.2916(1981).

22. Koudinov A.V., and Smondyrev M.A., Czech. J.Phys. 32, p.556(1982); Theor. Math. Phys., vol.56, p.357(1982) (in russian).
23. Dineykhan M., and Efimov G.V., Yad. Fiz. 56 ,p.89(1993).
24. Banerjee K., et al., Proc. Roy. Soc. London 360, p.575(1978).
25. Quigg C., and Rosner J., Phys. Rep., 56, p.206(1979);
    Richardson J., and Blankenbecier R., Phys. Rev. D19, p.496(1979);
    Dumont-Le Page M., et al., J. Phys. A13, p.1243(1980).
26. Dineykhan M., and Efimov G.V., Preprint, JINR, E4- 94-75 ,Dubna(1994).
27. Harris M., Phys. Rev. 125,p.1131(1962);
    Iafrate G.J., and Mendelsohn L.B., Phys.Rev., 182,p.244(1969);
    de Mayer H., et al., J. Phys. A18,L849(1985).
28. Rogerc F.J., et al., Phys. Rev. A1,p.1577(1970).
29. Belyaev V.B., Kartavtsev O.I., J. Comput. Phys. 59,p. 493(1985).
30. Gerry C.C., and Laub J., Phys. Rev. A30,p.1229(1984);
    Lam C.S., and Varshni Y.P., Phys.Rev. A6, p.139(1972);
    Becher A., Ann. Phys., 108, p.49(1977);
    Dutt R., et al., J. Phys. A18, p.1379(1985);
    Sever R., Tezcan C., Phys. Rev. A36,p.1045(1987).
31. Basdevant J.L., Boukraa S., Z.Phys. C, 28, p.413(1985);
    Goldfrey S. Phys. Rev., D31, p.2375(1985).
32. Eichten E., Feinberg F., Phys. Rev., D23, p.2724(1981);
    Sebastian J., Phys. Rev., D26, p.2295(1982).
33. Martin A., Z. Phys. C, 32,p.359(196).
34. Gupta S.N. et. all., Phys. Rev., D26, p.3305(1982).
    Godfrey S., Isgur N., Phys. Rev., D32, p.189(1983).
    Golangelo P., Nardulli G. and Pietroni M., Preprint, BARI TH/90-70; Phys. Lett., B220, p.265(1989).
35. Fock V.A. Izv. Aked. Nauk. SSSR ser. Fiz. 18, p.161(1954)[Norsk. Vidensk. Selsk. Forh. 31,p.138(1958)].
36. Belyaev V.B. et al., Zh. Eksp. Teor. Fiz.,37,p.1652(1959) [Sov. J. JETP 37,p.1171(1960)];
    Macek J. Jour. Phys. B1,p.831(1968);
    Vinitsky S.I. and Ponomarev L.I., Sov. J.Part. Nucl. 13,p.557(1982);
    Few Body Problems in Physics. Ed H.W.Fearing(North-Holland Publishing Co.,1990);
    Gusev V.V. et al.: Few-Body Systems. 9,p.137(1990).
37. Bhatia A.K.,Drachman R.J. Phys. Rev., A30,p.2138(1984);
    Hu C.Y. Phys. Rev., A32, p.1245(1985).
38. Frolov A.M., Efros V.D., Pis'ma Zh.Eksp. Teor.Fiz. 39,p.449(1984)[JETP Lett. 39,p.545(1984)]; J. Phys. B 18, p.265(1985);
    Frolov A.M., Yad. Fiz.44, p.1367(1986)[Sov.J.Nucl. Phys. 44,p.88(1986)].
39. Halpern A. Phys. Rev. Lett. 13, p.660(1964);
    Vinitsky S.I. et al., Sov. Phys. JETP, 64,p.417(1986);
    Phys. Lett., B196, p.272(1987).
40. Martin A., et al., Phys. Rev., A46, p.3697(1992);
    Martin A., Preprint CERN-TH 6376/92, Geneve (1992).
41. Thirring W., *A Course in Mathematical Physics* , V.3 (Springer-Verlag,1981).
42. Bhatia A.K. and Drachman R.J., Phys. Rev. A35, p.4051(1989);
    Papovic Z.S. and Vukajlovic F.R., Phys. Rev., A36, p.1936(1987).
43. Poshusta R.D., J. Phys. B18, p.1887(1985).
44. Rotenberg M. and Stein J., Phys. Rev. 182, p.1(1969).
45. Armour E.A.G. and Schrader D.M., Can. J. Phys. 60, p.581(1982).

46. Glaser V., et al.,in *Mathematical Problems in Theoretical Physics*, Proc.Int.Conf.Math.Phys. Lausanne(1979), ed.K.Osterwalder (Lectures Notes in Physics, vol.116, Springer-Verlag, Berlin,1980).

47. Cohen S., Hiskes J.R. and Riddell R.J., Phys. Rev. 119, p.1025(1960); Wind H., Chem J., Phys. 42, p.2371(1965); 43, p.2956(1965); Kolos W., Acta Phys. Acad. Sci. Hung. 27, p.241(1969); Beckel C.L., Hausen B.D. and Peek J.M., J. Chem. Phys. 53, p.3681(1970); Struensee M.C., Cohen J.S. and Pack R.T., Phys. Rev. A34, p.3605(1986); Fonseca A.C. and Pena A., Phys. Rev. A38, p.4967(1988).

48. Mills A.P., Phys. Rev. Lett. 46, p.717(1981).

49. Hill R.N., J.Math. Phys., 18, p.2316(1977).

50. Wightman A.S., Thesis, Princeton University (1949).

51. Gershtein S.S., Ponomarev L.I. Mesomolecular processes induced by μ^- and π^- mesons: In Muon Physics. Hughes V., Wu C.S. (eds), Vol 3, p.141 N.Y.(1975).

52. Aristov Yu.A., et al., Sov J. Nucl. Phys. 33, p.564(1981)

53. Matsuzaki I., Ishida K., Nagamine K., Hirata Y., and Kadono R. Muon Catalyzed Fusion 2, p.217(1988).

54. H. P. Von Arb et al., Muon Catalyzed Fusion 4, p.61(1989).

55. Gershtein S.S., Gusev V.V., Preprint IHEP 92-129, Protvino (1992).

56. Kravtsov   A.V.,   Mikhailov   A.I.,   Savichev   V.I.Preprint   1819, St.Petrsburg(1992).

57. Niinikoski T.O. Progres in polarized targets in high-energy physics with polarized beams and polarized targets. Jeseph C., Soffer J. (eds.) p.191 (1981) Besal Birhäuser.

58. Kravtsov A.V., Popov N.P., Solyakin G.E. Sov. J. Nucl. Phys. 35, p.876(1982).

59. Belyaev V.B., et al., Few-Body Systems, Suppl.6, p.332(1992).

60. Dineykhan M., and Efimov G.V., Preprint, JINR, E4-93-465, Dubna(1993).

61. Hara S., Ishihara T. Phys. Rev. A 39, p.5633(1989).

New Series m: Monographs

Lecture Notes in Physics

For information about Vols. 1–403
please contact your bookseller or Springer-Verlag